Biomedical Photoacoustic Imaging and Sensing Using Affordable Resources

Biomedical Photoacoustic Imaging and Sensing Using Affordable Resources

Editors

Mithun Kuniyil Ajith Singh
Wenfeng Xia

MDPI • Basel • Beijing • Wuhan • Barcelona • Belgrade • Manchester • Tokyo • Cluj • Tianjin

Editors

Mithun Kuniyil Ajith Singh
Research and Business
Development Department
CYBERDYNE, INC.
Rotterdam
Netherlands

Wenfeng Xia
School of Biomedical
Engineering and Imaging
Sciences
King's College London
London
United Kingdom

Editorial Office
MDPI
St. Alban-Anlage 66
4052 Basel, Switzerland

This is a reprint of articles from the Special Issue published online in the open access journal *Sensors* (ISSN 1424-8220) (available at: www.mdpi.com/journal/sensors/special_issues/BPISUAR).

For citation purposes, cite each article independently as indicated on the article page online and as indicated below:

LastName, A.A.; LastName, B.B.; LastName, C.C. Article Title. *Journal Name* **Year**, *Volume Number*, Page Range.

ISBN 978-3-0365-1199-3 (Hbk)
ISBN 978-3-0365-1198-6 (PDF)

Contents

About the Editors

Mithun Kuniyil Ajith Singh

Dr. Mithun Kuniyil Ajith Singh is an engineering scientist with extensive experience in preclinical and clinical photoacoustic and ultrasound imaging. He is presently working as a Research and Business Development Manager at CYBERDYNE, INC, the Netherlands. In his current role, he initiates and coordinates various scientific projects in collaboration with globally renowned research groups, especially focusing on the clinical translation of LED-based photoacoustic imaging technology. Mithun earned his PhD in 2016 from the University of Twente, the Netherlands, under the supervision of Professor Wiendelt Steenbergen. Mithun has published multiple international journal articles and book chapters mainly focused on accelerating the translation of photoacoustic imaging from bench to bedside. He is the editor of the first book on LED-based photoacoustic imaging, published by Springer Nature in 2020.

Wenfeng Xia

Dr. Wenfeng Xia is a Lecturer in the School of Biomedical Engineering & Imaging Sciences at King's College London, UK. He received a BSc in Electrical Engineering from Shanghai Jiao Tong University, China, and a MSc in Medical Physics from the University of Heidelberg, Germany, in 2005 and 2007, respectively. In 2013, he obtained his Ph.D from the University of Twente, Netherlands. From 2014 to 2018, he was a Research Associate in the Department of Medical Physics and Biomedical Engineering at University College London, UK. His research interests include non-invasive and minimally invasive photoacoustic imaging, and ultrasound-based medical devices tracking for guiding interventional procedures.

Preface to "Biomedical Photoacoustic Imaging and Sensing Using Affordable Resources"

The overarching goal of this book is to provide a current picture of the latest developments in the capabilities of biomedical photoacoustic imaging and sensing in an affordable setting, such as advances in the technology involving light sources, and delivery, acoustic detection, and image reconstruction and processing algorithms. This book includes 14 chapters from globally prominent researchers, covering a comprehensive spectrum of photoacoustic imaging topics from technology developments and novel imaging methods to preclinical and clinical studies, predominantly in a cost-effective setting. Affordability is undoubtedly an important factor to be considered in the following years to help translate photoacoustic imaging to clinics around the globe. This first-ever book focused on biomedical photoacoustic imaging and sensing using affordable resources is thus timely, especially considering the fact that this technique is facing an exciting transition from benchtop to bedside. Given its scope, the book will appeal to scientists and engineers in academia and industry, as well as medical experts interested in the clinical applications of photoacoustic imaging.

Mithun Kuniyil Ajith Singh, Wenfeng Xia
Editors

sensors

MDPI

Editorial

Biomedical Photoacoustic Imaging and Sensing Using Affordable Resources

Mithun Kuniyil Ajith Singh [1,*] and Wenfeng Xia [2,*]

1 Research and Business Development Division, CYBERDYNE INC., Stationsplein 45, A4.004, 3013 AK Rotterdam, The Netherlands

2 School of Biomedical Engineering & Imaging Sciences, King's College London, King's Health Partners, St. Thomas' Hospital, London SE1 7EH, UK

* Correspondence: mithun_ajith@cyberdyne.jp (M.K.A.S.); wenfeng.xia@kcl.ac.uk (W.X.)

check for updates

Citation: Kuniyil Ajith Singh, M.; Xia, W. Biomedical Photoacoustic Imaging and Sensing Using Affordable Resources. *Sensors* **2021**, *21*, 2572. https://doi.org/10.3390/s21072572

Received: 26 March 2021
Accepted: 2 April 2021
Published: 6 April 2021

Publisher's Note: MDPI stays neutral with regard to jurisdictional claims in published maps and institutional affiliations.

1. Introduction

The photoacoustic (PA) effect, also called the optoacoustic effect, was discovered in the 1880s by Alexander Graham Bell and has been utilized for biomedical imaging and sensing applications since the early 1990s [1]. In biomedical photoacoustic imaging, nanosecond-pulsed or intensity-modulated light illuminates tissue of interest, which when absorbed by intrinsic (such as hemoglobin, lipid) and extrinsic optical absorbers results in the generation of ultrasound (US) signals via thermoelastic expansion. These optically generated US signals can be detected on the tissue surface using conventional ultrasonic probes to generate the tissue's optical absorption maps with high spatiotemporal resolution. PA imaging thus offers advantages of both US imaging (imaging depth, spatiotemporal resolution) and conventional optical imaging techniques (spectroscopic contrast), making it an ideal modality for structural, functional, and molecular characterization of tissue in situ. Moreover, since both PA and US imaging rely on acoustic detection, it is feasible to share the probe and data acquisition system to perform naturally co-registered dual-mode PA and US imaging with complementary contrast. Since US imaging is ubiquitous in clinics, such a dual-mode approach is expected to facilitate accelerating the clinical translation of the PA imaging technique. Owing to all these advantages, PA imaging has been explored for myriads of preclinical and clinical applications and is undoubtedly one of the fastest-growing biomedical imaging modalities of recent times. Even though PA imaging is matured in lab settings, clinical translation of this promising technique is not happening at an expected pace. One of the important reasons behind this is the costs of pulsed light sources and acoustic detection hardware. Affordability is undoubtedly an important factor to be considered in the following years to help translate PA imaging to clinics around the globe. This first-ever Special Issue focused on biomedical PA imaging and sensing using affordable resources is thus timely, especially considering the fact that this technique is facing an exciting transition from benchtop to bedside.

The overarching goal of this Special Issue is to provide a current picture of the latest developments in the capabilities of PA imaging and sensing in an affordable setting, such as advances in the technology involving light sources, and delivery, acoustic detection, and image reconstruction and processing algorithms. This issue includes 13 papers (2 reviews, 2 letters, 1 communication and 8 full-length articles) which cover a comprehensive spectrum of research from technology developments and novel imaging methods to preclinical and clinical studies, predominantly in a cost-effective setting.

2. Review Articles

The Issue starts with a comprehensive review article on portable and affordable light-source-based PA tomography from Kuniyil Ajith Singh and Xia [2]. In this review, the authors focus on (1) basics of PA imaging, (2) cost-effective pulsed light sources for PA imaging and (3) important preclinical and clinical applications reported until now using affordable-light-source-based PA imaging. Because of tremendous developments in solid-state device technology, high-power LEDs have been explored heavily in recent years as illumination sources in PA imaging. In another interesting paper, for the first time, Zhu et al. comprehensively reviewed the use of LEDs in biomedical PA imaging, covering all technical details, preclinical and clinical applications reported until now [3].

3. Original Papers—Full-Length Articles, Communications, and Letters

The rest of the original papers in this issue are summarized in the following subsections based on the contents.

3.1. Instrumentation, Technology Developments

Conventional US probes are opaque and are not ideal for miniaturized PA imaging systems, especially for microscopic applications. In an exciting work from Chen et al., the authors developed an affordable transparent lithium niobate ($LiNbO_3$) US transducer with a 13-MHz central frequency and a 60% reception bandwidth for optical resolution PA microscopy [4]. The authors tested the system performance by imaging vasculature in chicken embryos and melanoma depth profiling using tissue phantoms. The system proposed in this work is expected to have a promising future in wearable and high-throughput imaging applications.

LED-based PA imaging is gaining more popularity in recent years because of its portability, affordability, and ease of use. However, due to the large beam divergence of LEDs compared to traditional laser beams, it is of paramount importance to quantify the angular dependence of LED-based illumination and optimize its performance for imaging superficial or deep-seated lesions. Kuriakose et al. reported on the development of a custom 3D printed LED array holder to be used along with a commercial LED-based PA imaging system (Acoustic X, CYBERDYNE INC, Tsukuba, Japan) and demonstrated the importance of changing LED illumination angle when used for superficial and deep-tissue applications [5].

Considering the tradeoffs between portability, cost, optical energy, and frame rate, it is critical to compare the PA imaging performance of LED and laser illuminations to help select a suitable source for a given biomedical imaging application. Agrawal et al. reported on the development of a setup for a head-to-head comparison of LED and laser-based PA imaging of vasculature [6]. With measurements on tissue-mimicking phantoms and human volunteers, authors concluded that LED-based PA imaging performs equally and sometimes even better than laser-based systems demonstrating its strong potential to be a mobile health care technology for diagnosing vascular diseases such as peripheral arterial disease and stroke in point-of-care and resource-limited settings.

In applications like wound screening and laser surgery guidance, conventional PA imaging systems that usually require US probes in contact with the tissue are not an ideal option. In a work by Lengenfelder et al., it is demonstrated for the first time that remote PA sensing by speckle analysis can be performed in the MHz sampling range by tracking a single speckle using a four-quadrant photo-detector [7]. By demonstrating PA sensing and endoscopic capabilities, this work demonstrated that single speckle sensing is, therefore, an easy, robust, contact-free photoacoustic detection technique and holds the potential for economical, and ultra-fast PA sensing.

3.2. Image Processing and Enhancement Techniques

Multispectral PA imaging can be used to non-invasively visualize and quantify tissue chromophores with high spatial resolution. Utilizing multiwavelength PA data, one can characterize the spectral absorption signature of prominent tissue chromophores, such as hemoglobin or lipid, by using spectral unmixing methods. Grasso et al. reported the feasibility of an unsupervised spectral unmixing algorithm to detect and extract the tissue chromophores without any a priori knowledge of the optical absorption spectra of the tissue chromophores or exogenous contrast agents and user interactions [8]. The authors validated this novel algorithm using simulations, phantom studies, and mouse in vivo experiments and demonstrated the feasibility of extracting and quantify the tissue chromophores in a completely unsupervised manner.

Conventionally, hand-held linear-array-based PA imaging systems operate in the reflection mode using a dark-field illumination scheme, where the illumination is on both sides of the elevation plane (short-axis) of the US probe. Uliana et al. reported and demonstrated a novel multiangle long-axis lateral illumination approach with several advantages [9]. Phantom, animal, and human in vivo results in this work demonstrate a remarkable improvement of the new illumination approach in light delivery for targets with a width smaller than the transducer's lateral dimension.

Portable and affordable light sources like pulsed laser diodes and LEDs have the potential in accelerating the clinical translation of PA imaging. However, pulse energy offered by these sources is often low when compared to solid-state lasers, thus resulting in a low signal-to-noise-ratio (SNR), especially in deep-tissue applications. Improvement of the SNR is of paramount importance in these cases. Thomas et al. proposed a continuous-wave laser-based pre-illumination approach to increase the temperature of the imaging sample and thus the PA signal strength from it [10]. In this work, using tissue-mimicking phantoms and common contrast agents, the authors showed the feasibility of enhancing the PA signal strength significantly.

3.3. *Preclinical and Clinical Applications*

Small animals are widely used as disease models in medical research, especially in the pharmaceutical industry. Since PA imaging can offer functional and molecular information, it is an ideal modality for small animal imaging. Kalloor Joseph et al. reported a portable and affordable approach for performing fast tomographic PA and US imaging of small animals [11]. In this work, the authors used LED-based PA and US tomographic imaging and showed its potential in liver fibrosis research.

Conventional PA imaging systems utilize expensive and bulky solid-state lasers with low pulse repetition rates; as such, their availability for preclinical cancer research is hampered. In an interesting study from Xavierselvan et al., the authors validated the capability of an LED-based PA and US imaging system for monitoring heterogeneous microvasculature in tumors (up to 10 mm in depth) and quantitatively compared the PA images with gold standard histology images [12]. The results of this work give a direct confirmation that LED-based PA and US imaging hold the potential to be a valuable tool in preclinical cancer research.

Hypoxia and hyper-vascularization are the hallmarks of cancer and oxygen saturation imaging is arguably the most important application of PA imaging. By illuminating tissue using two wavelengths, it is feasible to probe the oxygen saturation of tissue with microvasculature scale resolution. Bulsink et al. reported fluence-compensated oxygen saturation imaging using two wavelength LED-based PA imaging [13]. In this work, the authors demonstrated real-time fluence compensated oxygen saturation imaging in phantoms, small animals, and measurements on human volunteers.

Follicular unit extraction and follicular unit transplantation are used in most hair transplant procedures. In both cases, it is important for clinicians to characterize follicle density for treatment planning and evaluation. Hariri et al. utilized 2D and 3D LED-based PA imaging for measuring follicle density and angles across regions of varying density [14]. The authors validated the idea using experiments on small animals and also using measurements on healthy human volunteers.

4. Summary

PA imaging is growing at a tremendous pace and is expected to reach clinics soon. At this point, this promising technology is facing an exciting transition phase and thus this Special Issue with a focus on affordability is timely. Thirteen excellent papers in this

Special Issue from academia and industry represent a small sample that demonstrates the immense developments in the field. We hope that this Special Issue can provide the motivation and inspiration for further technological advances in this exciting field and accelerates the clinical translation of this promising biomedical imaging modality.

Author Contributions: M.K.A.S. and W.X.; writing—original draft preparation, review and editing. Both authors have read and agreed to the published version of the manuscript.

Acknowledgments: We would like to thank all the authors who have submitted their excellent works to this Special Issue, and reviewers who supported us in maintaining the high standard by providing constructive and timely feedback for manuscripts published in this Special Issue. Finally, we thank the editorial staff at MDPI, for helping us through the publication of all articles with incredibly fast turn-around time and excellent editorial process.

Conflicts of Interest: M.K.A.S. is employed by CYBERDYNE, INC. The authors have no other financial interests or conflict of interest to disclose. We clarify that in some of the reported studies in this Special Issue, CYBERDYNE, INC as a company was involved in technical developments or design of experiments either directly or through industry–academic collaborative research projects, but had no influence in the peer-review process or in the decision to publish the articles.

References

1. Manohar, S.; Razansky, D. Photoacoustics: A historical review. *Adv. Opt. Photonics* **2016**, *8*, 586–617. [CrossRef]
2. Kuniyil Ajith Singh, M.; Xia, W. Portable and affordable light source-based photoacoustic tomography. *Sensors* **2020**, *20*, 6173. [CrossRef] [PubMed]
3. Zhu, Y.; Feng, T.; Cheng, Q.; Wang, X.; Du, S.; Sato, N.; Yuan, J.; Kuniyil Ajith Singh, M. Towards clinical translation of LED-based photoacoustic imaging: A review. *Sensors* **2020**, *20*, 2484. [CrossRef] [PubMed]
4. Chen, H.; Agrawal, S.; Dangi, A.; Wible, C.; Osman, M.; Abune, L.; Jia, H.; Rossi, R.; Wang, Y.; Kothapalli, S.-R. Optical-resolution photoacoustic microscopy using transparent ultrasound transducer. *Sensors* **2019**, *19*, 5470. [CrossRef] [PubMed]
5. Kuriakose, M.; Nguyen, C.D.; Kuniyil Ajith Singh, M.; Mallidi, S. Optimizing irradiation geometry in LED-based photoacoustic imaging with 3D printed flexible and modular light delivery system. *Sensors* **2020**, *20*, 3789. [CrossRef] [PubMed]
6. Agrawal, S.; Kuniyil Ajith Singh, M.; Johnstonbaugh, K.; Han, C.D.; Pameijer, R.C.; Kothapalli, S.-R. Photoacoustic imaging of human vasculature using LED versus laser illumination: A comparison study on tissue phantoms and in vivo humans. *Sensors* **2021**, *21*, 424. [CrossRef] [PubMed]
7. Lengenfelder, B.; Hohmann, M.; Späth, M.; Scherbaum, D.; Weiß, M.; Rupitsch, S.J.; Schmidt, M.; Zalevsky, Z.; Klämpfl, F. Remote photoacoustic sensing using single speckle analysis by an ultra-fast four quadrant photo-detector. *Sensors* **2021**, *21*, 2109. [CrossRef] [PubMed]
8. Grasso, V.; Holthof, J.; Jose, J. An automatic unmixing approach to detect tissue chromophores from multispectral photoacoustic imaging. *Sensors* **2020**, *20*, 3235. [CrossRef] [PubMed]
9. Uliana, J.H.; Sampaio, D.R.T.; Fernandes, G.S.P.; Brassesco, M.S.; Nogueira-Barbosa, M.H.; Carneiro, A.A.O.; Pavan, T.Z. Multiangle long-axis lateral illumination photoacoustic imaging using linear array transducer. *Sensors* **2020**, *20*, 4052. [CrossRef] [PubMed]
10. Thomas, A.; Paul, S.; Mitra, J.; Singh, M.S. Enhancement of photoacoustic signal strength with continuous wave optical pre-illumination: A non-invasive technique. *Sensors* **2021**, *21*, 1190. [CrossRef] [PubMed]
11. Francis, K.J.; Booijink, R.; Bansal, R.; Steenbergen, W. Tomographic ultrasound and led-based photoacoustic system for preclinical imaging. *Sensors* **2020**, *20*, 2793. [CrossRef]
12. Xavierselvan, M.; Singh, M.K.A.; Mallidi, S. In vivo tumor vascular imaging with light emitting diode-based photoacoustic imaging system. *Sensors* **2020**, *20*, 4503. [CrossRef]
13. Bulsink, R.; Kuniyil Ajith Singh, M.; Xavierselvan, M.; Mallidi, S.; Steenbergen, W.; Francis, K.J. Oxygen saturation imaging using LED-based photoacoustic system. *Sensors* **2021**, *21*, 283. [CrossRef]
14. Hariri, A.; Moore, C.; Mantri, Y.; Jokerst, J.V. Photoacoustic imaging as a tool for assessing hair follicular organization. *Sensors* **2020**, *20*, 5848. [CrossRef]

sensors

Review

Portable and Affordable Light Source-Based Photoacoustic Tomography

Mithun Kuniyil Ajith Singh [1] and Wenfeng Xia [2,*]

1 Research and Business Development Division, CYBERDYNE INC., Stationsplein 45, A4.004, 3013 AK Rotterdam, The Netherlands; mithun_ajith@cyberdyne.jp
2 School of Biomedical Engineering& Imaging Sciences, King's College London, King's Health Partners, St Thomas' Hospital, London SE1 7EH, UK
* Correspondence: wenfeng.xia@kcl.ac.uk

Received: 8 October 2020; Accepted: 28 October 2020; Published: 29 October 2020

Abstract: Photoacoustic imaging is a hybrid imaging modality that offers the advantages of optical (spectroscopic contrast) and ultrasound imaging (scalable spatial resolution and imaging depth). This promising modality has shown excellent potential in a wide range of preclinical and clinical imaging and sensing applications. Even though photoacoustic imaging technology has matured in research settings, its clinical translation is not happening at the expected pace. One of the main reasons for this is the requirement of bulky and expensive pulsed lasers for excitation. To accelerate the clinical translation of photoacoustic imaging and explore its potential in resource-limited settings, it is of paramount importance to develop portable and affordable light sources that can be used as the excitation light source. In this review, we focus on the following aspects: (1) the basic theory of photoacoustic imaging; (2) inexpensive light sources and different implementations; and (3) important preclinical and clinical applications, demonstrated using affordable light source-based photoacoustics. The main focus will be on laser diodes and light-emitting diodes as they have demonstrated promise in photoacoustic tomography—the key technological developments in these areas will be thoroughly reviewed. We believe that this review will be a useful opus for both the beginners and experts in the field of biomedical photoacoustic imaging.

Keywords: photoacoustic imaging; photoacoustic computed tomography; light-emitting diodes; laser diodes

1. Introduction

Photoacoustic imaging (PAI) is an emerging biomedical imaging modality that is based on the detection of ultrasound (US) waves generated from tissue in response to the absorption of temporally varying optical energy [1–5]. Typically, in PAI, short-pulsed or temporally modulated light is delivered to a tissue surface, and the light then propagates diffusively through the tissue during which a portion of the optical energy is selectively absorbed by various light absorbing structures, such as endogenous tissue chromophores and exogenous contrast agents [2]. These endogenous tissue chromophores include DNA/RNA, oxy- and deoxy-hemoglobin, lipid, and melanin; exogenous contrast agents include small molecular dyes, organic nanostructures, metal and carbon nanoparticles, and genetically encoded chromophores [3]. The absorption of the time-varying optical energies leads to rapid and subtle local temperature rises, and subsequently, the generation of broadband US waves in the MHz frequency range. These initial pressure waves propagate outwards and can be received by US detectors at the tissue surface to form images of the absorbing structures. Since optical absorption spectra of common tissue chromophores (hemoglobin, melanin, lipid, etc.) are well known, it is possible to tune the light excitation wavelength and functionally characterize the tissue using multispectral PAI [1]. For example, one can quantitatively detect oxygen saturation inside a blood vessel with high spatial and temporal resolution at depths up to a few centimeters, which is unachievable using any other imaging modalities. PAI involves US detection and it is straight forward to implement this technique in conventional US imaging equipment, offering structural, functional, and molecular imaging details in a single image acquisition [6]. PAI is undoubtedly one of the fastest growing research-based medical imaging modalities in recent times. However, one of the key factors hindering the clinical translation of this promising technique is the requirement of bulky and expensive solid-state lasers for tissue illumination [7]. In this review, after covering the principles of PAI, we will focus on affordable light sources (light-emitting diodes and laser diodes) that are being explored as alternative excitation sources for photoacoustic (PA) tomography, demonstrate its applications, and also discuss the advantages and disadvantages of different illumination sources in PAI. To focus more on PA tomography (PAT) using light-emitting diodes (LEDs) and laser diodes (LDs), we exclude developments and applications demonstrating PA microscopy and endoscopy.

2. Principles of Photoacoustic Imaging

Optical tissue imaging utilizes spectroscopic features of light–tissue interactions, such as scattering, absorption, and polarization. Since many tissue molecules possess signature optical properties (especially absorption), this interaction is quite powerful in characterizing tissue at functional and molecular levels. Figure 1 shows the optical absorption spectra of prominent tissue chromophores, such as hemoglobin, melanin, lipid, and water. The rich spectroscopic contrast of these molecules and their relations to a spectrum of diseases make optical imaging very useful [1–3]. However, purely light-based imaging techniques suffer from poor resolution at depths larger than a few mm, due to high scattering in the tissue [1]. While in US imaging, the tissue is insonified using sound waves and echoes are used to generate acoustic reflectivity maps [8]. This is one of the most popular medical imaging modalities with several advantages, including a high portability, affordability, accessibility, and spatial and temporal resolution. Even though US imaging offers tissue anatomical information in real time, it can provide insufficient contrast for soft tissues, and insufficient sensitivity for differentiating malignant and benign abnormalities in deep tissue [6,9].

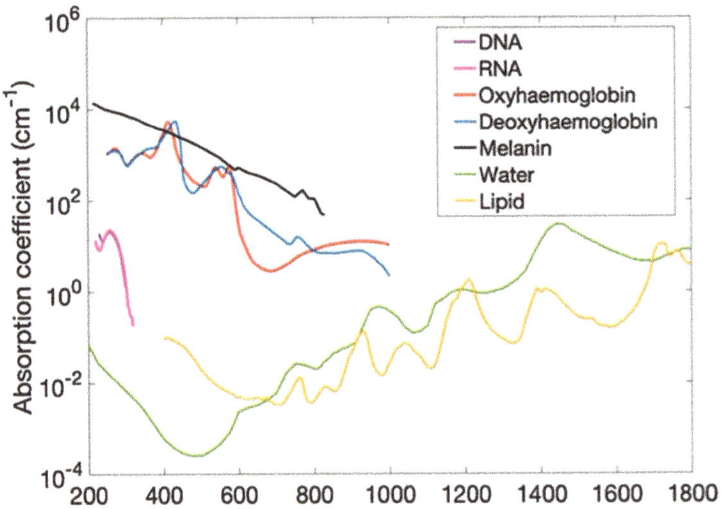

Figure 1. Absorption coefficient spectra (μa) of endogenous tissue chromophores, including DNA, RNA, oxyhemoglobin, deoxyhemoglobin (150 g L^{-1}), melanin, water, and lipid. Adapted with permission from T. Zhao, A. E. Desjardins, S. Ourselin, T. Vercauteren, W. Xia, Photoacoustics, Vol.16, Article ID100146, 2019; licensed under a Creative Commons Attribution (CC BY) license.

2.1. Generation of Photoacoustic Signals

The basic idea of the photoacoustic effect was described by Alexander Graham Bell long back in 1880 [10]. When pulsed light is shone on a sample that absorbs a fraction of the incident energy, the optical absorption will result in a temperature rise, leading to thermoelastic expansion of the absorbing object. This sudden pressure rise propagates as a sound wave, which then can be detected using conventional US transducers. By detecting the pressure waves, one can localize their sources (i.e., where the light was absorbed) and obtain important functional and molecular information about the studied sample [1]. The basic idea of PA imaging is schematized in Figure 2.

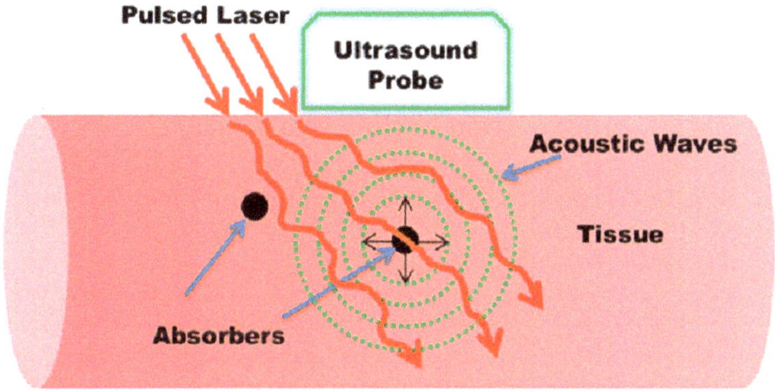

Figure 2. Basic principle of biomedical photoacoustic imaging. Reproduced with permission from Handheld Probe-Based Dual Mode Ultrasound/Photoacoustics for Biomedical Imaging. In: Olivo M., Dinish U. (eds) Frontiers in Biophotonics for Translational Medicine. Progress in Optical Science and Photonics, vol 3. Springer, Singapore (2016). Copyright 2016 Springer, Singapore.

To generate broadband PA signals (sound waves) in the sample efficiently and obtain high-resolution images, the light pulse must be shorter than both the thermal relaxation time τ_{th} (thermal confinement) and the stress relaxation time τ_s (stress confinement), respectively, defined by Equations (1) and (2) [1,11].

$$\tau_{th} = \frac{d_c^2}{\alpha_{th}} \quad (1)$$

where d_c is the characteristic dimension of the structure of interest and α_{th} is the thermal diffusivity.

$$\tau_s = \frac{d_c}{v_s} \quad (2)$$

where v_s is the speed of sound in the tissue (~1540 m/s).

Let us consider a 15 μm optically absorbing structure inside the tissue. In this case, τ_{th} and τ_s will be 1.4×10^{-3} s and 1×10^{-8} s, respectively. Thus, it is clear that the pulse width of the tissue illumination source must be shorter than 10 ns to spatially resolve this 15 μm structure in a PA image. It is also important to note that the detection bandwidth of the US probe also must be high to achieve such high-resolution images, which will be discussed in greater detail later in Section 2.4. The optically induced initial pressure distribution p_0 can be estimated as

$$p_0(r) = \Gamma\, \mu_a\, (r) F(r) \quad (3)$$

where r refers to a spatial location within the heated volume, μ_a is the optical absorption distribution of the absorbing structure, F is the local fluence, Γ is the Grüneisen coefficient, and a the dimensionless thermodynamic constant that defines the conversion efficiency of heat energy to pressure, and can be expressed as

$$\Gamma = \beta {v_s}^2 / C_p \quad (4)$$

where β is the isobaric volume expansion coefficient and C_p is the specific heat at a constant pressure. It can be seen that p_0 depends on a number of parameters. Although several studies have shown that Γ can be a source of PA contrast in tissue [12], PA image contrast is generally assumed to be dominated by the local optical absorption μ_a and the local fluence F, which is a product of optical absorption and scattering. Considering the strong light attenuation in the tissue, F decreases exponentially with tissue depth; however, optical fluence variations at a given depth in the tissue are usually much smaller compared to those of μ_a. As such, PA imaging is often considered as an imaging modality that is based on optical absorption of the tissue; however, it is important to note that the PA image signal is not directly proportional to μ_a.

2.2. Illumination Sources

The characteristics of the illumination sources mainly define the quality of the PA images and contribute to the high cost of the whole PAI system. PAI systems conventionally use solid-state lasers, such as Q-switched Nd: YAG lasers coupled with optical parametric oscillators (OPO). These lasers usually offer pulse widths in the range of 5–20 ns, which is very much suitable for high-resolution PAI. However, these bulky laser sources are very expensive and are not suitable for use in a clinical setting (for example in an operating room or an outpatient clinic) [13–17]. Recently, there has been significant research in the area of affordable light sources for PAI. Laser diodes (LDs) and light-emitting diodes (LEDs) are the most common PAI illumination sources, which offer high portability, affordability, and energy efficiency [13–17]. Solid-state lasers, LDs, and LEDs have completely different characteristics (pulse width, pulse repetition rate (PRR), optical output, and cost) and have their own advantages and disadvantages. Table 1 shows a comparison of the light sources used in PAI.

Table 1. Comparison of different light sources used in photoacoustic imaging. * Cost includes the driving electronics and may vary based on different features, number of wavelengths, etc. Integration to a US probe may also involve extra development cost. LD, Laser diode; LED: Light-emitting diode; DPSS: Diode-pumped solid-state; PRR, pulse repetition rate. Adapted with permission from Y. Zhu, T. Feng, Q. Cheng, X. Wang, S. Du, N. Sato, J. Yuan and M. Kuniyil Ajith Singh, Sensors, Vol.20, Article ID2484, 2020; licensed under a Creative Commons Attribution (CC BY) license.

	Energy (mJ)	PRR (Hz)	Pulse Width (ns)	Cost *	Advantages	Disadvantages
Solid-state lasers	5–120	10–200	<10	$70–200 K	Powerful, ~5 cm penetration depth, tunable wavelength	Bulky size, eye protection and laser safe rooms needed
LD	0.5–2.5	~1 K–6 K	30–200	~$10–25 K	Integration in a handheld probe feasible, high PRR	Limited penetration depth, eye protection and laser safe rooms needed, wavelength tuning not possible
LED	0.2	~200–16 K	30–100	$10–15 K	Integration in a handheld probe feasible, high PRR, wide wavelength range, no need of laser-safe rooms and eye-safety goggles	Limited penetration depth, wavelength tuning not possible
Q-switched DPSS laser	1	100 K	2–10	-	High PRR, low pulse width, Reasonably high optical energy per pulse	Less number of wavelengths (266 nm, 355 nm, 532 nm, 1064 nm) available and spectral tuning may be cumbersome
High-energy DPSS laser	200	200	10–30	-	High optical output per pulse, reasonably high PRR	Less number of wavelengths (266 nm, 355 nm, 532 nm, 1064 nm) available and spectral tuning may be cumbersome

2.3. Optical Absorption

As shown in Figure 1, intrinsic optical absorbers in the tissue possess the signature absorption characteristics and the spatial distributions of these absorbers can be quantified in PAI with a high spatial resolution and large imaging depths [1]. For example, PAI offers excellent contrast for blood. Since oxyhemoglobin and deoxyhemoglobin have different optical absorption spectra, careful selection of the excitation wavelengths (even two wavelengths with good absorption difference would suffice) can help in obtaining oxygen saturation maps of the tissue with an unprecedented resolution (micro-vasculature level) [16]. Apart from intrinsic absorbers, it is feasible to image exogenous contrast agents by selecting the light wavelengths based on the absorption peak of them [17]. Considering the low absorption of water and high absorption of blood, near infrared (NIR) wavelengths are most commonly used in PAI for achieving higher imaging depths.

2.4. Ultrasound Propogation and Detection

The velocity of the US waves in tissue is considered constant at 1540 m/s in general. Variations are typically less than 10%, and are usually not accounted for, unless the sample is highly heterogenous. Acoustic scattering in tissue is roughly three orders of magnitude lower than optical scattering and this is the key reason for a higher imaging depth in PAI compared to other optical imaging modalities that rely on optical focusing [1–5]. One important factor to consider acoustically is the attenuation in the tissue, which can be modelled as below.

$$\alpha = af^b \ [\mathrm{dB/cm}] \tag{5}$$

where a is a tissue dependent constant, b is a power law factor whose value is usually ranged between 1 and 2 in the soft tissue, and f is the frequency expressed in MHz. In general, for soft tissue, α is considered as 1 dB cm^{-1} MHz^{-1} on average; it can reach as high as 20 dB cm^{-1} MHz^{-1} for bone (assuming b = 1 in both cases). As such, the characteristics of the frequency-dependent acoustic attenuation are similar to those of a low pass filter so that higher frequency signals are attenuated more in tissue. Although the PA signals can be extremely broadband with frequency contents ranging from several tens of kHz to several tens of MHz, depending on the size of the object, frequency-dependent acoustic attenuation in tissue has limited the maximum frequency content that can arrive at the detector side and thus fundamentally limits the achievable spatial resolution. The spatial resolution of a PA tomography system also depends on other factors, including the frequency bandwidth of the US detector, the active surface area of the detector element, number of detector elements, detector aperture, and image reconstruction algorithms. Moreover, as the acoustic sensitivity of an US detector decreases with the increase in frequency (which largely determines the spatial resolution), there is a trade-off between the imaging depth and spatial resolution. For example, according to a recent study by Xia et al. [18], the measured axial resolution using resolution targets immersed in water for a PAI system with a linear array US probe (central frequency: 9 MHz; −6 dB bandwidth: 77%; active surface area: 5 mm × 0.3 mm; number of element: 128; and detector aperture: 38.4 mm) was 0.22 mm and remained consistent over a depth range of 13 to 36 mm; the lateral resolution depended on the spatial location, ranging from 0.35 mm to 0.76 mm over the same depth range.

Linear array piezoelectric probes are commonly used as US detectors for PAI similar to those used for conventional US imaging. As such, real-time interleaved US and PA imaging can be performed with the same US detector and DAQ electronics [6]. US detection in PAI mode is the same as that in the US mode, except that in this case no US transmissions are performed. This dual modality approach offers many advantages involving complementary contrast and easiness in clinical translation. US imaging is a well-accepted imaging modality and it would be easier for clinicians to accept PAI as an additional technique along with conventional US imaging [7]. Typical frequencies for US imaging are in the range of 1–25 MHz. For utilizing the full benefits of PA imaging, it is important to develop new US probes with a higher bandwidth and sensitivity as the PA signals are usually broadband

and their amplitudes are much lower than the pulse echo signals in US imaging. There have been significant developments in this area recently [6]. Acoustic waves can be detected using single-element transducers (and employing scanning) or using an array of elements as in conventional US probes (planar, cylindrical, or spherical) to enable 2D or 3D imaging.

2.5. Image Reconstruction

PAI involves the detection of acoustic signals generated by optical absorption, and image reconstruction strategies are thus stemmed from both optical and US imaging. Basic PA image reconstruction aims to retrieve the spatial distribution of the initial PA pressure or locations of the absorbed optical energy [11]. This is usually termed the acoustic inversion problem and uses various algorithms that originate from the US and sonar world. From the distribution of the initial PA pressure, it is desired to reconstruct a distribution of the optical absorption coefficient, especially when multiple wavelengths are used for PA excitation; however, this is a non-linear problem due to the wavelength-depended optical attenuation of the tissue, which is termed the optical inversion problem. Reconstruction of the optical absorption coefficient is usually termed quantitative PA imaging [11].

There has been extensive research in the area of PA image reconstruction and different methods have been reported. Time reversal, Fourier domain analysis, analytic back-projection, radon transfer, and model-based algorithms are some of the most commonly applied methods. Kuchment et al. reported a detailed review about the different PA-based image reconstruction methods [19]. Based on the type of acoustic detection probes (linear arrays, curved arrays, etc.) and the computational capability, one can choose the right image reconstruction algorithm. Considering the ease in implementation and speed offered, back-projection is the most commonly used image reconstruction technique for processing PA data [1]. Figure 3 illustrates the concept of back-projection algorithms when the PA detection geometry is linear/planar. In this case, each detector element at a specific location records the PA data using the speed of sound (c) and time of flight (t), and then the recorded time-resolved signals are back-projected over a spherical surface of radius R = ct into the imaging volume [1].

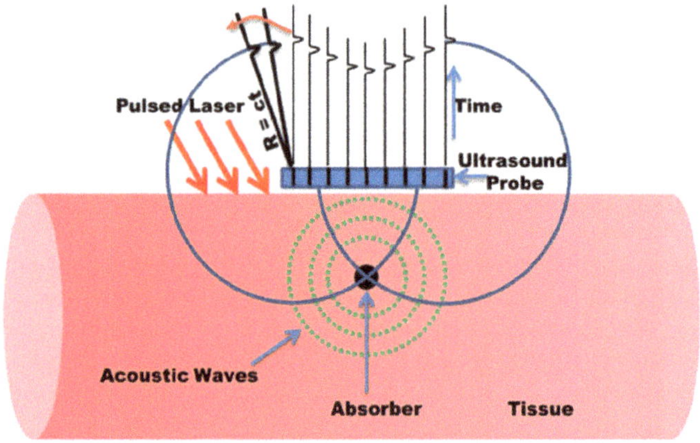

Figure 3. Back-projection PAI reconstruction for a planar detection geometry. Reproduced with permission from Handheld Probe-Based Dual Mode Ultrasound/Photoacoustics for Biomedical Imaging. In: Olivo M., Dinish U. (Eds) Frontiers in Biophotonics for Translational Medicine. Progress in Optical Science and Photonics, volume 3. Springer, Singapore (2016). Copyright 2016 Springer, Singapore.

This method is comparable to the conventional delay-and-sum beamformer, which is the most common algorithm employed in US imaging using phased arrays. It is important to note that a US-based delay-and-sum reconstruction algorithm can be used for PAT data by reducing the time-of-flight to half (PA data traverse half the time in tissue when compared to US signals). This offers the possibility to switch between these two image reconstruction modes at a high speed, which is a key requirement in handheld dual-mode PA/US imaging systems. However, the image quality offered by such an approach is not up to the mark. To solve this issue, there have been tremendous developments in the area of frequency domain algorithms, offering superior image quality with computing speeds suitable for systems suitable for real-time imaging [20,21].

3. LED-Based Photoacoustic Tomography

In recent years, there have been significant developments in the use of high-power LEDs in biomedical PAI, especially for superficial imaging applications [13,22]. In this section, we review the developments in this area after discussing briefly the characteristics of the LEDs and its advantages and disadvantages.

3.1. *High-Power LEDs Suitable for Photoacoustic Tomography*

An LED is a semiconductor device (P–N junction) that emits light when an electrical current passes through it. Free electrons are the majority charge carriers in N-type semiconductors. On the other hand, holes are the main charge carriers in P-type semiconductors. A P–N junction is formed when the N-type and P-type semiconductors are joined together. When a voltage is applied across the P–N junction diode (forward bias), holes (from P-substrate) and electrons (from N-substrate) move towards the junction, resulting in plenty of electrons and holes in the junction region. This fosters a strong electron–hole radiative recombination and consequent emission of photons. In case of LEDs, the radiative recombination process is dependent on spontaneous emission in which electrons in a high energy state (E_2) move down to a lower energy state (E_1) and combine with the available holes. The difference in energy between these two states ($E_g = E_2 - E_1$) results in spontaneous emission of photons in random directions [23].

3.2. *High-Power LEDs—Technical Aspects*

3.2.1. Emission Wavelength

When considering the biomedical imaging applications, a key advantage of LEDs over LDs is the availability of a wide range of wavelengths. As shown in Figure 4, LEDs are available even in the visible wavelengths in which the optical absorption of hemoglobin is quite high [23].

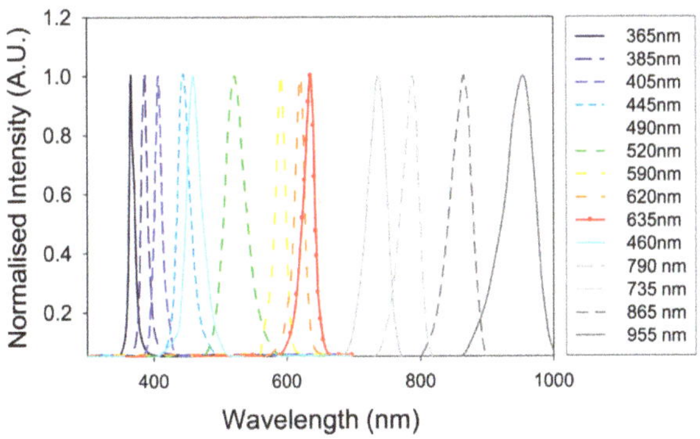

Figure 4. Optical spectra for a range of commercially available LEDs. Reproduced with permission from T. J. Allen and P. C. Beard, Biomedical Optics Express, Vol.7, Article ID1260, 2016; licensed under a Creative Commons Attribution (CC BY) license.

Materials used for designing the semiconductor (P–N junction) and its energy bandgap (E_g) are the factors that define the emission wavelength of the LEDs. Thus, one can tune the emission wavelength (visible to NIR range suitable for PAI) of LEDs by carefully selecting the semiconductor material [23]. For example, NIR wavelengths, the most suitable wavelengths for deep-tissue PAI can be developed by using aluminum gallium arsenide (AlxGa1-xAs). By changing the mole fraction (x) of the aluminum (Al) in AlxGa1-xAs, it is straightforward to tune the wavelength range from 624 nm to 920 nm. For obtaining wavelengths in the visible range (570–650 nm), aluminum gallium indium phosphide ((AlxGa1-x)0.5In0.5P) is the right candidate for the semiconductor material. In this case, by increasing the mole fraction (x) of the aluminum (Al), one can reduce the emission wavelength. For even shorter wavelengths (440–550 nm), it is recommended to use indium gallium nitride (InGaN) as the semiconductor material. An increase in indium (In) will result in shifts of emission from shorter to longer wavelengths. The materials used to design LEDs with different wavelengths are summarized in Table 2. With the availability in a wide range of wavelengths, LED illumination is very well suited for multispectral PAI [22].

Table 2. Summary of the materials used to design LEDs with different wavelengths.

Wavelength (nm)	440–550	570–650	624–920
Material	InGaN	AlGaInP	AlGaAs

3.2.2. Overdriving LEDs

LEDs are designed to be used for continuous wave (CW) operation and its rated current is valid in this mode. However, it is possible to operate LEDs in the pulsed mode by driving it with higher current pulses of lower duty cycles (less than <0.1%). This way, one can safely increase the optical output and use these affordable devices for PAI [22,24]. Even though LED manufacturers do not provide specifications for pulsed operation, recent studies have demonstrated that it is safe to drive LEDs using a higher current (ten times their rated current) at a duty cycle lower than 0.1% without any noticeable damage [24].

To operate LEDs in the pulsed mode, special electronic drivers are required. Pulsed LED drivers are most commonly composed of a capacitor, which is used as a storage element that discharges through the LED when a fast-switching device (metal oxide semiconductor field-effect transistors (MOSFETs) are used commonly) is activated. Figure 5 shows a schematic of a typical LED driver.

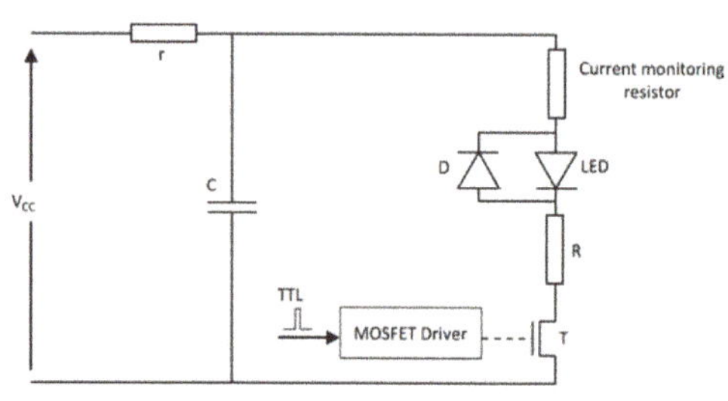

Figure 5. Schematic of a typical LED driver for the pulse operation mode. Vcc is the voltage provided by the power supply, T is the transistor used to switch the LED on and off, C is the storage capacitor, R is the limiting resistor, r is the charging resistor, D is a diode, and MOSFET is a metal oxide semiconductor field-effect transistor. Reprinted with permission from High-Power Light Emitting Diodes; An Alternative Excitation Source for Photoacoustic Tomography. In: Kuniyil Ajith Singh M. (eds) LED-Based Photoacoustic Imaging. Progress in Optical Science and Photonics, vol 7. Springer, Singapore. Copyright 2020 Springer, Singapore.

3.2.3. Optical Output Power

In its default continuous wave mode, typical high-power LEDs can provide optical output up to a few watts, which is dependent on the wavelength and also the size of the elements [24]. It is essential to use the LED elements within its allowed current rating (usually around 1 A). For avoiding heating issues, high power LED elements are usually mounted on heatsinks. However, it is critical to consider the amount of heat generated, especially when the goal is to integrate a light source and US probe in a single housing for handheld PA and US imaging. Photographs of two representative high-power LEDs are shown in Figure 6, where Figure 6a shows a device with a large emitting area (9 mm^2) and Figure 6b shows a multi-wavelength device composed of 4 LED elements, each with a smaller emitting area (1 mm^2) and at a different wavelength and mounted on a metal-core printed circuit (MCPC) board for efficient heat removal.

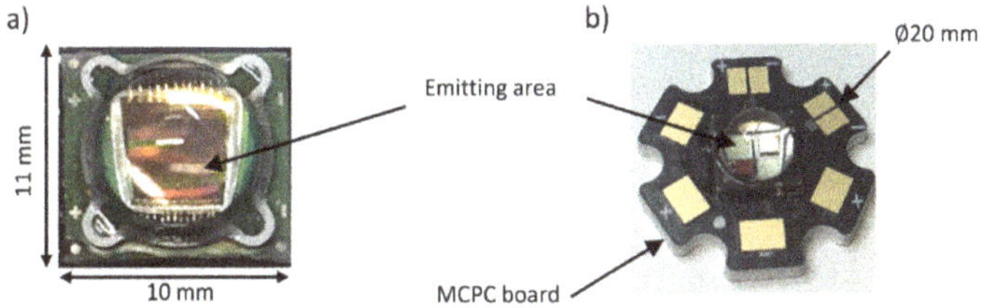

Figure 6. Photographs of high-power LEDs: (**a**) high-power LED (SST-90) with an emitting area of 9 mm^2; (**b**) high-power multi-wavelength LED (LZ4-00MC00, LedEngin, Inc., California, USA), composed of 4 LEDs emitting at 452, 520, 520, and 618 nm, each with a 1 mm^2 emitting area and mounted on a metal-core printed circuit (MCPC) board. These devices are encapsulated in spherical glass lenses. Reprinted with permission from High-Power Light Emitting Diodes; An Alternative Excitation Source for Photoacoustic Tomography. In: Kuniyil Ajith Singh M. (Eds) LED-Based Photoacoustic Imaging. Progress in Optical Science and Photonics, vol 7. Springer, Singapore. Copyright 2020 Springer, Singapore.

In recent years, the possibility of using multiple LED elements as an array have also been explored to generate a higher optical output required for PAT. Figure 7 shows a photograph of an LED array with 750 nm and 850 nm LED elements arranged in an interleaved manner (left panel, 36 elements per row, 4 rows in total) and also the integrated PA and US probe in which the LED arrays are fixed on a conventional linear array US probe (right panel). These high-power LED arrays are commercialized by CYBERDYNE INC (Tsukuba, Japan) for research use [25].

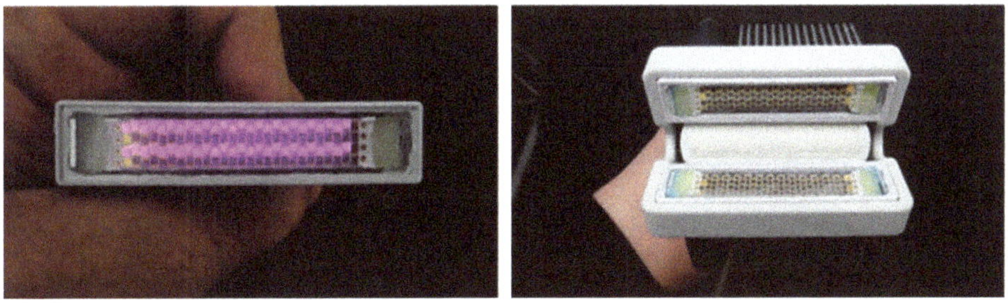

Figure 7. Photograph of an LED array developed by CYBERDYNE INC with four rows of LED elements, in which Rows 1 and 3 are 850 nm elements, and Rows 2 and 4 are 750 nm elements (**left**). In this picture, the 850 nm elements are activated and captured using an IR camera. Photograph of a LED-based PA/US probe developed by CYBERDYNE INC, in which two LED arrays (750/850 nm) are placed on both sides of a linear array US probe (7 MHz) (**right**).

3.2.4. Pulse Repetition Rate (PRR)

The typical PRR of conventional solid-state lasers are in the range of 10–200 Hz. For obtaining good signal-to-noise ratios (SNRs) in deep tissue PAI, it is usually required to perform signal averaging over multiple imaging frames, leading to low frame rates in a solid-state laser-based PAI system. On the other hand, PRRs of LEDs are in the KHz range and this will help to average more frames to improve SNR, without compromising the temporal resolution. Moreover, the large PRR makes high-speed PAT a possibility. LED arrays at 850 nm with a PRR of 16 KHz have been reported by Zhu et al. recently and they demonstrated dynamic PAI using commercially available LED arrays and conventional linear array US probes [26].

3.2.5. Pulse Width

The pulse width of LEDs are tens of nanoseconds (30–100 ns), whereas that of the solid-state lasers is usually less than ten nanoseconds. The temporal pulse width imposes a limit on the spatial resolution of the PAI systems [1]. For example, if an LED element with a 70 ns pulse width is used as the excitation source, the finest spatial resolution that is potentially achievable can be roughly estimated as 105 μm (= 70 ns × 1500 μm/μs). However, conventional US probes (5–7 MHz probes) have reception bandwidth limits and thus further limits its axial resolution to approximately 200–300 μm. One may generate broad bandwidth PA signals when the excitation pulse width is low (for example, 5 ns as in a solid-state laser), but detection sensitivity beyond the US bandwidth is usually very low and the spatial resolution is thus limited by the bandwidth of the US probe. It has been demonstrated that the pulse widths of LEDs are suitable for deep tissue imaging using conventional US probes [13]. A recent study also has shown that tuning the pulse width of the LEDs is feasible and this would be helpful if transducers with different center frequencies and bandwidths are used for the detection [27].

3.2.6. Spatial Divergence of LEDs

Conventional high-power LEDs have larger emission angles when compared to other sources. The spatial divergence of commercially available high-power LED arrays (CYBERDYNE INC, Tsukuba, Japan) is approximately +/−60° [28,29], which is acceptable in PAT where a relatively large illumination area is usually required. However, because of the large divergence and larger emission area (~1 mm^2), it is difficult to collimate the light and couple it to optical fibers for applications like minimally invasive PA imaging [9,30–32]. In terms of eye/skin safety, high spatial divergence in combination with a low optical output makes LEDs a safer alternative to solid-state lasers for non-invasive PAI [20].

3.3. *Technical Developments in LED-Based Photoacoustic Tomography*

The first report on the use of LEDs as an illumination source in PAI was from Jansen in 2011 [33]. In this work, a 627 nm LED element was used (Luxeon LXHL_PD09), which has been measured to yield approximately 250 mW of light output when supplied with 1 A DC current. Using a special electronic driver, the LED was supplied with 60-ns current pulses with a peak value of 40 A, resulting in a pulse energy of 400 nJ per pulse with a pulse width of 60 ns and a PRR of 200 Hz. Light focusing was performed to generate the fluence required for generating a PA response. To obtain PA signals from a non-realistic gelatin-based phantom, 50,000 A-lines were averaged. Owing to significant developments in the field of solid-state technology, there have been significant developments in LED technology (optical output, PRR, and pulse width), which resulted in the step-by-step development of LED-based PAI technology after 2011.

In 2013, Allen and Beard worked on this further and demonstrated that LEDs can be used as illumination sources in biomedical PA imaging [34]. In this work, they demonstrated the potential of using LEDs as an alternative excitation source in multispectral PA imaging. Recently, the same group achieved an imaging depth of 1.5 cm in tissue mimicking phantoms when using LEDs as an illumination source in PA imaging [24]. In this work, they obtained an LED output power of 10 μJ/pulse by overdriving 620 nm elements and averaging thousands of frames to improve the SNR. Figure 8a shows a schematic of the experimental setup used by them. Figure 8b shows a raw RF signal obtained and 8c shows a reconstructed image. With the same PRR of 200 Hz, as in the previous study, they achieved a frame rate that is 1000 times better, thanks to the higher optical output achieved using a MOSFET-based electronic driver.

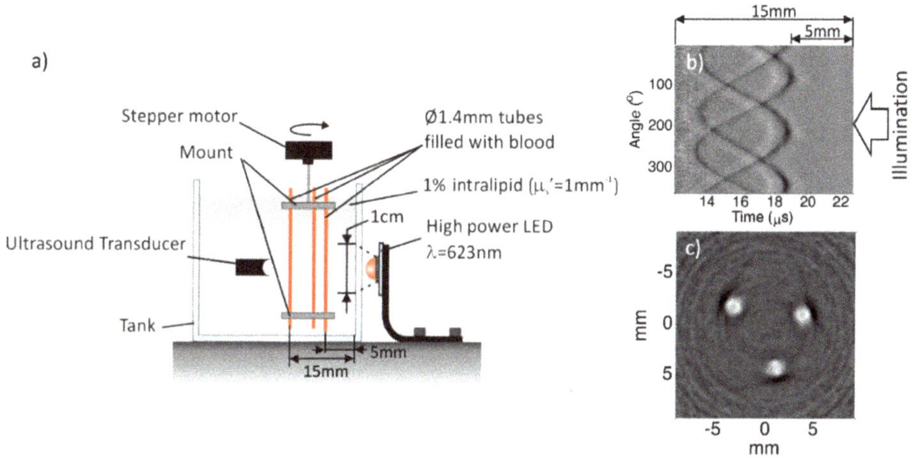

Figure 8. (**a**) Experimental setup; (**b**) RF PA signals of three 1.4 mm tubes filled with human blood immersed in water mixed with intralipid; (**c**) final reconstructed PA image. Optical output energy attained = 9 μJ. Number of frames averaged: 5000. Reproduced with permission from T. J. Allen and P. C. Beard, Biomedical Optics Express, Vol.7, Article ID1260, 2016; licensed under a Creative Commons Attribution (CC BY) license.

Although these pioneering works demonstrated the feasibility of using LEDs as illumination sources in PAI, no in vivo results have been reported. The main reason behind this is the optical energy of the LEDs, which is magnitudes lower than that of a solid-state laser or even some LDs. In 2015, Agano et al. demonstrated that it was feasible to combine hundreds of high-power LED elements in a rectangular package and could be pulsed simultaneously to perform biomedical PAI [35–37]. Using this novel technology, a high-power LED-array-based PA imaging system (AcousticX) was commercialized by a Japanese company (PreXion Corporation, later the technology was acquired by CYBERDYNE, INC, Japan).

With AcousticX, a single LED element provided an output energy of 0.024 μJ per pulse, with a pulse duration of 70 ns and 1 A DC current. By developing LED elements with a double stack structure, arranging them in an array, and applying 20 times the rated current, a light output of 200 μJ per pulse at a wavelength of 850 nm was achieved. Two of these arrays were kept on both sides of a linear array US probe for performing real-time PA and US imaging. The repetition rate of this first commercial LED-based PAI system was 4 KHz.

The system was thoroughly characterized by several research groups for its spatial resolution and imaging depth, and subsequently its capability for functional, structural, and molecular imaging was demonstrated [26,38–42]. Figure 9 shows a photograph of the system and the PA and US images of a human volunteer's wrist, which was acquired in a real-time handheld operation, demonstrating the capability of obtaining a complementary contrast (structural details from the US image and vasculature details from PA image) in a single measurement [18].

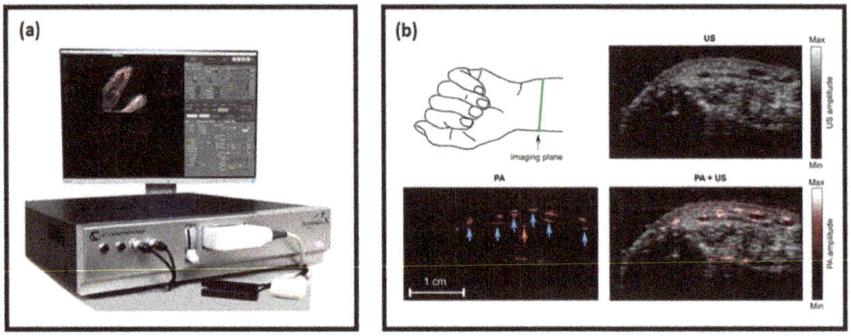

Figure 9. (**a**) Photograph of the LED-based PA and US imaging system—AcousticX; (**b**) US (gray colormap), PA (hot colormap), and US/PA overlay cross-section images of a human volunteer's wrist acquired at a frame rate of 10 Hz. Blood vessels are marked using blue arrows in the PA image. Adapted with permission from W. Xia, M. Kuniyil Ajith Singh, E. Maneas, N. Sato, and A. E. Desjardins, Sensors, Vol.18, Article ID1394, licensed under a Creative Commons Attribution (CC BY) license.

In another study [7], utilizing the high temporal resolution of LED-based PAI, it was also possible to scan through an area of interest to generate 3D maps of the vasculature on a human volunteer's foot dorsum, as shown in Figure 10. The total time required to generate such a high-resolution 3D PA image with a large field-of-view was less than 15 s, including scanning and image reconstruction and rendering [7].

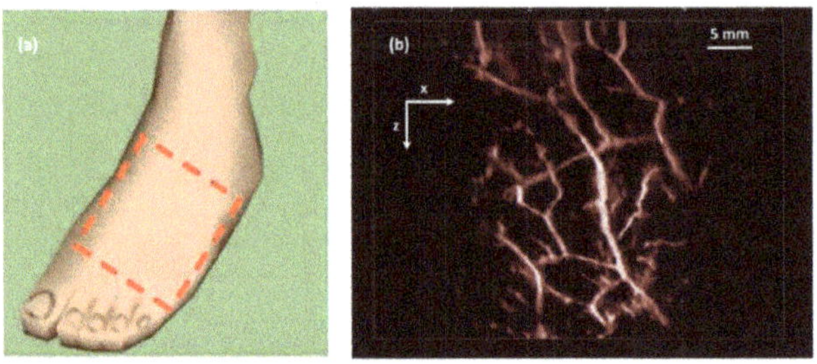

Figure 10. (**a**) Dashed box showing the imaging area on a human volunteer's foot dorsum; (**b**) 3D maximum intensity projection PA image of the area marked in (**a**), clearly visualizing the vasculature network. Adapted with permission from. Reprinted with permission from Clinical Translation of Photoacoustic Imaging—Opportunities and Challenges from an Industry Perspective. In: Kuniyil Ajith Singh M. (Eds) LED-Based Photoacoustic Imaging. Progress in Optical Science and Photonics, vol 7. Springer, Singapore. Copyright 2020 Springer, Singapore.

The system has been recently used extensively by several research groups in a number of studies to explore its potential preclinical and clinical applications [43–51], which are summarized in Table 3. All these reports clearly demonstrated the potential of LED-based PAI in structural, functional, and molecular imaging applications in a preclinical and a clinical setting. Even though the optical output power of the LEDs or LED arrays are not comparable to a solid-state laser, it is encouraging that LED-based PAI is able to consistently achieve an imaging depth of around 5–10 mm. The main reason for this is the high PRR (~16 KHz), which allows averaging over a large number of image frames while maintaining a good temporal resolution; one can easily achieve video rates even after averaging hundreds of frames. Considering the portability, affordability, safety, and ease-of-use, LED-based PAI holds strong potential in clinical translation of PA imaging as an additional modality along with conventional pulse-echo US imaging.

Table 3. Summary of the preclinical and clinical applications of LED-based PAI. Adapted with permission from Y. Zhu, T. Feng, Q. Cheng, X. Wang, S. Du, N. Sato, J. Yuan and M. Kuniyil Ajith Singh, Sensors, Vol.20, Article ID2484, 2020; licensed under a Creative Commons Attribution (CC BY) license. ICG, indocyanine green.

Target	Application		Depth (mm)	Contrast Agent	Wavelength (nm)
Medical needles, Vasculature	Guidance of minimally invasive procedures with peripheral tissue targets [18]	Phantom and ex vivo studies	38	N/A	850
Vasculature	Imaging of human placental vasculature [48]		7	N/A	850
Tumor	Imaging of intraocular tumors [26]		10	N/A	850
Vasculature	Non-invasive monitoring of angiogenesis [51]	Animal in vivo	10	N/A	850
Ulcer	Noninvasive imaging of pressure ulcers [47]		10	N/A	690
Oxygen saturation	Oxygen saturation imaging in rheumatoid arthritis [39]		5	N/A	750/850
Molecular	Detection and monitoring of reactive oxygen and nitrogen species [49]		10	CyBA	850
Tumor/Contrast agents	Imaging of tumor using contrast enhancement [44]		10	NC	850
Cells/Contrast agents	Imaging of molecular-labelled cells [38]		10	DiR	850
Vasculature	Imaging of peripheral microvasculature and function [26]	Healthy human	10	N/A	690/850
Vasculature	Simultaneous imaging of veins and lymphatic vessels [40]		10	ICG	940/820
Finger joints	Full view tomography of finger joints [28]		5	N/A	850
Finger joints	Imaging of inflammatory arthritis [42]	Patient	5	N/A	850
Skin	Imaging of port wine stain [43]		10	N/A	850

4. Laser Diode-Based Photoacoustic Tomography

Just like LEDs, LDs are also used as an excitation source in PAT, as an alternative to bulky and expensive lasers [52]. With high optical energy (when compared to LEDs) and compactness, pulsed LDs are suitable for performing deep-tissue PAT. In this section, after discussing some of the technical details, specifications, and development of LD-based PAT, we shortly introduce the applications that demonstrate the use of this technique. For the technical aspects, we will focus on the key differences when compared to LEDs.

4.1. High-Power Pulsed Laser Diodes

LD is a semiconductor laser device in which the laser beam is produced at the interface region in a P–I–N (positive–interface–negative) diode region. They are electrically pumped semiconductor laser sources. They convert the input electric energy into light energy, in comparison with conventional lasers in which the input light energy from a flash lamp is converted into laser output [14]. The energy efficiency of these bulky lasers is not high, and this results in an enormous amount of heat generation (water cooling, etc., is thus required, which increases the footprint of the device). On the other hand, the energy efficiency of an LD is very high and thus a minimal amount of heat will be generated.

In recent years, there have been multiple studies about the development and use of pulsed LDs in PAT [14,15,53].

4.2. Pulsed Laser Diodes—Technical Aspects

4.2.1. Emission Wavelengths

Conventionally, continuous wave laser diodes are available in a wide range of wavelengths (visible to NIR range). However, pulsed LDs are available only in the NIR range, most probably because of the lower optical energy generated when overdriven with the short pulses required in PAI. In the NIR range, pulsed LDs offer a far higher optical energy when compared to LEDs and thus are ideal sources of illumination in multispectral PAI of deep tissue.

Even though the fundamental process of light generation of LEDs and LDs are the same, LDs generate stimulated emissions and are considered as laser sources with a high spatial coherence. Just as in LEDs, semiconductor materials and their energy band gaps are the deciding factors for the emission wavelengths in LDs. Interestingly, there are some semiconductor compounds in which the bandgap energy can be altered by varying the details of the composition. For example, to achieve shorter wavelengths, one can increase the bandgap energy by increasing the aluminum content (increased x) in AlxGa1-xAs. Table 4 gives a summary of the semiconductor compounds used for developing LDs with different wavelengths. Even though the visible wavelength range is listed here, these are not commonly available in the pulsed mode.

Table 4. Summary of the materials used to design LDs with different wavelengths.

Wavelength (nm)	630–670	720–850	900–1100
Material	AlGaInP/GaAs	AlGaAs/GaAs	InGaAs/GaAs

4.2.2. Pulsed Laser Diode Drivers

Even though continuous wave modulations have also been explored to drive LDs, the most common way is to pulse modulate to generate the short light pulses required for PAI [15,54]. A typical LD driver circuit is shown in Figure 11 and the basic idea of driving is similar to that of an LED driver detailed in Section 3.2.2 (charging and discharging of a capacitor).

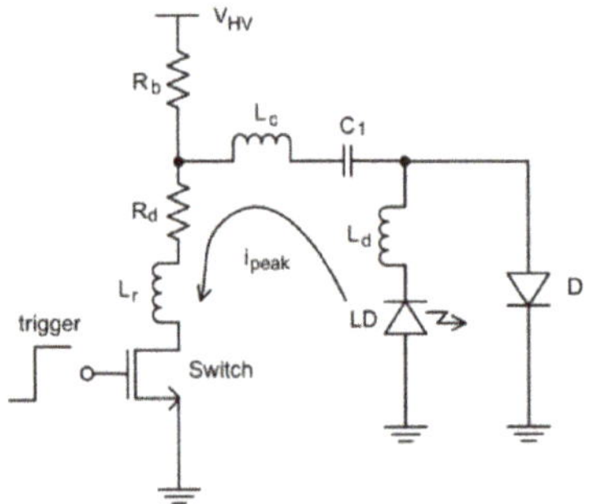

Figure 11. Typical laser driver circuit. Reproduced with permission from H. Zhong, T. Duan, H. Lan, M. Zhou, F. Gao, Sensors, Vol.18, Article ID2264, 2018; licensed under a Creative Commons Attribution (CC BY) license.

The pulse current flows from the LD to the switching device, as illustrated by the arrow in Figure 11. In order to attain the lower pulse width and boost the peak current, the parasitic inductance (L_r, L_c and L_d) should be as small as possible [15,55]. Usually the time constant for these circuits are fixed and tuning of the pulse width is not an option. It is worth mentioning that LEDs on the other hand has already proven to generate light pulses of different pulse widths (30–100 ns). To the best of our knowledge, the most efficient LD driver in the literature so far was reported in 2016, with a size of 40 × 50 mm^2, a PRR up to 10 kHz, and a pulse energy of 1.7 mJ per pulse with a pulse width of 40 ns [56]. In this work, Canal et al. reported the possibility of integrating a high power LD in its own driver electronic board to save space (Figure 12) and thus opening up the possibility to develop handheld PA probes.

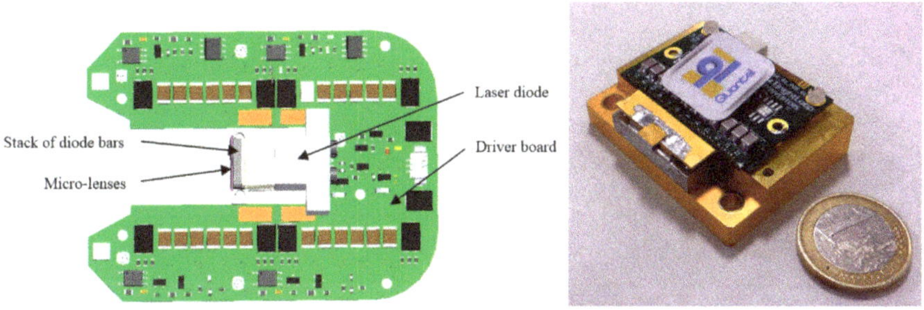

Figure 12. Schematic diagram of the diode laser source showing implementation of the laser diode within the driver board (**left**). An ultra-short pulse laser source integrating a mini-laser diode providing 1.7 mJ in pulses as short as 40 ns from Quantel, France (**right**). Adapted with permission from Proc. SPIE 9887, Biophotonics: Photonic Solutions for Better Health Care V, 98872B (2016). Copyright 2016 Society of Photo-Optical Instrumentation Engineers (SPIE).

4.2.3. Optical Output Power

Conventional pulsed LDs generate laser pulses of relatively low energies over a range of few hundreds of nanojoules to a few microjoules per pulse. However, similar to LEDs, with high PRR LDs (~10 KHz), one can average multiple frames and improve the SNR and imaging depth without much impact on the temporal resolution. Recent developments in semiconductor technology also had positive impacts on the field of LDs. Using diode-stacking technology and ultracompact drivers, a maximum output energy of 1.7 mJ per pulse was reported in 2016 [56]. Combined with a PRR of 10 KHz and a pulse width of 40 ns, these powerful LDs hold strong potential in deep-tissue high-resolution PAI. Figure 13 shows a schematic comparing diode bars with single and multiple active regions.

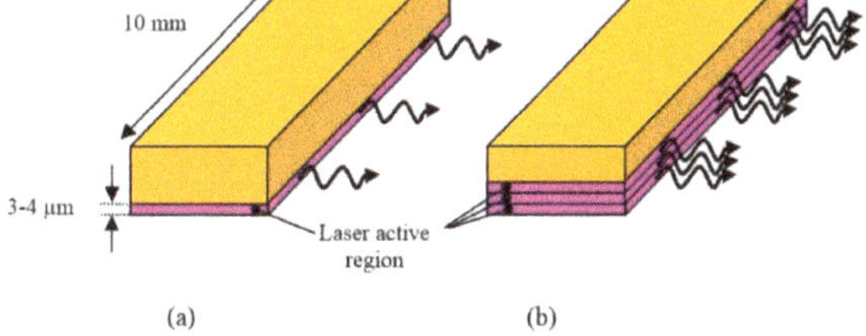

Figure 13. Schematic diagram of (**a**) a standard diode bar and (**b**) a multiple active region diode bar. Adapted with permission from Proc. SPIE 9887, Biophotonics: Photonic Solutions for Better Health Care V, 98872B (2016). Copyright 2016 Society of Photo-Optical Instrumentation Engineers (SPIE).

4.2.4. Pulse Repetition Rate

Typically, pulsed LDs can be driven at repetition rates of 1–10 KHz, which is far higher than the PRR of conventional lasers. To the best of our knowledge, the highest PRR reported for an LD is 10 KHz, which is high enough to be useful for compensating for a low output pulse energy [56]. In this regard, there is flexibility to average multiple frames, maintaining the real-time imaging capability just as in the case of LEDs. Since LDs are coherent light sources as lasers, it is also important to consider the maximum permissible exposure when using higher PRR to ensure light safety, especially in a clinical setting [53].

4.2.5. Pulse Width

Pulse widths of commercially available pulsed LDs are in the range of 30–200 ns [53]. This range is higher than that of a conventional solid-state laser, which is capable of delivering 5–10 ns pulses. However, as discussed before, if stress and thermal confinement can be met, such a low pulse width is not a necessity for deep-tissue PAT. For example, when using 100 ns LDs as an illumination source, the maximum PA signal frequency expected will be around 10 MHz, which is well within the detection bandwidths of US transducers suitable for imaging deep-tissue structures. However, for high-resolution imaging of shallow structures using high-frequency US probes, the pulse widths of the LDs may not be sufficient; also, it is important to mention that the pulse energy of the LD-generated light pulses will drop when the pulse width is reduced. Pulse width-tunable LDs are yet to be reported, which may be because of the technical difficulties in the driver circuit.

4.2.6. Spatial Divergence of LD

The spatial divergence of LDs is larger than that of solid-state lasers but smaller than that of LEDs in general, with angles of up to 40 degrees in the axis perpendicular to the diode arrays ("fast axis") and 10 degrees in the parallel axis ("slow axis") [57]. It has been shown that one can efficiently reshape the divergent beam from LDs to be used in an integrated PA and US probe [57]. However, tight optical focusing with these high-power LDs is not feasible due to their multimode nature, which limits their application in optical-resolution PA microscopy.

4.3. Technical Developments in LD-Based Photoacoustic Tomography

To the best of our knowledge, the first report on using LDs as a PA illumination source was from Allen and Beard back in 2005 [58]. In this work, they used a pulsed LD (P GAF5S24 from EG&G; wavelength: 905 nm; and pulse duration: 200 ns) driven by a commercially available electronic driver (PCO−7120 from DEI). In this proof-of-concept work, they used a non-realistic phantom (ink-filled cell) and the light was delivered to it using a multimode fiber coupled to the diode. Apart from showing the feasibility of generating PA signals using an LD, they also studied the impact of pulse duration in PAI. In another work immediately after this, Kolkman et al. demonstrated that it was feasible to generate PA signals from superficial blood vessels in a human volunteer [59]. They used a commercially available pulsed laser diode module (iRLS, Laser Components GmbH, Germany) and compared its PAI efficiency with a conventional solid-state laser. The LD used was of 905 nm wavelength with a pulse duration of 112 ns and a PRR of 5 kHz, and offered a maximum optical energy of 23 μJ per pulse. Acoustic detection was performed using a double-ring sensor that can generate 1D depth images (A-scans), which then can be used to generate 2D images by rendering multiple A-lines together. Using this setup, for the first time they demonstrated that LDs could image blood vessels in vivo. However, in this encouraging study, the time for one scan (above 3 min) and the imaging depth (1 mm) was not optimal.

In late 2016, Allen and Beard improved their 905 nm LD-based PAT system and showed that it was feasible to image a tissue-mimicking blood vessel phantom effectively [60]. This work confirmed that LDs can very well be an alternative to solid-state lasers for superficial PA imaging applications. In this system, the maximum PRR was 5 kHz and the pulse width was tunable from 50–500 ns. Pulse width settings of 65 ns (pulse energy: 24 μJ) and 500 ns (pulse energy: 184 μJ) were used to demonstrate how pulse width affects PA image quality because of stress confinement requirements. A 3.5 MHz single-element PZT focused transducer was used for acoustic detection and 5000 PA frames were averaged to generate one image. Even though slightly slow because of limitations in electronics and data transfer, this work was a solid demonstration of LD-based PAT. Key issues to resolve for using LD-based PAT for in vivo applications were pulse energy and time of acquisition and processing.

After these pioneering works on the use of LDs in PAI, there have been significant developments in the field of combining PAI with conventional pulse-echo US imaging. In this regard, by 2014, Daoudi et al. reported an integrated US and PA imaging probe with a high-power LD (Quantel, France) and all optical components integrated inside one housing along with a commercially available 7 MHz linear array US probe (ESAOTE Europe) [57]. The wavelength of the LD was 805 nm, which emitted 130 ns pulses with an optical energy of 0.56 mJ per pulse. The PRR was 10 kHz, opening up the possibility to average multiple frames to improve the SNR without sacrificing the frame rate. Figure 14 shows a photograph of this dual-modality PA and US system. As one can see, the probe is quite portable and well suited for a real-time handheld operation in a clinical setting.

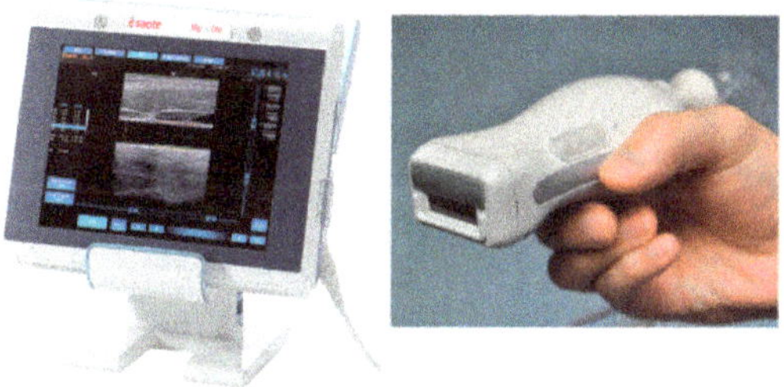

Figure 14. Portable imaging scanner combining photoacoustics and ultrasound. Left is the ultrasound scanner system and right is the picture of the probe integrating the laser module and ultrasound transducer array. Reprinted with permission from K. Daoudi, P.J. van den Berg, O. Rabot, A. Kohl, S. Tisserand, P. Brands, and W. Steenbergen, Optics Express, Vol.22, pp. 26365–26374, 2014; licensed under OSA's "Copyright Transfer and Open Access Publishing Agreement" (OAPA). Copyright 2014 Optical Society of America.

A key feature of this probe design is that all optical components, including the deflecting prism, diffractive optical elements, diode stack, micro-cylindrical lenses, and aluminum cooling rim, were packaged inside a single casing along with a 128-element US probe (Figure 15a). Along with reporting the design of the probe, they performed a detailed characterization of the system and demonstrated the feasibility of in vivo real-time imaging on a human finger. An imaging depth of 4 mm was achieved in the human finger measurement at a frame rate of 20 Hz (Figure 15b,c). Higher imaging depths (10–15 mm) were achieved in phantom studies, but at a higher PRF, which cannot be used in human experiments because of laser safety regulations. Radiant exposure of 1.3 mJ/cm^2 on the skin with an illumination spot size of 18.2 × 2.3 mm^2 was achieved in this prototype, resulting in the possibility of imaging micro-vasculature with unprecedented contrast and resolution. The same system was later used for multiple preclinical applications and early clinical pilot studies [61–66]. Adding multiple diode stacks with different wavelengths can make this system a power tool with excellent structural, functional, and molecular imaging capability.

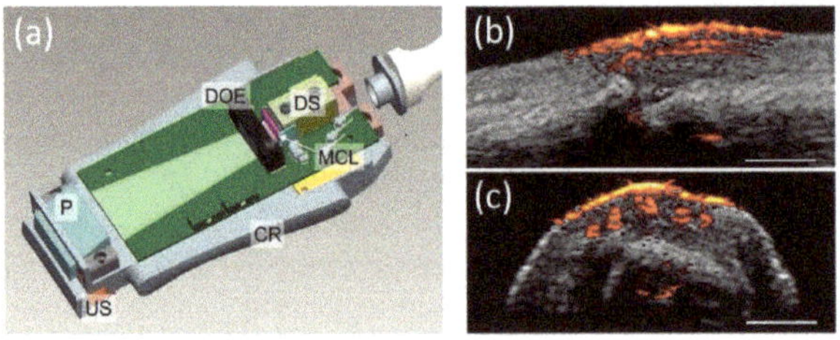

Figure 15. (**a**) A schematic of the handheld PA and US probe. US: ultrasound array transducer; P: deflecting prism; DOE: diffractive optical elements; DS: diode stack; MCL: micro-cylindrical lenses; CR: aluminum cooling rim. Photoacoustic/ultrasound images of a human proximal interphalangeal joint in the (**b**) sagittal and (**c**) transverse planes. Adapted with permission from K. Daoudi, P.J. van den Berg, O. Rabot, A. Kohl, S. Tisserand, P. Brands, and W. Steenbergen, Optics Express, Vol.22, pp. 26365–26374, 2014; licensed under OSA's "Copyright Transfer and Open Access Publishing Agreement" (OAPA). Copyright 2014 Optical Society of America.

In late 2015, Upputuri et al. demonstrated the possibility of PAT using LD excitation and scanning of a single-element US transducer [67]. Using a powerful LD from Quantel (wavelength: 803 nm; pulse energy: 1.4 mJ; and PRR: 7 KHz) and a conventional single-element US transducer rotating around the object, they achieved to obtain an image every 3 s and they also demonstrated an imaging depth of 2 cm in phantom studies, which is commendable. They also compared the results with an OPO-based laser system in similar settings. This system was upgraded by them with multiple single-element transducers and used in multiple preclinical applications recently [68–70]. Figure 16 shows a schematic of the PLD-PAT system designed for in vivo small animal brain imaging and Figure 17 shows images of brain vasculature in a 95 g female rat acquired non-invasively with PAT at different scan times.

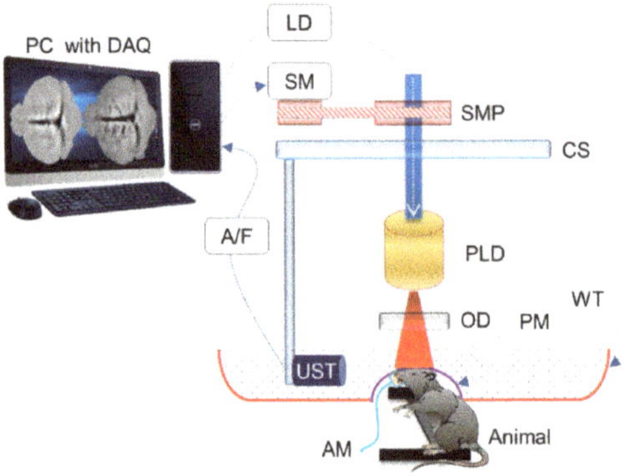

Figure 16. Schematic of the PLD-PAT system for in vivo small animal brain imaging: PLD, pulsed laser diode; OD, optical diffuser; CS, circular scanning plate; SMP, stepper motor pulley unit; UST, ultrasound transducer; A/F, amplifier/filter unit; LD, laser driver unit; SM, stepper motor; PC, personal computer; WT, water tank; DAQ, data acquisition card; AM, anesthesia machine; PM, transparent polythene membrane. Reproduced with permission from P. K. Upputuri and M. Pramanik, Journal of Biomedical Optics, Vol.22, Article ID090501, 2017; licensed under a Creative Commons Attribution (CC BY) license.

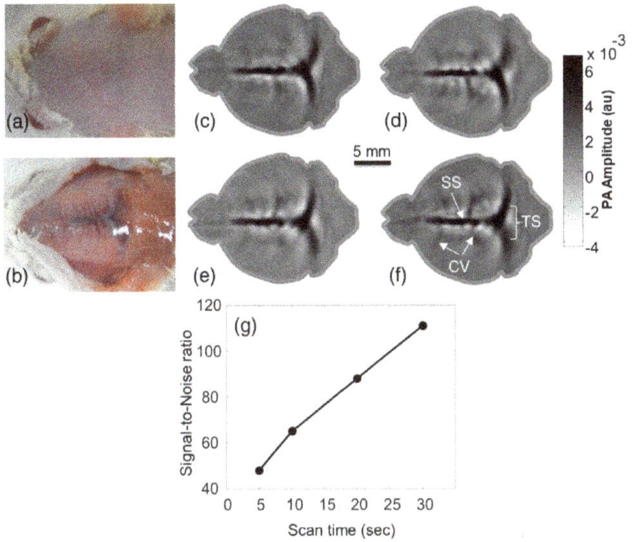

Figure 17. Images of brain vasculature in a 95 g female rat acquired non-invasively with a laser diode-based photoacoustic tomography system at different scan times: photograph of rat brain before (**a**) and after (**b**) removing the scalp. In vivo brain images at a (**c**) 5-s, (**d**) 10-s, (**e**) 20-s, and (**f**) 30-s scan time. (**g**) SNR of the in vivo images as a function of scan time. SS, sagittal sinus; TS, transverse sinus; CV, cerebral veins. Adapted with permission from P. K. Upputuri and M. Pramanik, Journal of Biomedical Optics, Vol.22, Article ID090501, 2017; licensed under a Creative Commons Attribution (CC BY) license.

Out of these PLD-PAT works, a recent study from Rajendran et al. [69] is commendable. In this work, the authors used their LD-PAT setup for detecting changes in the sagittal sinus due to intra-cranial hypotension in a rat model. A key advantage of this LD-PAT setup is the cost reduction because of (1) not using US probes with multiple elements; (2) not using multichannel DAQ systems; and (3) replacement of bulky and expensive lasers with LDs. In another recent work from the same group, Upputuri et al. [71] demonstrated a pulsed LD-based PA temperature sensing system for monitoring tissue temperature in real time. The system takes advantages of a laser diode with a high repetition rate (7000 Hz), a near-infrared wavelength (803 nm), and a relatively high energy (1.42 mJ/pulse). Results gave a direct confirmation that this LD-based PAI system is capable of providing local temperature information at a high temporal resolution of 1 ms and high sensitivity of 0.31 °C.

Even though no LD-based PAT system is available commercially, several research studies (Table 5) have validated the potential of LD-based PAT in multiple preclinical and clinical applications, in which encouraging imaging depths were achieved despite low pulse energies when compared with lasers. Pulsed LDs are powerful, portable, and cost-effective, and it is believed that LD-based PAT could be a complementary modality to conventional clinical US imaging [72].

Table 5. Summary of the preclinical and clinical applications of LD-based PAI. ICG, indocyanine green.

Target	Application		Depth (mm)	Contrast Agent	Wavelength (nm)
Vasculature	Detection of intraplaque hemorrhage in carotid artery [62]	Phantom and ex vivo studies	20	N/A	808
Vasculature	Dynamic imaging studies (for example in cardiovascular medicine) [73]		20	N/A	803
Red blood cells	Non-invasive blood flow imaging [66]		7	N/A	805
Vasculature	Detection of liver fibrosis [64]	Animal in vivo	5	N/A	808
Cortical vasculature	Brain imaging [68]		5	ICG	803
Cerebro-spinal fluid volume level	Detection of venous sinus distension by measuring intra-cranial hypertension [69]		5	N/A	803
Vasculature and perfusion	Vascular/dermal pathologies [63]	Healthy human	5	N/A	805
Vasculature	Detection of intraplaque hemorrhage in carotid artery [65]		15	N/A	808
Finger joints	Imaging of rheumatoid arthritis [57]		5	N/A	808
Finger joints	Imaging of rheumatoid arthritis [61]	Patient	5	N/A	808

5. Discussion

In this review, we focused on the basics of PAI and the use of affordable light sources (LEDs and LDs) as illumination sources in PAT. After a short introduction to the physical phenomenon behind biomedical PAI, we elaborated on the key specifications and technological developments in the area of high-power LEDs and LDs and detailed their use as illumination sources in PAT. It is encouraging that even with low optical energies, LED and LD-based PAT systems have already demonstrated their potential in a wide range of functional (oxygen saturation imaging, blood flow imaging, etc.) and molecular imaging applications (tracking contrast agents, pharmacokinetic studies, etc.), thanks to the high PRR (maximum reported PRR for LED: 16 KHz; LD: 10 KHz) and possibility to average over a large number of frames to improve the SNR [13–15]. With these affordable light source-based PAT systems, several in vivo preclinical and clinical pilot studies have reported imaging depths of above 8–10 mm, at frame rates unachievable for conventional laser-based systems [13–15]. Apart from the technical aspects in developing an LED and LD-based PAT system, we also shortly introduced the wide variety of preclinical applications and clinical pilot studies reported using these. Divergence of the LEDs and their large beam size when using an array of LDs make these semiconductors not the optimum choice for PA microscopy applications where a tight light focus is an important requirement [23]. However, there are some reports on the use of LEDs and LDs for microscopic and shallow-depth imaging applications, too [74–79]. We did not include these details in this review as our main focus was on tomographic setups and applications.

PAI is already matured in the research setting and has demonstrated its unprecedented potential in a wide range of biomedical imaging applications. However, the clinical translation of this promising technology is not happening at the expected pace. One of the important reasons for this is the requirement of bulky, slow, and expensive pulsed lasers. In the last decade, there have been significant developments in the semiconductor device technology and use of portable and affordable devices like LEDs and LDs became popular as pulsed light sources in PAT. Q-switched solid-state lasers with water-cooling are indeed not an optimal choice for use in a clinical setting because of their sizes, power consumption, costs, and also skin/eye safety aspects (users and patients must wear eye safety glasses and systems must be installed in a laser-safe room). The PRRs of these lasers are also slow and real-time imaging can be challenging, especially when frame averaging is a requirement for improving SNR. Considering the portability, affordability, and dynamic imaging capability aspects, both LEDs and LDs hold strong potential in replacing solid-state lasers, especially for superficial and sub-surface imaging applications, such as in rheumatology, dermatology, and cardiovascular medicine. Handheld US imaging is one of the most popular medical imaging modalities and it is seamless to implement PAT in conventional clinical pulse-echo US equipment. In such handheld systems, integration of LEDs and LDs in a single housing along with a US probe will be a straightforward development and an important aspect to consider in these integrated probes is the mechanism for absorbing the heat generated because of high-speed pulsing. In terms of eye and skin safety, LDs with its coherent nature is similar to that of a conventional laser and the requirements of eye-safety goggles and laser-safe rooms remain important. Even though the fundamental light generation processes for LEDs and LDs are similar, LEDs do not generate stimulated emissions and they possess a broad optical bandwidth and their spatial coherence is low. Considering these factors, LEDs are not practically considered as laser sources and thus could be potentially used as an eye/skin-safe light sources for PAI given the low optical fluence, and the system could be installed in non-laser safe rooms, too, making it an ideal choice for using in a resource-limited clinical setting [13].

Even though LEDs and LDs have shown their potential in biomedical PAT, the pulse energy of these diodes is still a bottleneck when compared to a solid-state laser, and this limits their use only to superficial imaging applications. The maximum pulse energy reported by an LED source (when used in an array form) is 200 μJ at an 850 nm wavelength. When two such arrays are used on both sides of an US probe, the total light output will be 400 μJ per pulse. In such a setting, the maximum imaging depth achieved in vivo was reported to be around 10 mm, with a combined US and PA frame rate above 10 Hz. On the other hand, LDs are more powerful and a pulse energy above 2 mJ is achievable. However, in an animal/human in vivo situation, so far no LD-based PAI studies has reported an imaging depth above 5–6 mm. An exception is the study from Jaeger et al., in which a carotid artery at a depth of 15 mm was visualized using an integrated probe with an 808 nm LD and a 7 MHz linear array probe. This was achieved by applying an algorithm called DCA (deformation compensated averaging) to reduce the clutter and thereby improving the SNR. It is not straightforward to use a clinical US system for PA detection as highly sensitive and broadband detection with high amplification is very important to achieve high imaging depths. In a nutshell, the maximum imaging depth currently achievable using LED and LED-based PAI systems is less than 2 cm, making it difficult to target deep-tissue imaging applications. Imaging depth in PAI is dependent on several factors, including illumination (type of illumination source, optical output power, wavelength, illumination area, etc.), acoustic detection strategies (sensitivity, directionality, spatial distributions of the US detectors, and noise performance of DAQ, etc.), and image processing and reconstruction algorithms. The conventional solid-state laser-based PAT has already achieved an excellent imaging depth above 6 cm in phantom studies and 3–4 cm in in vivo clinical pilot studies [80]. It is worth mentioning that a higher imaging depth (4 cm) was achieved using curved US probes in a computed tomographic setup [81,82]. Using clinical linear array US probes, to the best of our knowledge, an imaging depth above 2 cm has not been demonstrated in vivo even when using bulky and powerful solid-state lasers as illumination sources. We believe that further advancements in semiconductor device technology and lighting industry in general will improve the pulse energy offered by LDs and LEDs in the near future, thus making these devices well-suited for PAT applications. Theoretically, it is feasible to improve the SNR and thereby imaging depth by increasing the PRR and thus offering more room to increase frame averaging. However, heat generation (especially in fully integrated US/PA probes) and also laser safety will be serious concerns when the PRR is extremely high. It is also well known that when the N frames are averaged, the SNR will be improved only by $\sqrt{N}$. Several US-received side strategies and AI-based methods have also been reported for improving the image quality [65,83–91]. We foresee that developments in the area of semiconductor devices and low noise electronics, combined with advanced image reconstruction and enhancement algorithms, will accelerate the clinical translation of affordable light sources based PAI. Since SNR is a key problem to solve, use of novel coded excitation schemes [24,92] and implementation of clutter and reflection artifact reduction algorithms [65,93] also will have a significant impact in clinical translation of LED and LD-based PAT. When compared to OPO-based lasers, one of the key disadvantages of LEDs and LDs is the difficulty in obtaining multiple illumination wavelengths for multispectral PAI [13]. When multiple diode elements with different wavelengths are embedded in an arrayed form, the pulse energy will be further reduced, which then has a serious impact on the imaging depth. Dual-wavelength LED-based PAT for applications like oxygen saturation imaging and differentiation of veins and lymphatic vessels are already reported [25,40]. However, in a complex situation with more than two optical absorbers in the tissue, OPO-based laser systems are still preferred because of the wide range of wavelengths available and fast tuning capability.

The pulse duration of the excitation light source is one of the key factors that has an impact on PA signal generation and image quality. It is well known that a short pulse width is important for satisfying the stress confinement criteria and generating broadband PA signals suitable for high-resolution images. Typical pulsed lasers offer a pulse width in the range of 3–10 ns, which then can generate broadband PA signals from the optically absorbing objects. For example, a 3.5 ns light pulse can generate PA signals with frequencies up to 300 MHz, depending on the size of the targets. Such high frequencies from deep tissue will be heavily attenuated and will not reach the US probe. If one has to detect such signals from superficial tissue, it is important to use high frequency US probes with a super-high reception bandwidth. Even though this is feasible technically, applications will be limited to PA microscopy. When using a conventional mid-frequency US probe suitable for deep tissue imaging, the US detection efficiency with frequencies above 12–14 MHz is extremely low [13]. Considering these factors, we believe that such a low pulse width is not an absolute necessity, provided that the stress confinement is met. As discussed before, a very low pulse widths will be only interesting for PA microscopy imaging applications with imaging depth less than 1–2 mm.

In this review, we mainly focused on LEDs and LDs as these are the most common and well-explored affordable PAT light source, especially in the NIR wavelength range, which is the most suitable range of light wavelengths for deep tissue imaging applications [14,15,53]. However, in recent years there have been several other reports too on affordable illumination strategies in PAT. In 2016, Wong et al. reported on the use of a single Xenon flash lamp as a light source in PAT [94]. In this promising work, the PRR of the Xenon lamp could be controlled between 10–100 Hz and 3 mJ of energy was carried in pulses with a 1 µs pulse width. Using a 0.5 MHz ring-shape detector, the authors demonstrated the feasibility of the imaging tissue, mimicking phantoms (imaging depth: 3.5 cm) and a whole mouse body in vivo. Recently, there have been promising reports on the use of diode-pumped solid-state lasers (DPSSLs) as a light source in PAT. PDSSLs are portable and energy efficient, and thus a promising light source for PAI in a resource-limited setting. Wang et al. used a compact high-power DPSSL (Montfort Laser GmbH Inc., Germany) for deep tissue single wavelength in vivo PAT imaging [95]. This laser source has a miniature size of 13.2 × 14.0 × 6.5 cm^3, a weight of 1.6 kg, and an average power output of 4 W, with a high pulse energy of up to 80 mJ at the wavelength 1064 nm with a PRR up to 50 Hz. Using this DPSSL-based system, the authors successfully imaged murine whole-body vascular structures and cardiac functions in vivo, and mapped the arm, palm, and breast vasculatures of living human subjects. One of the key advantages of these DPSSLs are the possibility of fiber coupling and consequent focusing, which is very much important for PA microscopy imaging applications (this is not possible using LED and LEDs). Recently, Jeng et al. reported on the use of a compact 700–900 nm tunable DPSSL with a pulse energy of around 1 mJ over the range of wavelengths and a PRR of 1 KHz [96]. In this work, the authors demonstrated automatic laser-fluence compensation in spectroscopic PAT and inter-wavelength motion correction using US speckle tracking, which has never been shown before in real-time systems. The 50-Hz video rate PAUS system was demonstrated in vivo using a murine model of drug delivery monitoring. As an energy-efficient and portable illumination source, pulsed fiber lasers were also explored as a PAT illumination source. Using a fiber laser emitting at 1060 nm, with a maximum pulse energy of 0.8 mJ and a Fabry Perot ultrasound scanner, Allen et al. demonstrated the potential of 3D PAT with a realistic blood vessel phantom and the palm vasculature of a human volunteer [97]. Frequency domain PAI systems using continuous-wave modulated laser diodes and LEDs have also been developed recently for several biomedical applications [98,99].

Since short-pulsed light can have a serious impact on the eyes/skin, it is very important to keep the optical pulse energy within the maximum permissible exposure (MPE) limit by following, e.g., the American National Standards Institute (ANSI). Even though the LEDs used in PAI generate pulsed light, they are not considered as laser sources and at present there are no safety standards defined in this regard, to the best of our knowledge. Light from LEDs is optically broad and non-coherent, making it considerably safer to use in a clinical setting as compared to lasers in general (divergence of LEDs is also high, and no safety issues are expected unless the light source is kept very close to the eyes). As a high PRR pulsed laser source, it is very important to consider the MPE limits of the LDs. One may get an initial impression that portable devices like LDs with a low optical power will be completely safe for the eyes and skin. Unfortunately, this is not the case because the light density on the skin will be high when the repetition rate is very high (a large number of light pulses over a period of time). ANSI has clear definitions for this and the safety limits for the skin depend on the light wavelength, pulse width, duration of the exposure, and illumination area [51]. For example, let us consider the case of 800-nm wavelength (λ), the MPE limit for the skin over an exposure time (t) of 0.5 s (assumption) is given by $1.1 \times 10^{(2(\lambda-700)/1000)} \times t^{0.25}$ J/cm^2 = ~1.47 J/cm^2. For an LD with a PRR of 2 KHz, the MPE per pulse becomes ~0.735 mJ/cm^2 (1.47/2000). If the LD can deliver optical energy of 1 mJ per pulse, the light beam must be expanded in such a way that the total area of illumination is about 1.36 cm^2 (1/0.735). From this, it is clear that for increasing the PRR for the same exposure time of 0.5 s, the MPE per pulse must be lower than 0.735 mJ/cm^2. In a nutshell, the total pulse energy density will be lower when one increases the PRR, subsequently reducing the possibility of averaging and thus SNR reduction. In clinical PAI, it is of paramount importance to carefully select the pulse energy, PRR, illumination area, and time of exposure for keeping up with the MPE safety limits [52].

6. Conclusions

Compact, fast, affordable, and energy-efficient light sources are important requirements for accelerating the clinical translation of PAI. LEDs and LDs fit very well into this category and have been explored extensively in recent years as an illumination source in PAI. This review, which is focused on affordable light source-based PAI, is timely considering the fact that this promising technology is facing an exciting transition from bench to bedside.

In this paper, we first covered the basic theory of PAI, clearly conveying the key advantages of the technique, including optical contrast and acoustic resolution/imaging depth. In the first section, we also shortly introduced the different light sources used in PAI with its specifications, including pulse energy, PRR, pulse width, and cost. After this, the principles of light generation and basic performance characteristics of both LEDs and LDs were discussed with a focus on the advantages and disadvantages when they are used as a light source in PAI. The historical developments (from single-point measurements to in vivo imaging and to commercialization) of both LED- and LD-based PAI systems were also discussed, listing out the key preclinical and clinical application demonstrations.

Even though LEDs and LDs possess some disadvantages, such as low optical energy, lack of spectral tuning capability, and long pulse widths, they are portable, affordable, and energy-efficient light sources. Apart from point-of-care biomedical imaging, LEDs and LDs would be an ideal choice for wearable PA equipment and may find a plethora of applications in the field of therapeutic drug monitoring. Integration of LED- or LD-based PAI to a clinical US scanner will have an easier clinical acceptance when compared to laser-based PAI. We expect that dual-mode US and PA equipment utilizing affordable light sources will have a significant impact on bedside diagnostic imaging, accelerating the translation of this technology from research labs to clinics.

Author Contributions: Conceptualization, M.K.A.S. and W.X.; writing—original draft preparation, M.K.A.S.; writing—review and editing, W.X.; supervision, W.X.; project administration, W.X.; funding acquisition, W.X. All authors have read and agreed to the published version of the manuscript.

Funding: This work was supported by the Wellcome Trust [203148/Z/16/Z, WT101957] and the Engineering and Physical Sciences Research Council (EPSRC) [NS/A000027/1, NS/A000049/1].

Conflicts of Interest: Mithun Kuniyil Ajith Singh is employed by CYBERDYNE, INC. The authors have no other financial interests or conflict of interest to disclose. We clarify that in some of the reported studies in this review article, CYBERDYNE, INC as a company was involved in technical developments or design of experiments either directly or through industry-academic collaborative research projects.

References

1. Beard, P. Biomedical photoacoustic imaging. *Interface Focus* **2011**, *1*, 602–631. [CrossRef] [PubMed]
2. Manohar, S.; Razansky, D. Photoacoustics: A historical review. *Adv. Opt. Photonics* **2016**, *8*, 586–617. [CrossRef]
3. Wang, L.V. Multiscale photoacoustic microscopy and computed tomography. *Nat. Photonics* **2009**, *3*, 503–509. [CrossRef]
4. Omar, M.; Aguirre, J.; Ntziachristos, V. Optoacoustic mesoscopy for biomedicine. *Nat. Biomed. Eng.* **2019**, *3*, 354–370. [CrossRef]
5. Wang, L.V.; Yao, L.V.W.J. A practical guide to photoacoustic tomography in the life sciences. *Nat. Methods* **2016**, *13*, 627–638. [CrossRef]
6. Singh, M.K.A.; Steenbergen, W.; Manohar, S. Handheld Probe-Based Dual Mode Ultrasound/Photoacoustics for Biomedical Imaging. In *Frontiers in Biophotonics for Translational Medicine*; Progress in Optical Science and Photonics; Springer: Singapore, 2015; Volume 3, pp. 209–247. [CrossRef]
7. Singh, M.K.A.; Sato, N.; Ichihashi, F.; Sankai, Y. Clinical Translation of Photoacoustic Imaging—Opportunities and Challenges from an Industry Perspective. In *LED-Based Photoacoustic Imaging*; Progress in Optical Science and Photonics; Springer: Singapore, 2020; Volume 7, pp. 379–393. [CrossRef]
8. Szabo, T.L. *Diagnostic Ultrasound Imaging: Inside Out*; Elsevier BV: Amsterdam, The Netherlands, 2014.
9. Zhao, T.; Desjardins, A.E.; Ourselin, S.; Vercauteren, T.; Xia, W. Minimally invasive photoacoustic imaging: Current status and future perspectives. *Photoacoustics* **2019**, *16*, 100146. [CrossRef]
10. Bell, A.G. On the production and reproduction of sound by light. *Am. J. Sci.* **1880**, *118*, 305–324. [CrossRef]
11. Lutzweiler, C.; Razansky, D. Optoacoustic Imaging and Tomography: Reconstruction Approaches and Outstanding Challenges in Image Performance and Quantification. *Sensors* **2013**, *13*, 7345–7384. [CrossRef]
12. Tian, C.; Xie, Z.; Fabiilli, M.L.; Wang, X. Imaging and sensing based on dual-pulse nonlinear photoacoustic contrast: A preliminary study on fatty liver. *Opt. Lett.* **2015**, *40*, 2253–2256. [CrossRef]
13. Zhu, Y.; Feng, T.; Cheng, Q.; Wang, X.; Du, S.; Sato, N.; Yuan, J.; Singh, M.K.A. Towards Clinical Translation of LED-Based Photoacoustic Imaging: A Review. *Sensors* **2020**, *20*, 2484. [CrossRef]
14. Erfanzadeh, M.; Zhu, Q. Photoacoustic imaging with low-cost sources; A review. *Photoacoustics* **2019**, *14*, 1–11. [CrossRef] [PubMed]
15. Zhong, H.; Duan, T.; Lan, H.; Zhou, M.; Gao, F. Review of Low-Cost Photoacoustic Sensing and Imaging Based on Laser Diode and Light-Emitting Diode. *Sensors* **2018**, *18*, 2264. [CrossRef] [PubMed]
16. Li, M.; Tang, Y.; Yao, J. Photoacoustic tomography of blood oxygenation: A mini review. *Photoacoustics* **2018**, *10*, 65–73. [CrossRef]
17. Mallidi, S.; Luke, G.P.; Emelianov, S. Photoacoustic imaging in cancer detection, diagnosis, and treatment guidance. *Trends Biotechnol.* **2011**, *29*, 213–221. [CrossRef]
18. Xia, W.; Singh, M.K.A.; Maneas, E.; Sato, N.; Shigeta, Y.; Agano, T.; Ourselin, S.; West, S.J.; Desjardins, A.E. Handheld Real-Time LED-Based Photoacoustic and Ultrasound Imaging System for Accurate Visualization of Clinical Metal Needles and Superficial Vasculature to Guide Minimally Invasive Procedures. *Sensors* **2018**, *18*, 1394. [CrossRef]
19. Kuchment, P.; Kunyansky, L. Mathematics of thermoacoustic tomography. *Eur. J. Appl. Math.* **2008**, *19*, 191–224. [CrossRef]

20. Jaeger, M.; Schüpbach, S.; Gertsch, A.; Kitz, M.; Frenz, M. Fourier reconstruction in optoacoustic imaging using truncated regularized inverse k -space interpolation. *Inverse Probl.* **2007**, *23*, S51–S63. [CrossRef]
21. Niederhauser, J.J.; Jaeger, M.; Lemor, R.; Weber, P.; Frenz, M. Combined ultrasound and optoacoustic system for real-time high-contrast vascular imaging in vivo. *IEEE Trans. Med. Imaging* **2005**, *24*, 436–440. [CrossRef]
22. Singh, M.K.A. *LED-Based Photoacoustic Imaging*; Springer Nature Singapore Pte Ltd.: Singapore, 2020.
23. Allen, T.J. High-Power Light Emitting Diodes; An Alternative Excitation Source for Photoacoustic Tomography. In *LED-Based Photoacoustic Imaging*; Progress in Optical Science and Photonics; Springer: Singapore, 2020; Volume 7, pp. 23–43. [CrossRef]
24. Allen, T.J.; Beard, P.C. High power visible light emitting diodes as pulsed excitation sources for biomedical photoacoustics. *Biomed. Opt. Express* **2016**, *7*, 1260–1270. [CrossRef]
25. Singh, M.K.A.; Sato, N.; Sankai, Y.; Ichihashi, F. In vivo demonstration of real-time oxygen saturation imaging using a portable and affordable LED-based multispectral photoacoustic and ultrasound imaging system. *Photons Plus Ultrasound: Imaging Sens. 2019* **2019**, *10878*, 108785N. [CrossRef]
26. Zhu, Y.; Xu, G.; Yuan, J.; Jo, J.; Gandikota, G.; Demirci, H.; Agano, T.; Sato, N.; Shigeta, Y.; Wang, X. Light Emitting Diodes based Photoacoustic Imaging and Potential Clinical Applications. *Sci. Rep.* **2018**, *8*, 1–12. [CrossRef]
27. Singh, M.K.A. Effect of light pulse width on frequency characteristics of photoacoustic signal—An experimental study using a pulse-width tunable LED-based photoacoustic imaging system. *Int. J. Eng. Technol.* **2018**, *7*, 4300–4303.
28. Francis, K.J.; Boink, Y.E.; Dantuma, M.; Singh, M.K.A.; Manohar, S.; Steenbergen, W.; Joseph, F.K. Tomographic imaging with an ultrasound and LED-based photoacoustic system. *Biomed. Opt. Express* **2020**, *11*, 2152–2165. [CrossRef]
29. Francis, K.J.; Boink, Y.E.; Dantuma, M.; Singh, M.K.A.; Manohar, S.; Steenbergen, W. Light Emitting Diodes Based Photoacoustic and Ultrasound Tomography: Imaging Aspects and Applications. In *LED-Based Photoacoustic Imaging*; Progress in Optical Science and Photonics; Springer: Singapore, 2020; Volume 7, pp. 245–266. [CrossRef]
30. Xia, W.; Nikitichev, D.I.; Mari, J.M.; West, S.J.; Pratt, R.; David, A.L.; Ourselin, S.; Beard, P.C.; Desjardins, A.E. Performance characteristics of an interventional multispectral photoacoustic imaging system for guiding minimally invasive procedures. *J. Biomed. Opt.* **2015**, *20*, 086005. [CrossRef]
31. Mari, J.M.; Xia, W.; West, S.J.; Desjardins, A.E. Interventional multispectral photoacoustic imaging with a clinical ultrasound probe for discriminating nerves and tendons: An ex vivo pilot study. *J. Biomed. Opt.* **2015**, *20*, 110503. [CrossRef] [PubMed]
32. Singh, M.K.A.; Parameshwarappa, V.; Hendriksen, E.; Steenbergen, W.; Manohar, S. Photoacoustic-guided focused ultrasound for accurate visualization of brachytherapy seeds with the photoacoustic needle. *J. Biomed. Opt.* **2016**, *21*, 120501. [CrossRef] [PubMed]
33. Hansen, R.S. Using high-power light emitting diodes for photoacoustic imaging. *SPIE Med. Imaging* **2011**, *7968*, 79680. [CrossRef]
34. Allen, T.J.; Beard, P.C. Light emitting diodes as an excitation source for biomedical photoacoustics. In Proceedings of the Photons Plus Ultrasound: Imaging and Sensing 2013, San Francisco, CA, USA, 4 March 2013; p. 85811F. [CrossRef]
35. Agano, T.; Sato, N.; Nakatsuka, H.; Kitagawa, K.; Hanaoka, T.; Morisono, K.; Shigeta, Y. Comparative experiments of photoacoustic system using laser light source and LED array light source. In Proceedings of the Photons Plus Ultrasound: Imaging and Sensing 2015, San Francisco, CA, USA, 11 March 2015; p. 93233X. [CrossRef]
36. Agano, T.; Sato, N.; Nakatsuka, H.; Kitagawa, K.; Hanaoka, T.; Morisono, K.; Shigeta, Y. Attempts to increase penetration of photoacoustic system using LED array light souce. In Proceedings of the Photons Plus Ultrasound: Imaging and Sensing 2015, San Francisco, CA, USA, 11 March 2015; p. 93233Z. [CrossRef]
37. Agano, T.; Sato, N. Photoacoustic Imaging System using LED light source. In Proceedings of the 2016 Conference on Lasers and Electro-Optics (CLEO), San Jose, CA, USA, 5–10 June 2016; The Optical Society: Washington, DC, USA, 2016; p. ATh3N.5.
38. Hariri, A.; LeMaster, J.; Wang, J.; Jeevarathinam, A.S.; Chao, D.L.; Jokerst, J.V. The characterization of an economic and portable LED-based photoacoustic imaging system to facilitate molecular imaging. *Photoacoustics* **2018**, *9*, 10–20. [CrossRef]

39. Joseph, F.K.; Xavierselvan, M.; Singh, M.K.A.; Mallidi, S.; Van Der Laken, C.; Van De Loo, F.; Steenbergen, W. LED-based photoacoustic imaging for early detection of joint inflammation in rodents: Towards achieving 3Rs in rheumatoid arthritis research. In Proceedings of the Photons Plus Ultrasound: Imaging and Sensing 2020, San Francisco, CA, USA, 17 February 2020; p. 112400M. [CrossRef]
40. Singh, M.K.A.; Agano, T.; Sato, N.; Shigeta, Y.; Uemura, T. Real-time in vivo imaging of human lymphatic system using an LED-based photoacoustic/ultrasound imaging system. In Proceedings of the Photons Plus Ultrasound: Imaging and Sensing 2018, San Francisco, CA, USA, 19 February 2018; p. 1049404. [CrossRef]
41. Singh, M.K.A.; Shigeta, Y.; Hanaoka, T.; Agano, T.; Sato, N. High-speed photoacoustic imaging using an LED-based photoacoustic imaging system. In Proceedings of the Photons Plus Ultrasound: Imaging and Sensing 2018, San Francisco, CA, USA, 19 February 2018; p. 104943N. [CrossRef]
42. Jo, J.; Xu, G.; Zhu, Y.; Burton, M.; Sarazin, J.; Schiopu, E.; Gandikota, G.; Wang, X. Detecting joint inflammation by an LED-based photoacoustic imaging system: A feasibility study. *J. Biomed. Opt.* **2018**, *23*, 110501–110504. [CrossRef]
43. Cheng, Q.; Qian, M.; Wang, X.; Zhang, H.; Wang, P.; Wen, L.; Pan, J.; Gao, Y.; Wu, S.; Zhang, M.; et al. Diagnosis and Treatment Monitoring of Port-Wine Stain Using LED-Based Photoacoustics: Theoretical Aspects and First In-Human Clinical Pilot Study. In *LED-Based Photoacoustic Imaging*; Progress in Optical Science and Photonics; Springer Nature Singapore Pte Ltd.: Singapore, 2020; pp. 351–377.
44. Xavierselvan, M.; Mallidi, S. LED-Based Functional Photoacoustics—Portable and Affordable Solution for Preclinical Cancer Imaging. In *LED-Based Photoacoustic Imaging*; Progress in Optical Science and Photonics; Springer: Singapore, 2020; Volume 7, pp. 303–319. [CrossRef]
45. Agrawal, S.; Fadden, C.; Dangi, A.; Yang, X.; AlBahrani, H.; Frings, N.; Zadi, S.H.; Kothapalli, S.-R. Light-Emitting-Diode-Based Multispectral Photoacoustic Computed Tomography System. *Sensors* **2019**, *19*, 4861. [CrossRef]
46. Francis, K.J.; Booijink, R.; Bansal, R.; Steenbergen, W. Tomographic Ultrasound and LED-Based Photoacoustic System for Preclinical Imaging. *Sensors* **2020**, *20*, 2793. [CrossRef]
47. Hariri, A.; Chen, F.; Moore, C.; Jokerst, J.V. Noninvasive staging of pressure ulcers using photoacoustic imaging. *Wound Repair Regen.* **2019**, *27*, 488–496. [CrossRef]
48. Maneas, E.; Aughwane, R.; Huynh, N.; Xia, W.; Ansari, R.; Singh, M.K.A.; Hutchinson, J.C.; Sebire, N.J.; Arthurs, O.J.; Deprest, J.; et al. Photoacoustic imaging of the human placental vasculature. *J. Biophotonics* **2019**, *13*, e201900167. [CrossRef]
49. Hariri, A.; Zhao, E.; Jeevarathinam, A.S.; LeMaster, J.; Zhang, J.; Jokerst, J.V. Molecular imaging of oxidative stress using an LED-based photoacoustic imaging system. *Sci. Rep.* **2019**, *9*, 11378. [CrossRef]
50. Xia, W.; Maneas, E.; Huynh, N.T.; Singh, M.K.A.; Brown, N.M.; Ourselin, S.; Gilbert-Kawai, E.; West, S.J.; Desjardins, A.E. Imaging of human peripheral blood vessels during cuff occlusion with a compact LED-based photoacoustic and ultrasound system. In Proceedings of the Photons Plus Ultrasound: Imaging and Sensing 2019, San Francisco, CA, USA, 27 February 2019; p. 1087804. [CrossRef]
51. Zhu, Y.; Lu, X.; Dong, X.; Yuan, J.; Fabiilli, M.L.; Wang, X. LED-Based Photoacoustic Imaging for Monitoring Angiogenesis in Fibrin Scaffolds. *Tissue Eng. Part C Methods* **2019**, *25*, 523–531. [CrossRef]
52. Kalva, S.K.; Pramanik, M. *Photoacoustic Tomography with High-Energy Pulsed Laser Diodes*; SPIE Spotlight Series Book; SPIE-Intl Soc Optical Eng: Bellingham, WA, USA, 2020; Volume SL56, ISBN 9781510636620. Available online: https://spie.org/Publications/Book/2566566?SSO=1 (accessed on 28 October 2020).
53. Fatima, A.; Kratkiewicz, K.; Manwar, R.; Zafar, M.; Zhang, R.; Huang, B.; Dadashzadeh, N.; Xia, J.; Avanaki, K. (Mohammad) Review of cost reduction methods in photoacoustic computed tomography. *Photoacoustics* **2019**, *15*, 100137. [CrossRef]
54. Choi, S.S.S.; Mandelis, A. Review of the state of the art in cardiovascular endoscopy imaging of atherosclerosis using photoacoustic techniques with pulsed and continuous-wave optical excitations. *J. Biomed. Opt.* **2019**, *24*, 1–15. [CrossRef]
55. Nissinen, J.; Kostamovaara, J. A 4 a peak current and 2 ns pulse width CMOS laser diode driver for high measurement rate applications. In Proceedings of the 2013 ESSCIRC (ESSCIRC), Bucharest, Romania, 16–20 September 2013; Institute of Electrical and Electronics Engineers (IEEE): Bucharest, Romania, 2013; pp. 355–358.
56. Canal, C.; Laugustin, A.; Kohl, A.; Rabot, O. Portable multiwavelength laser diode source for handheld photoacoustic devices. *Biophotonics: Photonic Solutions for Better Health Care V* **2016**, *9887*, 98872. [CrossRef]

57. Daoudi, K.; Berg, P.V.D.; Rabot, O.; Kohl, A.; Tisserand, S.; Brands, P.; Steenbergen, W. Handheld probe integrating laser diode and ultrasound transducer array for ultrasound/photoacoustic dual modality imaging. *Opt. Express* **2014**, *22*, 26365–26374. [CrossRef]
58. Allen, T.J.; Cox, B.T.; Beard, P.C. Generating photoacoustic signals using high-peak power pulsed laser diodes. In Proceedings of the Photons Plus Ultrasound: Imaging and Sensing 2005: The Sixth Conference on Biomedical Thermoacoustics, Optoacoustics, and Acousto-optics, San Jose, CA, USA, 25 April 2005; SPIE-Intl Soc Optical Eng: Bellingham, WA, USA, 2005; Volume 5697, pp. 233–243.
59. Kolkman, R.G.M.; Steenbergen, W.; Van Leeuwen, T.G. In vivo photoacoustic imaging of blood vessels with a pulsed laser diode. *Lasers Med Sci.* **2006**, *21*, 134–139. [CrossRef] [PubMed]
60. Allen, T.J.; Beard, P.C. Pulsed near-infrared laser diode excitation system for biomedical photoacoustic imaging. *Opt. Lett.* **2006**, *31*, 3462–3464. [CrossRef]
61. Berg, P.J.V.D.; Daoudi, K.; Moens, H.J.B.; Steenbergen, W. Feasibility of photoacoustic/ultrasound imaging of synovitis in finger joints using a point-of-care system. *Photoacoustics* **2017**, *8*, 8–14. [CrossRef]
62. Arabul, M.U.; Heres, M.; Rutten, M.C.M.; Van Sambeek, M.R.; Van De Vosse, F.N.; Lopata, R.G.P. Toward the detection of intraplaque hemorrhage in carotid artery lesions using photoacoustic imaging. *J. Biomed. Opt.* **2016**, *22*, 41010. [CrossRef]
63. Heres, H.M.; Arabul, M.U.; Rutten, M.C.M.; Van De Vosse, F.N.; Lopata, R.G.P. Visualization of vasculature using a hand-held photoacoustic probe: Phantom and in vivo validation. *J. Biomed. Opt.* **2017**, *22*, 41013. [CrossRef]
64. Berg, P.J.V.D.; Bansal, R.; Daoudi, K.; Steenbergen, W.; Prakash, J. Preclinical detection of liver fibrosis using dual-modality photoacoustic/ultrasound system. *Biomed. Opt. Express* **2016**, *7*, 5081–5091. [CrossRef]
65. Jaeger, M.; Schwab, H.-M.; Almallouhi, Y.; Canal, C.; Song, M.; Sauget, V.; Sontrop, D.; Mulder, T.; Roumen, P.; Humblet, A.; et al. Deformation-Compensated Averaging for Deep-Tissue LED and Laser Diode-Based Photoacoustic Imaging Integrated with Handheld Echo Ultrasound. In *LED-Based Photoacoustic Imaging*; Progress in Optical Science and Photonics; Springer: Singapore, 2020; Volume 7, pp. 47–78. [CrossRef]
66. Berg, P.J.V.D.; Daoudi, K.; Steenbergen, W. Pulsed photoacoustic flow imaging with a handheld system. *J. Biomed. Opt.* **2016**, *21*, 026004. [CrossRef]
67. Upputuri, P.K.; Pramanik, M. Performance characterization of low-cost, high-speed, portable pulsed laser diode photoacoustic tomography (PLD-PAT) system. *Biomed. Opt. Express* **2015**, *6*, 4118–4129. [CrossRef]
68. Kalva, S.K.; Upputuri, P.K.; Pramanik, M. High-speed, low-cost, pulsed-laser-diode-based second-generation desktop photoacoustic tomography system. *Opt. Lett.* **2018**, *44*, 81–84. [CrossRef]
69. Rajendran, P.; Sahu, S.; Dienzo, R.A.; Pramanik, M. In vivo detection of venous sinus distension due to intracranial hypotension in small animal using pulsed-laser-diode photoacoustic tomography. *J. Biophotonics* **2020**, *13*, e201960162. [CrossRef]
70. Upputuri, P.K.; Pramanik, M. Dynamic in vivo imaging of small animal brain using pulsed laser diode-based photoacoustic tomography system. *J. Biomed. Opt.* **2017**, *22*, 1. [CrossRef]
71. Upputuri, P.K.; Das, D.; Maheshwari, M.; Yaowen, Y.; Pramanik, M. Real-time monitoring of temperature using a pulsed laser-diode-based photoacoustic system. *Opt. Lett.* **2020**, *45*, 718–721. [CrossRef]
72. Upputuri, P.K.; Pramanik, M. Fast photoacoustic imaging systems using pulsed laser diodes: A review. *Biomed. Eng. Lett.* **2018**, *8*, 167–181. [CrossRef]
73. Sivasubramanian, K.; Pramanik, M. High frame rate photoacoustic imaging at 7000 frames per second using clinical ultrasound system. *Biomed. Opt. Express* **2016**, *7*, 312–323. [CrossRef]
74. Hariri, A.; Fatima, A.; Mohammadian, N.; Mahmoodkalayeh, S.; Ansari, M.; Bely, N.; Nasiriavanaki, M. Development of low-cost photoacoustic imaging systems using very low-energy pulsed laser diodes. *J. Biomed. Opt.* **2017**, *22*, 75001. [CrossRef]
75. Zeng, L.; Piao, Z.; Huang, S.; Jia, W.; Chen, Z. Label-free optical-resolution photoacoustic microscopy of superficial microvasculature using a compact visible laser diode excitation. *Opt. Express* **2015**, *23*, 31026–31033. [CrossRef]
76. Wang, T.; Nandy, S.; Salehi, H.S.; Kumavor, P.D.; Zhu, Q. A low-cost photoacoustic microscopy system with a laser diode excitation. *Biomed. Opt. Express* **2014**, *5*, 3053–3058. [CrossRef]
77. Zeng, L.; Liu, G.; Yang, D.; Ji, X. Portable optical-resolution photoacoustic microscopy with a pulsed laser diode excitation. *Appl. Phys. Lett.* **2013**, *102*, 053704. [CrossRef]

78. Erfanzadeh, M.; Kumavor, P.D.; Zhu, Q. Laser scanning laser diode photoacoustic microscopy system. *Photoacoustics* **2018**, *9*, 1–9. [CrossRef]
79. Dai, X.; Yang, H.; Jiang, H. In vivo photoacoustic imaging of vasculature with a low-cost miniature light emitting diode excitation. *Opt. Lett.* **2017**, *42*, 1456–1459. [CrossRef]
80. Steinberg, I.; Huland, D.M.; Vermesh, O.; Frostig, H.E.; Tummers, W.S.; Gambhir, S.S. Photoacoustic clinical imaging. *Photoacoustics* **2019**, *14*, 77–98. [CrossRef]
81. Kruger, R.A.; Kuzmiak, C.M.; Lam, R.B.; Reinecke, D.R.; Del Rio, S.P.; Steed, D. Dedicated 3D photoacoustic breast imaging. *Med Phys.* **2013**, *40*, 113301. [CrossRef]
82. Lin, L.; Hu, P.; Shi, J.; Appleton, C.M.; Maslov, K.; Li, L.; Zhang, R.; Wang, L.V. Single-breath-hold photoacoustic computed tomography of the breast. *Nat. Commun.* **2018**, *9*, 1–9. [CrossRef]
83. Abu Anas, E.M.; Zhang, H.K.; Kang, J.; Boctor, E. Enabling fast and high quality LED photoacoustic imaging: A recurrent neural networks based approach. *Biomed. Opt. Express* **2018**, *9*, 3852–3866. [CrossRef]
84. Mozaffarzadeh, M.; Hariri, A.; Moore, C.; Jokerst, J.V. The double-stage delay-multiply-and-sum image reconstruction method improves imaging quality in a LED-based photoacoustic array scanner. *Photoacoustics* **2018**, *12*, 22–29. [CrossRef]
85. Singh, M.K.A.; Sivasubramanian, K.; Sato, N.; Ichihashi, F.; Sankai, Y.; Xing, L. Deep learning-enhanced LED-based photoacoustic imaging. In Proceedings of the Photons Plus Ultrasound: Imaging and Sensing 2020, San Francisco, CA, USA, 17 February 2020; p. 1124038. [CrossRef]
86. Farnia, P.; Najafzadeh, E.; Hariri, A.; Lavasani, S.N.; Makkiabadi, B.; Ahmadian, A.; Jokerst, J.V. Dictionary learning technique enhances signal in LED-based photoacoustic imaging. *Biomed. Opt. Express* **2020**, *11*, 2533–2547. [CrossRef]
87. Manwar, R.; Hosseinzadeh, M.; Hariri, A.; Kratkiewicz, K.; Noei, S.; Avanaki, K. Photoacoustic Signal Enhancement: Towards Utilization of Low Energy Laser Diodes in Real-Time Photoacoustic Imaging. *Sensors* **2018**, *18*, 3498. [CrossRef]
88. Chandramoorthi, S.; Thittai, A.K. Ultrasound Receive-Side Strategies for Image Quality Enhancement in Low-Energy Illumination Based Photoacoustic Imaging. In *LED-Based Photoacoustic Imaging*; Progress in Optical Science and Photonics; Springer: Singapore, 2020; Volume 7, pp. 79–112. [CrossRef]
89. Chandramoorthi, S.; Thittai, A.K. Enhancing Image Quality of Photoacoustic Tomography Using Sub-Pitch Array Translation Approach: Simulation and Experimental Validation. *IEEE Trans. Biomed. Eng.* **2019**, *66*, 3543–3552. [CrossRef]
90. Sowmiya, C.; Thittai, A.K. Noise Reduction in Inherently low-SNR PLD-based PAT images. In Proceedings of the TENCON 2019—2019 IEEE Region 10 Conference (TENCON), Kochi, India, 17–20 October 2019; Institute of Electrical and Electronics Engineers (IEEE): Kochi, India, 2019; pp. 106–108.
91. Zhang, H.K. Democratizing LED-Based Photoacoustic Imaging with Adaptive Beamforming and Deep Convolutional Neural Network. In *LED-Based Photoacoustic Imaging*; Progress in Optical Science and Photonics; Springer: Singapore, 2020; Volume 7, pp. 183–202. [CrossRef]
92. Xia, W.; Ginsberg, Y.; West, S.J.; Nikitichev, D.I.; Ourselin, S.; David, A.L.; Desjardins, A.E. Coded excitation ultrasonic needle tracking: An in vivo study. *Med Phys.* **2016**, *43*, 4065. [CrossRef]
93. Singh, M.K.A.; Jaeger, M.; Frenz, M.; Steenbergen, W. In vivo demonstration of reflection artifact reduction in photoacoustic imaging using synthetic aperture photoacoustic-guided focused ultrasound (PAFUSion). *Biomed. Opt. Express* **2016**, *7*, 2955–2972. [CrossRef]
94. Wong, T.T.W.; Zhou, Y.; Garcia-Uribe, A.; Li, L.; Maslov, K.; Lin, L.; Wang, L.V. Use of a single xenon flash lamp for photoacoustic computed tomography of multiple-centimeter-thick biological tissue ex vivo and a whole mouse body in vivo. *J. Biomed. Opt.* **2016**, *22*, 041003. [CrossRef]
95. Wang, D.; Wang, Y.; Wang, W.; Luo, D.; Chitgupi, U.; Geng, J.; Zhou, Y.; Wang, L.; Lovell, J.F.; Xia, J. Deep tissue photoacoustic computed tomography with a fast and compact laser system. *Biomed. Opt. Express* **2016**, *8*, 112–123. [CrossRef] [PubMed]
96. Jeng, G.-S.; Li, M.-L.; Kim, M.W.; Yoon, S.J.; Pitre, J.J.; Li, D.S.; Pelivanov, I.; O'Donnell, M. *Real-Time Spectroscopic Photoacoustic/Ultrasound (PAUS) Scanning with Simultaneous Fluence Compensation and Motion Correction for Quantitative Molecular Imaging*; bioRxiv: New York, NY, USA, 2019.
97. Allen, T.J.; Alam, S.; Zhang, E.; Laufer, J.; Richardson, D.J.; Beard, P.C. Use of a pulsed fibre laser as an excitation source for photoacoustic tomography. In Proceedings of the Photons Plus Ultrasound: Imaging and Sensing 2011, San Francisco, CA, USA, 28 February 2011; p. 78991V. [CrossRef]

98. Lashkari, B.; Choi, S.S.S.; Dovlo, E.; Dhody, S.; Mandelis, A. Frequency-Domain Photoacoustic Phase Spectroscopy: A Fluence-Independent Approach for Quantitative Probing of Hemoglobin Oxygen Saturation. *IEEE J. Sel. Top. Quantum Electron.* **2015**, *22*, 127–136. [CrossRef]
99. Lashkari, B.; Mandelis, A. Comparison between pulsed laser and frequency-domain photoacoustic modalities: Signal-to-noise ratio, contrast, resolution, and maximum depth detectivity. *Rev. Sci. Instrum.* **2011**, *82*, 094903. [CrossRef] [PubMed]

Publisher's Note: MDPI stays neutral with regard to jurisdictional claims in published maps and institutional affiliations.

sensors

MDPI

Review

Towards Clinical Translation of LED-Based Photoacoustic Imaging: A Review

Yunhao Zhu [1,2,†], Ting Feng [3,4,†], Qian Cheng [4], Xueding Wang [2], Sidan Du [1], Naoto Sato [5], Jie Yuan [1,2,*] and Mithun Kuniyil Ajith Singh [6,*]

1 Department of Electronic Science and Engineering, Nanjing University, Nanjing 210023, China; yunhaoz@umich.edu (Y.Z.); coff128@nju.edu.cn (S.D.)
2 Department of Biomedical Engineering, University of Michigan, Ann Arbor, MI 48109, USA; xdwang@umich.edu
3 Department of Electronic and Optical Engineering, Nanjing University of Science and Technology, Nanjing 210094, China; fengting@njust.edu.cn
4 Institution of Acoustics, Tongji University, Shanghai 200092, China; Q.cheng@tongji.edu.cn
5 Research and Development Division, CYBERDYNE INC, Tsukuba 3050818, Japan; sato_naoto@cyberdyne.jp
6 Research and Business Development Division, CYBERDYNE INC, 3013 Rotterdam, The Netherlands
* Correspondence: yuanjie@nju.edu.cn (J.Y.); mithun_ajith@cyberdyne.jp (M.K.A.S.)
† These authors contributed equally.

Received: 29 March 2020; Accepted: 22 April 2020; Published: 27 April 2020

Abstract: Photoacoustic imaging, with the capability to provide simultaneous structural, functional, and molecular information, is one of the fastest growing biomedical imaging modalities of recent times. As a hybrid modality, it not only provides greater penetration depth than the purely optical imaging techniques, but also provides optical contrast of molecular components in the living tissue. Conventionally, photoacoustic imaging systems utilize bulky and expensive class IV lasers, which is one of the key factors hindering the clinical translation of this promising modality. Use of LEDs which are portable and affordable offers a unique opportunity to accelerate the clinical translation of photoacoustics. In this paper, we first review the development history of LED as an illumination source in biomedical photoacoustic imaging. Key developments in this area, from point-source measurements to development of high-power LED arrays, are briefly discussed. Finally, we thoroughly review multiple phantom, ex-vivo, animal in-vivo, human in-vivo, and clinical pilot studies and demonstrate the unprecedented preclinical and clinical potential of LED-based photoacoustic imaging.

Keywords: photoacoustic; LED; clinic; optical imaging

1. Introduction

Photoacoustic imaging (PAI) holds strong potential in providing structural, functional and molecular information on tissue, with scalable resolution and imaging depth [1]. Since the optical scattering is high in biological tissue, ballistic optical microscopic techniques cannot provide any useful information beyond a depth of 1 mm. PAI overcomes this difficulty since it involves acoustic detection and sound scattering in tissue is orders of magnitude lower than that of light. In PAI, short-pulsed light is irradiated on the tissue, and endogenous optical absorbers in the tissue absorb light resulting in a temperature rise [2]. This transient temperature rise results in thermoelastic expansion and produces light-induced ultrasound (US) waves, which then can be detected by US detectors placed on the skin surface for reconstructing an optical absorption map with acoustic resolution [3]. In a clinical context, PAI is easy to combine with US imaging and is capable of providing anatomical, functional, molecular, and metabolic information by utilizing the signature optical absorption contrast of the vasculature, hemodynamics, oxygen metabolism, biomarkers, and gene expression. Utilizing the useful information provided by PAI, a plethora of clinical applications have been explored in vascular biology [4–7], oncology [8,9], neurology [2,10,11], ophthalmology [12,13], dermatology [14,15], gastroenterology [16–20], osteology [21–24], and cardiology [25,26].

In laser-based PAI, where the tissue of interest is illuminated by a pulsed laser beam, the optical energy used usually ranges from tens to hundreds of mJ per pulse, with a typical pulse duration of 5–10 ns. Most of the commercial and research lab-made PAI systems utilize Q-switched Nd:YAG pumped OPO (optical parametric oscillator), Ti: Sapphire or dye laser systems. However, because of their high cost, larger footprint and strict requirement for eye-safety goggles and laser-safe rooms, these laser sources are not suitable for a clinical environment. Furthermore, the repetition rate of most high-power laser sources is relatively low (~10 Hz), which limits the imaging speed, especially when the signal-to-noise ratio (SNR) is not sufficient and frame averaging is a necessity. In recent years, laser diodes (LD) and light emitting diodes (LEDs) have been heavily explored to be used as an illumination source in PAI, resulting in portable, affordable and clinically translatable PAI systems [6,7,27–37]. LD offers a higher pulse repetition rate (PRR) (typical 2–4 KHz), average power around 6 W, and optical energy of around 0.56–2.5 mJ per pulse, but is only available at wavelengths greater than 750 nm [38,39]. Additionally, for an LD-based PAI system, laser-safe rooms and goggles are requirements, just as in the case of conventional laser sources. On the other hand, LEDs, which are available in a wide wavelength range (e.g., 470, 520, 620, 660, 690, 750, 820, 850, 940 and 980 nm) provide lower optical energy in the range of μJ per pulse, but at a higher repetition rate (~16 KHz) offering the possibility to average more frames without compromising on temporal resolution. Compared to fixed pulse widths in lasers, the optical pulse width of an LED/LD source can be tuned based on the required spatial resolution and imaging depth [39,40].

The pulse width of LED/LD sources is tens of nanoseconds, whereas that of solid-state lasers could be less than ten nanoseconds. The temporal pulse width imposes a limit to the spatial resolution of the imaging system. For example, the 35-ns pulse width of the 850-nm LED corresponds to a spatial resolution of 52.5 μm (=35 ns × 1500 μm/μs). The LD usually offers more energy than LEDs. An LD light source's (Quantel, Bozeman, MT) pulse width, for example, can be tuned from 30 ns to 200 ns, with the pulse energy correspondingly changing from 1–4 mJ. As a limitation, the LED array can only reach up to 0.200 mJ per pulse (highest reported optical output for 850 nm LED arrays). LD, even though it can offer higher pulse energy, is the same as class-IV lasers in terms of optical coherence and subsequent eye/skin safety issues. Owing to its portability, affordability, imaging speed and safety aspects, LED-based PAI holds potential in real-time functional and structural characterization of tissue in various superficial and sub-surface imaging applications and also to accelerate the clinical translation of PAI. Typically used laser, LD and LED performance are listed in Table 1.

Table 1. Comparison of pulsed laser, laser diodes and LED [6,7,27–40].

	Energy (mJ)	PRR (Hz)	Pulse Width (ns)	Cost *	Advantages	Disadvantages
Laser	5 ~120	~10	<10	$70–200 K	Powerful, ~5 cm penetration depth, tunable wavelength	Bulky size, eye protection and laser safe rooms needed
LD	0.5–2.5	~1 K–6 K	30–200	~$10–25 K	Integration in a handheld probe feasible, high PRR	Limited penetration depth, eye protection and laser safe rooms needed, wavelength tuning not possible
LED	0.2	~200–16 K	30–100	$10–15 K	Integration in a handheld probe feasible, high PRR, no need for laser-safe rooms or eye-safety goggles	Limited penetration depth, wavelength tuning not possible

* Cost includes the driving electronics and may vary based on different features, number of wavelengths etc. Integration to a US probe may also involve extra development cost.

In this paper, we review the progress of LED-based PAI technology and its potential preclinical and clinical applications.

2. Fundamental Development of LED-Based PAI Technology

PAI has already demonstrated its unparalleled potential in multiple preclinical and clinical applications and is quite mature in a research setting. At this point, this technology is facing an exciting transition from bench to bedside and LD- and LED-based systems are being explored heavily because of their portability, affordability, and ease of use in a clinical setting. LDs operating in pulsed mode have been investigated by different research groups for multiple point-of-care applications [36–38]. However, typical commercial pulsed LDs are available only in the near-infrared wavelength range, and combining multiple wavelengths in a handheld setting is a cumbersome process [41,42]. On the other hand, the LEDs could be fabricated to operate in a 400 nm to 1000 nm wavelength range (not continuously) with reasonable optical energy by developing arrays of multiple elements and overdriving them [43]. Within this wavelength range, PAI could provide high contrast for melanin, hemoglobin, and fat to an extent, making LEDs one of the ideal illumination sources for multispectral PAI of tissue up to a depth of 1–1.5 cm.

An LED is a semiconductor device based on a p–n junction diode. A p-n junction diode is a two-terminal semiconductor device, which allows the electric current in only one direction and blocks the electric current in the opposite or reverse direction. If the diode is forward biased by applying a voltage, it allows the electric current flow. A small increase in voltage results in a significant change in current flow. Holes (from the P-type material) and electrons (from the N-type material) flowing across the junction promote strong electron–hole radiative recombination, resulting in the emission of a large number of photons [44]. Typically, LEDs are designed for continuous wave (CW) operation, but it is also feasible to drive them with pulsed current. In the pulsed mode, output energy of LEDs is dependent on the peak current, and this can be far higher than the CW rated current, especially if the duty cycle is kept low (<0.1%) to avoid any thermal damage. Since LEDs, when overdriven in a pulsed mode, can generate significantly higher optical output than in conventional CW operation, these types of overdriven pulsed-mode LEDs are often referred to as high-power LEDs. However, operating an LED at excessively high drive currents may lead to faster ageing due to heat generation, and this can even cause immediate failure of the LEDs. The quantum efficiency of the device will also drop with increasing current [45]. Considering this, it is important to design and develop efficient, safe electronic drivers and heat sinks to use LEDs in high-power mode.

About a decade ago, Hansen first proposed the use of LEDs working at a 627-nm wavelength as an inexpensive and compact excitation source for biomedical PAI [46]. In this proof-of-concept work, he demonstrated the feasibility of using LEDs as a light source in PAI for the first time. The basic idea of creating pulsed high-power LED is that, when the low-power LED is overdriven, they can deliver optical output that is far higher than their normal specifications in CW-operation [47–50]. The LED they used was Luxeon LXHL_PD09 which has been measured to yield approximately 250 mW of light output when supplied with 1 A DC current. A derivative of the MOSFET-based circuit presented by Alton and Raji was employed as a driver for generating pulsed current [51]. When the LED was supplied with 60-ns current pulses with peak value of 40 A, it was able to provide pulse energy of 400 nJ per pulse with pulse width of 60 ns, and light focusing was performed to generate the radiant energy required for generating a photoacoustic (PA) response. Also, 50,000 A-line signals were averaged to detect PA response from a non-realistic phantom. Based on the PRR of the proposed LED (200 Hz), this system requires 250 s to acquire an image, which was not good enough for imaging tissue in real-time. Owing to the advances in solid-state device technology and efforts of different research groups in the last decade, there were significant improvements in the performance of LEDs (improvement of pulse energy, PRR, etc.), which consequently resulted in the step-by-step development and commercialization of an LED-based PAI system that is comparable to a laser-based machine.

In 2013, Allen et al. proposed the use of high-power LEDs (CBT−120 from Luminus) working at 400-nm to 65-nm wavelengths as an illumination source in biomedical PAI [43]. In this work, the pulse energy of the LED is increased to 22 µJ per pulse with a pulse width of 500 ns, by overdriving the LED elements by 10 times their rated current. The driver they used was made based on MOSFET, which is described in work of Chaney et al. [51]. As a result of increase in light energy, with the same PRR of 200 Hz, they were able to significantly reduce the frame averaging (by 1000 times), which is commendable.

In 2016, Allen et al. further improved their system performance by using high-power LED (SST−90 from Luminus) elements working at 400-nm to 650-nm wavelengths, driven by a commercial electronic driver (PCO−7120, Directed Energy, Inc., Loveland, CO, USA), which provided 9 µJ per pulse, with a pulse duration of 200 ns, a peak current of 50 A and a PRR of 500 Hz when overdriving LEDs by 20 times their nominal current [33]. They confirmed that the duty cycle was 0.01%, still below the 1% which has been previously reported as safe (no noticeable damage to the device) [45]. In this work, they first imaged a realistic tissue-mimicking phantom by averaging 5000 image frames and using a wide field illumination strategy. The best imaging depth they achieved was 15 mm in 1% intralipid (Figure 1).

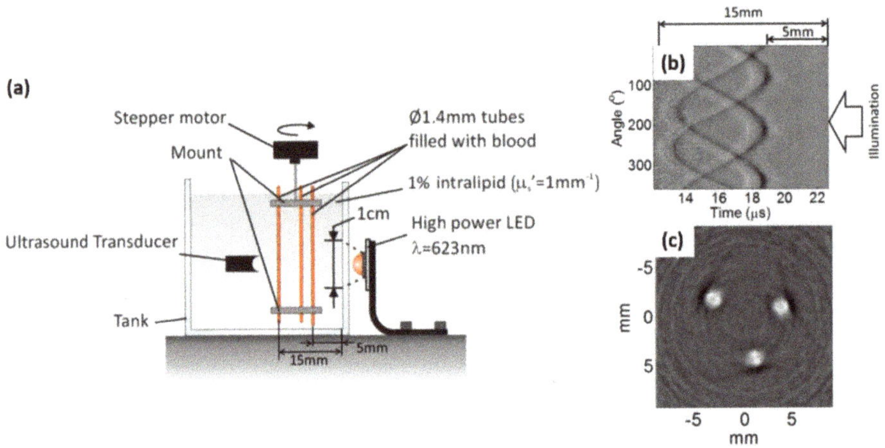

Figure 1. (**a**) PAI setup. (**b**) Time-resolved PA signals of three 1.4 mm tubes filled with human blood (35% haematocrit) and immersed in 1% Intralipid (µs′ = 1 mm^{-1}). (**c**) Reconstructed PA image. Energy = 9 µJ, averaging image frames = 5000. Reprinted from Ref. [33].

The pioneering works mentioned above laid the foundation for LED-based PAI, especially demonstrating the feasibility of using visible light for PAI, which is not possible using lasers or LDs. Oxygen saturation imaging is one of the most important applications of PAI, and it is of paramount importance to have illumination wavelengths suitable for this. Considering the absorption peak for deoxy-hemoglobin (690 nm) and oxy-hemoglobin (850 nm), it was critical to develop LEDs in these wavelength ranges too. Considering this, there has been significant efforts from different researchers in this direction, which are detailed below.

In 2016 and 2017, Agano et al. for the first time proposed the use of high-power LED arrays working at 850 nm with optical energy of 200 µJ per pulse and with a pulse duration of ~70–100 ns. They optimized the SNR by increasing the optical output, by developing highly noise-efficient front end electronics with multiple stages of amplification (~106 dB), and also by matching the US transducer's frequency characteristics with LED light pulse width [34,35,52]. To increase the optical output, they developed LED arrays and further optimized them using hybrid techniques. The single LED element provided output energy of 0.024 µJ per pulse, with a pulse duration of 70 ns and 1 A DC-current.

By developing LED elements with double stack structure, arranging them in an array, and applying 20 times the rated current, they were able to achieve 200 μJ per pulse at a wavelength of 850 nm. The repetition rate of this first commercial LED-based PAI system (AcousticX) was 4 KHz. In 2018, Zhu et al. improved the PRR of LEDs to 16 KHz and used the same system for demonstrating its dynamic structural and functional imaging capabilities [6]. The performance of the mentioned LED-based PAI technology is summarized in Table 2.

Table 2. Summary of the Fundamental development of LED-based PAI technology.

Year	Authors	Pulse Width (ns)	Peak Current (A)	Pulse Energy (mJ)	Repetition Rate (Hz)	Wavelength (nm)
2011	Hansen et al. [46]	60	40	0.0004	200	627
2013	Allen et al. [43]	500	200	0.0022	200	623
2016	Allen et al. [33]	200	50	0.0009	500	623
2016–2017	Agano et al. [34,35,52]	70 *	15–20	0.15–0.2 **	4000	850
2018	Zhu et al. [6]	70 *	20	0.2 **	4000–16,000	850

* Pulse width is tunable from 35–150 ns and above measurements are with 70 ns; ** value is measured from the arrays of the LED.

The AcousticX system was thoroughly characterized for its imaging depth, spatial resolution, frame rate, and oxygen saturation imaging accuracy in multiple studies. When using a 9-MHz US probe and 850-nm LED arrays, Xia et al. reported mean axial and lateral resolutions of 220 μm and 460 μm respectively, which was similar for both US and PA imaging. They also reported an imaging depth of 2.8 cm at an interleaved US and PA frame rate of 30 Hz. Imaging depth was further improved to 3.8 cm after averaging more frames, resulting in a final frame rate of 1.5 Hz. In another study, Hariri et al. reported mean axial and lateral resolution of 268 μm and 570 μm respectively, when using a combination of a 7-MHz US probe and 850-nm LED arrays. They also obtained an imaging depth of 3.2 cm (frame rate = 15 Hz) in their phantom study. In this molecular imaging study, they also measured the sensitivity of the system in detecting commonly used molecular contrast agents. The limits of detection for ICG, MB, and DiR were reported to be 9 μM, 0.75 mM, and 68 μM, respectively. Oxygen saturation imaging is one of the most important applications of PAI. Using a two-wavelength (750/850 nm) approach, Kalloor Joseph et al. evaluated the potential of AcousticX in oxygen saturation imaging using phantom and in-vivo small animal imaging experiments. Based on 28 human-blood-based in-vitro measurements, a standard error of 8.4% was observed between actual oxygen saturation values and the oxygen saturation image formed by the system. In the same work, they showed repeatability and reproducibility of oxygen saturation imaging using an in-vivo mouse oxygen breathing challenge experiment.

From the beginning of 2017, there has been a tremendous push in this area and multiple studies using LED-based PAI were reported. In the next section, we will review the potential preclinical and clinical applications demonstrated using AcousticX, the commercially available LED-based PA and US imaging system.

3. Preclinical and Clinical Applications of LED-Based PAI

Several preclinical and clinical studies utilizing LED-based PAI are reviewed in this section to provide a perspective for future clinical translation.

3.1. *Phantom and Ex-Vivo Studies*

3.1.1. Guidance on Minimally Invasive Procedures with Peripheral Tissue Targets

Precise and efficient device guidance is critical for minimally invasive vascular access procedures. US imaging is commonly used in clinics for this purpose, but the visualization of medical needles and tissue targets are often challenging [53]. PAI with high contrast for metallic needles and vasculature holds strong potential for guiding these needle-based vascular access procedures. However, the bulky and expensive solid-state laser that is required for tissue illumination is hindering the transition of PAI from bench to clinic, for this point-of-care application [54–56]. In 2018, Xia et al. used AcousticX and demonstrated the potential of LED-based PAI and US imaging in guiding minimally invasive procedures [32]. Their results demonstrated that LED-based PAI enabled needle visualization with SNRs that were 1.2 to 2.2 times higher than those obtained with US imaging, over insertion angles of 26 to 51 degrees. In the reported phantom study, an imaging depth close to 4 cm was achieved as shown in Figure 2a–c. In this work, they also demonstrated that LED-based PAI can visualize the superficial vasculature of the finger and wrist of a human volunteer in real time, along with conventional pulse-echo US. Based on these promising results, it can be concluded that LED-based PAI combined with US imaging holds potential in guiding minimally invasive procedures with peripheral tissue targets.

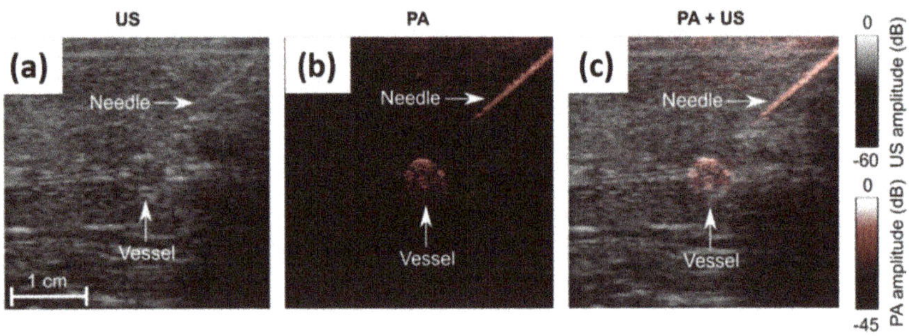

Figure 2. (**a**) US, (**b**) PA and (**c**) US-PA overlay images showing a medical needle inserted towards a vessel-mimicking phantom embedded in chicken tissue, respectively. Here, the uppermost 5 mm, which contained the US gel, is not shown. Reprinted from Ref. [32].

3.1.2. Imaging of Human Placental Vasculature

Accurate guidance is important not only for vascular access procedures, but also for minimally invasive fetal surgeries, for example, for the procedure performed to treat twin-to-twin transfusion syndrome (TTTS) [57]. TTTS is caused by imbalanced blood flow between twin vascular connections (anastomoses) in the placenta. If not treated in time, it will bring high morbidity and risk of death [58]. In current clinical practice of the TTTS procedure, in-vivo imaging of the placenta is performed using white-light fetal endoscopy and external B-mode US imaging. Both these methods do not provide sufficient contrast to visualize small anastomotic vessels below the surface of the chorionic placenta [59]. In 2018, Maneas et al. employed the LED-based PA imaging for wide field imaging of the human placental vasculature ex vivo, taking advantage of high optical absorption contrast offered by oxy- and deoxy-hemoglobin [60,61]. Previous studies have shown that PA signals from placental vessels can be detected using external clinical US probes [62,63]. Here, they used an LED-based PAI system (AcousticX) to image chorionic (fetal) superficial and subsurface vasculature in normal and TTTS-treated human placentas [64]. The LED-based PAI system based on a clinical US probe enabled fast interleaved 2D PA and US imaging (Figure 3). This work gives direct confirmation that volumetric LED-based PAI of the human placenta can generate detailed 3D maps of surface and subsurface vasculature up to a depth of around 7 mm. The authors also foresee that interventional PA imaging for visualizing the chorionic surface and superficial placental blood vessels with high spatial resolution may be valuable for minimally invasive fetal treatments.

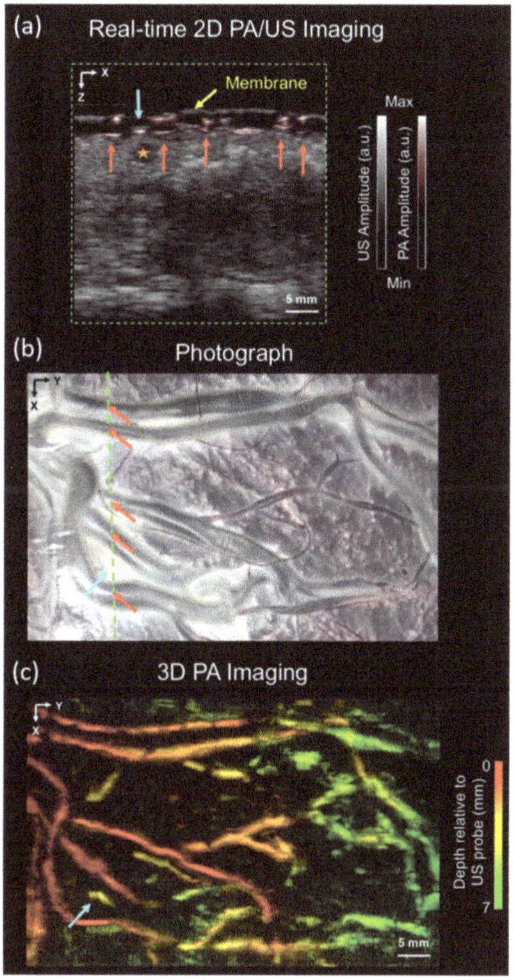

Figure 3. Wide-field PA-US images of the chorionic placental vasculature in an untreated part of a TTTS photocoagulated placenta. (**a**) 2D PA and US overlay image, acquired from a region corresponding to the green line in (**b**), which is a photograph of the imaged placenta. (**c**) 3D PA image displayed as a maximum intensity projection (MIP) of the reconstructed image volume. Reprinted from Ref. [64].

3.1.3. Imaging of Intraocular Tumors

PA imaging has shown great potential in diagnosis and characterization of cancer, especially in applications like head and neck tumor imaging [65–70]. As a head and neck cancer, intraocular tumors are relatively rare, but life-threatening [70,71]. Xu et al. have demonstrated the ability of laser-based PA imaging to characterize intraocular tumors through molecular composition and structural heterogeneity [72]. Zhu et al. recently explored the feasibility of using LED-based PAI to image intraocular tumors in a complete human eyeball. Figure 4a,b shows 2D US and PA images of the eyeball of a choroidal melanoma tumor, respectively. US is shown in grayscale and PA is shown in pseudo-color. Figure 4c is the PA/US overlay image, offering complementary information. PA results show that LED-based PA imaging has sufficient penetration depth to cover the entire tumor volume. Figure 4d shows a 3D PA image of the eyeball. In this work, single-wavelength LED arrays were used to visualize vasculature. LED arrays with multiple wavelengths can be potentially used to observe individual molecular components in the future. These results reveal the potential of LED-based PAI in the broad area of cancer diagnosis and staging.

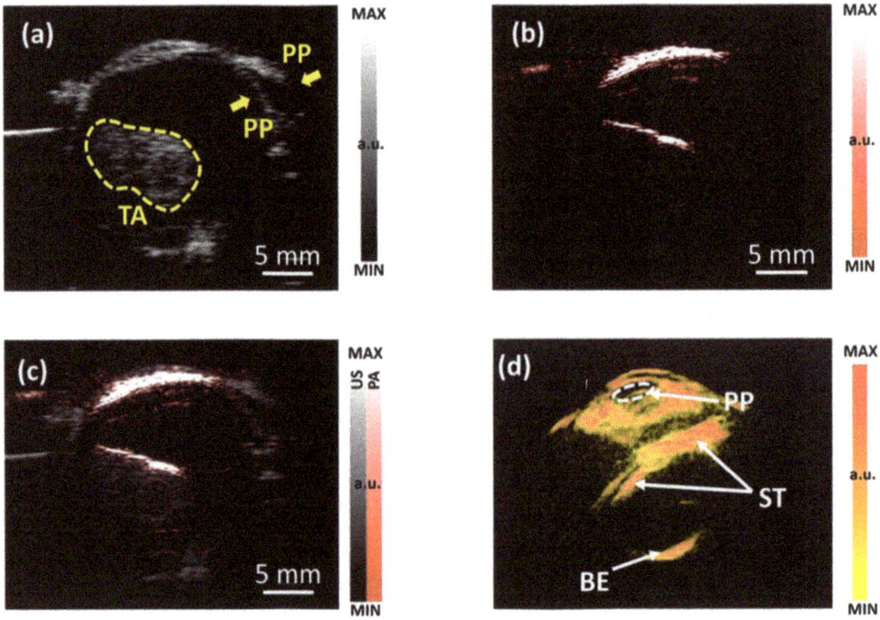

Figure 4. (**a**) 2D B-scan US image and (**b**) PA image of an ocular globe ex vivo with a tumor inside (confirmed in clinic). (**c**) PA and US combined image. (**d**) Perspective view of a 3D PA image of ocular globe with a choroidal melanoma tumor. PP-pupil. TA-tumor area. ST-the surface of the tumor. BE-the back of the eye. Reprinted from Ref. [6].

3.2. In-Vivo Preclinical Small Animal Imaging Studies

3.2.1. Non-Invasive Monitoring of Angiogenesis

Vascularization of engineered constructs is required to integrate an implant within the host blood supply. The ability to non-invasively monitor neovascularization of an implanted construct is ultimately critical for translation. Laser speckle contrast analysis (LASCA), a widely used imaging technique within regenerative medicine, has high spatial resolution, but offers limited imaging depth and is only sensitive to perfused blood vessels [73–75]. In 2019, Zhu et al. used LED-based PA and US imaging to potentially solve this challenge in regenerative medicine [7]. They used an LED-based PA-US dual-mode system to image and monitor angiogenesis for 7 days in fibrin-based scaffolds, which were subcutaneously implanted in mice. Scaffolds, with or without basic fibroblast growth factor (bFGF), were imaged on days 0 (i.e., post implantation), 1, 3, and 7 with both LASCA and PA-US imaging systems (Figure 5). Quantified perfusion measured by LASCA and PA imaging was compared with histologically determined blood vessel density on day 7. Vessel density corroborated changes in perfusion measured by both LASCA and PA. PAI enabled delineation of differences in neovascularization in the upper and the lower regions of the scaffold. Overall, this study has demonstrated that PAI could be a noninvasive and highly sensitive method for monitoring deep-seated vascularization in regenerative applications.

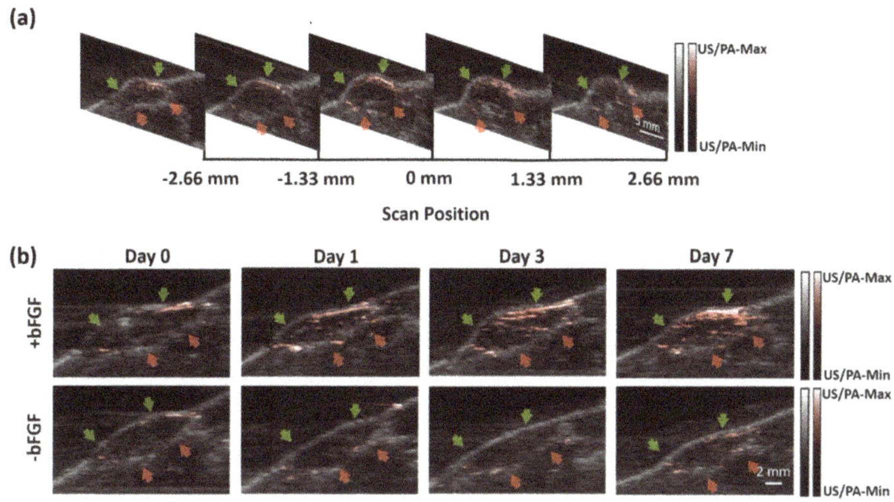

Figure 5. Longitudinal LED-based PA-US imaging of two subcutaneous implants. Green and red arrows indicate the upper and lower edges of the scaffold, as determined through the B-mode US. (**a**) A series of two-dimensional US-PA images from a +bFGF scaffold on day 7 at different scan positions. Note that only images within the range of −2.5 to 2.5 mm are used for MIP image. (**b**) A series of longitudinal MIP US-PA images of +bFGF and −bFGF scaffolds from the same mouse. PA intensity represented in red has the greatest difference on day 7. Reprinted from Ref. [7].

3.2.2. Noninvasive Imaging of Pressure Ulcers

Hariri et al. used PAI as a noninvasive method for detecting early tissue damage that cannot be visually observed. They used a mouse model of pressure ulcers by implanting subdermal magnets in the dorsal flank and periodically applying an external magnet to the healed implant site. The magnet-induced pressure was applied in cycles, and the extent of ulceration was dictated by the number of cycles. They evaluated these ulcers with LED-based PAI. Figure 6 shows baseline (top) and stage I ulcers (bottom) via US, PA, and US/PA overlay images obtained using LED excitation. The insets in Figure 6B–C are photographs of mice without and with pressure ulcers, respectively. They used a LED-based PAI system to detect early stage (stage I) pressure ulcers and observed a 2.5-fold increase in PA signal. Importantly, they confirmed the capacity of this technique to detect dysregulated skin even before stage I ulcers have erupted. They also observed significant changes in PA intensity during healing, suggesting that this approach can monitor therapy. These findings were confirmed with histology (not shown here). These results suggest that this PA-based approach might have clinical value for monitoring skin diseases, including pressure ulcers.

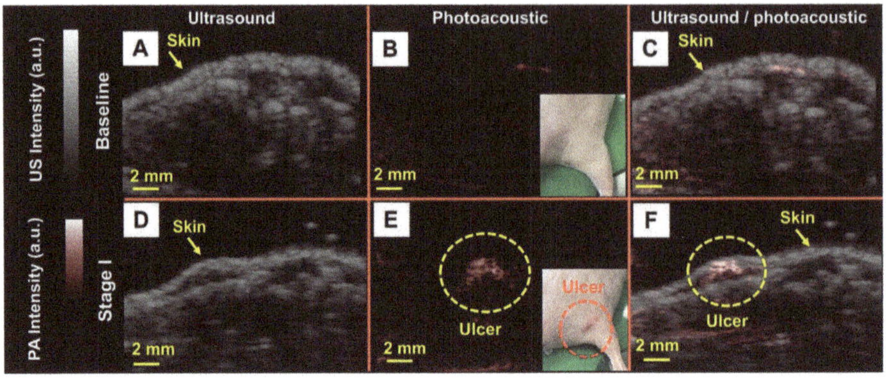

Figure 6. LED-based PA evaluation of pressure ulcers at stage I. (**A**) B-mode US image at baseline conditions when no pressure has been applied. (**B**) B-mode PA image at baseline at the same position as panel A. Minor PA signal is observed from the epidermis. The photographic inset shows the mouse in absence of ulcer. (**C**) B-mode PA/US overlay at baseline conditions. (**D**) B-mode US image at stage I. (**E**) B-mode PA image at stage I at the same position as panel D. We observed a 2.5-fold increase in PA intensity compared to baseline. The photographic inset shows the stage I ulcer. (**F**) B-mode PA/US overlay at stage I. The image depth is 1 cm and the scale bars are 2 mm. Reprinted with permission from Ref. [76].

3.2.3. Oxygen Saturation Imaging in Rheumatoid Arthritis Diagnosis

Multiwavelength LED-based PAI is a useful tool for functionally characterizing tissue in different clinical applications, for example, to obtain oxygen saturation. Hypoxia in the joints are biomarkers of Rheumatoid Arthritis (RA). The ability to accurately estimate the oxygen concentration makes multiwavelength PAI a potential tool for early detection of RA. Joseph et al. used in vivo animal study to find the capability of measuring the oxygen saturation using this system [77]. First, ex-vivo PA oxygen saturation imaging using human blood was validated against oximeter readings and further verified with in-vivo animal studies. The PA oxygen saturation estimation correlates with oximeter readings, which is confirmed with in-vivo studies. In the case of oxygen saturation, an approximately 5-mm imaging depth was achieved. This imaging depth is considered sufficient for RA imaging of the finger joint.

Results from the blood oxygen saturation imaging are shown in Figure 7. Figure 7a–c shows an oxygen saturation map of two tubes, one with normal blood (oxygen concentration approximately 65%) and another with blood having 18%, 43% and 87% oxygen concentration respectively. This correlates with the estimated oxygen saturation generated by the imaging system. Further, in-vivo imaging of the thigh muscle of the mouse is shown in Figure 7d–f. In the case of the mouse breathing normal air, an average oxygen concentration of 61.1% was observed and with 100% oxygen it was 105.3% using the system. This shows that the system can provide a fairly accurate estimate of oxygen saturation with in-vivo imaging. Through their results, they have shown that early stage rheumatoid arthritis changes such as synovial angiogenesis and hypoxia can be imaged using the LED-based PA imaging system. The results give a direct confirmation that multispectral LED-based PA holds potential in early detection and staging of RA in animal studies.

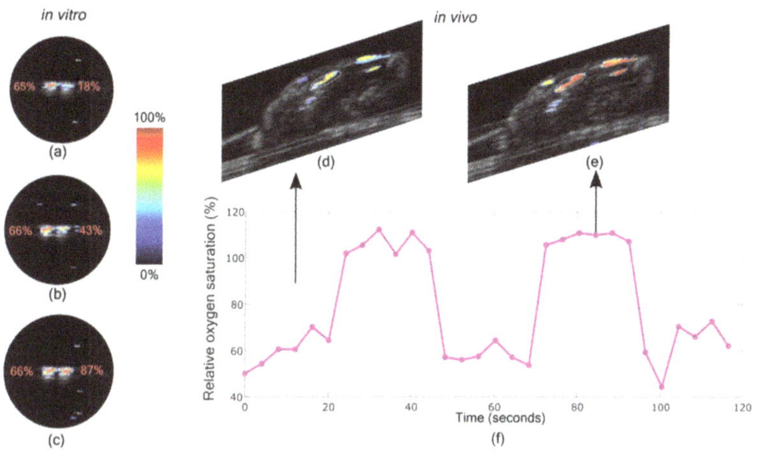

Figure 7. (**a–c**) The in-vitro study oxygen saturation map, with three different oxygen concentrations in the tube on the right, keeping oxygen unchanged in the left tube. The oxygen concentrations values in left and right tube: (**a**) 65% and 18%, (**b**) 66% and 43%, (**c**) 66% and 87%, (**d**) PA and (**e**) US imaging of mouse knee. In-vivo oxygen saturation images of mouse thigh muscle during high and low cycles of oxygen concentration. (**f**) Average oxygen saturation in the region of interested plotted over time. Reprinted from Ref. [77].

3.2.4. Molecular Imaging: Detection and Monitoring of Reactive Oxygen and Nitrogen Species

Reactive oxygen and nitrogen substances (RONS) regulate important functions in living systems. Endogenous RONS contribute to signal transduction, smooth muscle relaxation, and blood pressure regulation [66,67]. RONS disorders can cause diseases, such as cancer, and RONS detection can be used to diagnose and treat infections and various diseases [68–70].

In 2019, Hariri et al. reported on molecular imaging of RONS using near-infrared absorbing small molecules (CyBA) and an LED-based PA imaging system [29]. They evaluated CyBA's ability to measure inflammation in mice. Figure 8a–d shows the US/PA images from the injection location of CyBA and zymosan at 0, 10, 20, and 60 min, respectively (Figure 8). These figures show an increasing trend in PA signals due to the diffused probe (CyBA) at the region of interest (ROI), marked using a yellow circle. Figure 8e represents the quantitative analysis of CyBA's in-vivo PA imaging assessment. These results indicated a gradual increase in PA intensity, and a ~3.2-fold increase was quantified 90 min after CyBA injection. It also shows the flat PA intensity trend of separated zymosan and the dye.

As we know, the LED output energy is about 1000-fold lower than that of conventional Nd:YAG lasers. However, it turns out to be a positive factor here: the higher power generated by Nd:YAG lasers bleach dyes, but not the light from LEDs. Therefore, their work not only revealed the sensitivity of PA in detecting RONS, but also emphasized the practicality of LED-based PAI in the clinic for molecular imaging applications involving dyes. In this work, they demonstrated that advantages of LED as an illumination source may help to accelerate the clinical translation of PAI. It is foreseen that LED-based PAI and reported probes can be used for clinical monitoring of RONS, especially for keloid diagnosis and for drug toxicity studies [78].

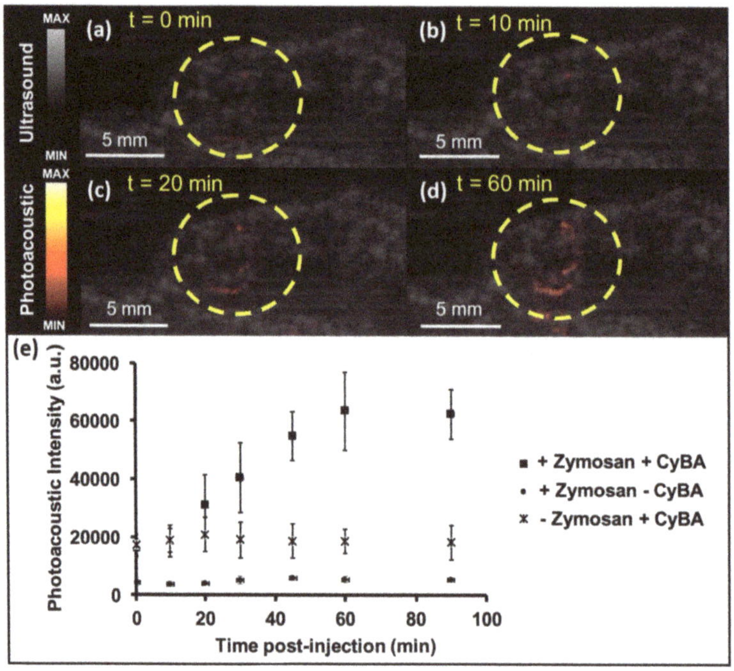

Figure 8. In-vivo PA evaluation of CyBA. US-PA image at (**a**) baseline, (**b**) 10, (**c**) 20, and (**d**) 60 min after CyBA injection. (**e**) Quantitative analysis of PA intensity as a function of time post-injection of CyBA. Reprinted from Ref. [29].

3.2.5. Imaging of Tumor Vasculature Using Contrast Enhancement

Contrast enhancement is prevalently used in vasculature imaging. Xavierselvan et al. tested the LED-based PAI system for its ability to image the vasculature in the tumor using contrast enhancement [79]. For their study, they used subcutaneous head and neck tumor (FaDu) xenografts in nude mice. When the tumor size reached about 100 mm^3, the tumors were imaged using the AcousticX system. The image shows heterogenous vascular density in the tumor which demonstrates the ability of the LED-based PAI system to obtain vascular information from tissues that are more than 1 cm deep.

Through their study, they found LED-based PAI was a more affordable option for various research groups to avail the opportunity of utilizing the technology to understand nanoparticle or drug uptake non-invasively at high resolution. Figure 9 demonstrates the capability of the LED-based PAI system in imaging a naphthalocyanine (NC) dye, which has strong absorption in the NIR region (~absorption peak at 860 nm). Before and after 100 µL of NC dye was injected intratumorally into the FaDu tumor interstitium, the distribution of dye inside the tumor was imaged respectively. The NC dye has distributed into almost all parts of the tumor, but a strong PA signal was received from the top of tumor, which is potentially due to these areas receiving stronger light energy.

3.2.6. Imaging of Molecular-Labelled Cells

Hariri et al. have previously used PA imaging for stem cell imaging [80,81]. In a recent work [27], they used labeled cells to understand the in-vivo molecular imaging performance of AcousticX. They used DiR, which is an effective contrast agent for cellular imaging. The cells used here were human mesenchymal stem cells (HMSCs; Lonza, PT−2501, NJ, USA).

Figure 10 shows PA images before and after injection of DiR, DiR + HMSC, and HMSC, respectively. The needle generates a strong PA signal and simultaneous US acquisition, offering detailed structural information along with the functional details from DiR-labeled cells. A strong PA signal in the presence of DiR is visible in Figure 10D. A signal increase is also visible when HMSC is labelled using DiR, as shown in Figure 10H. Unlabeled HMSCs were also injected as control but there was no increase in PA signal. This study showed the potential of LED-based PA in cellular imaging.

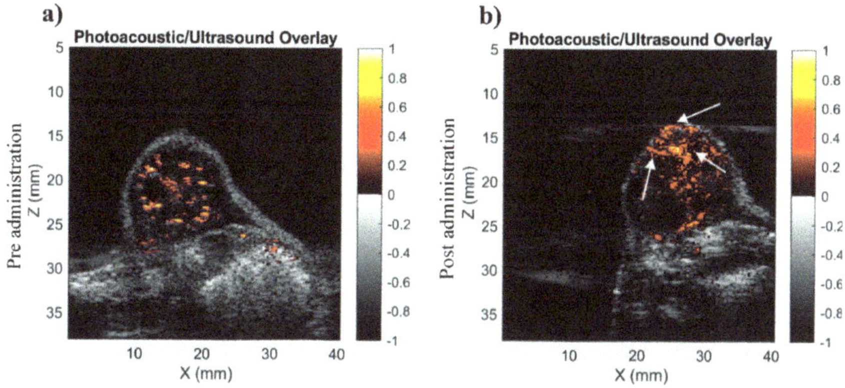

Figure 9. Contrast enhancement in LED-based PAI using exogeneous contrast agents. Combined overlay of PA and US image of subcutaneous FaDu tumor in mice before (**a**) and after (**b**) the NC dye administration. PA images were acquired using 850 nm LED light source (areas with greater contrast is shown with white arrows). Reprinted with permission from Ref. [79].

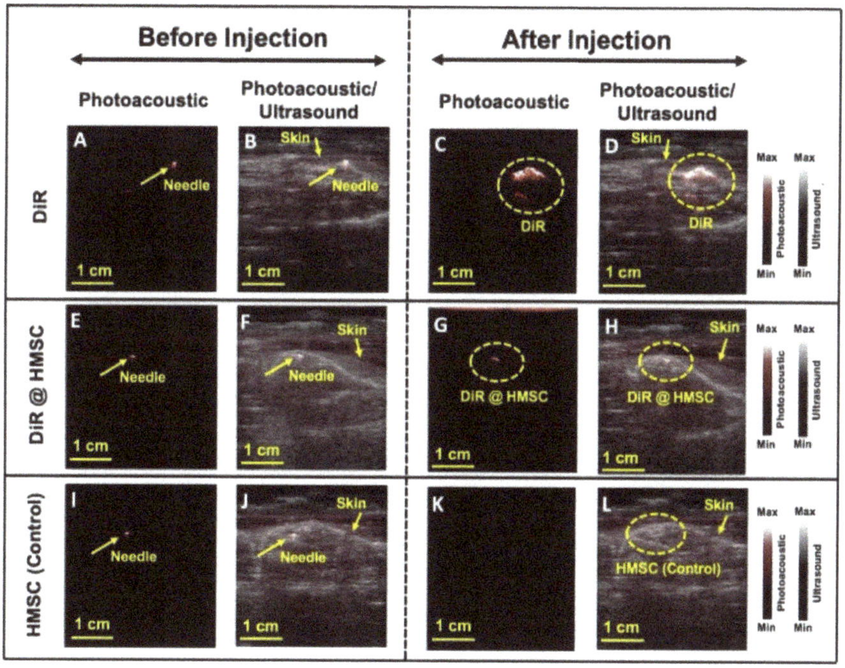

Figure 10. Images were separated in three groups: injecting DiR (**A–D**), DiR + HMSC (**E–H**), and HMSC (**I–L**). Each group contains PA and overlaid PA-US images before and after injection. Needles marked in images are subcutaneously injected on spinal cord area before DiR injection. The needle has a strong PA signal. Reprinted from Ref. [27].

3.3. In-Vivo Human Volunteer Studies

3.3.1. Imaging of Peripheral Microvasculature and Function

Hypoxia is an important biomarker that reflects the occurrence and development of many diseases, such as cancer [82]. Multispectral PA imaging has been demonstrated to detect the relative hemoglobin oxygen saturation and hypoxia in biological samples in vivo in a non-invasive manner by detecting the spectral differences between oxygenated hemoglobin and deoxyhemoglobin [83–85]. Figure 11a shows LED-based PA and US B-scan images of a human finger along the sagittal section, which is from our recent work [6]. To show the arterial pulsation (shown by the arrow), four frames of the movie are shown. In this work, we also explored the feasibility of LED-based PA imaging to measure blood oxygen content on human finger blood vessels using a pair of two-wavelength LED bars (850 nm and 690 nm), as shown in Figure 11b [6].

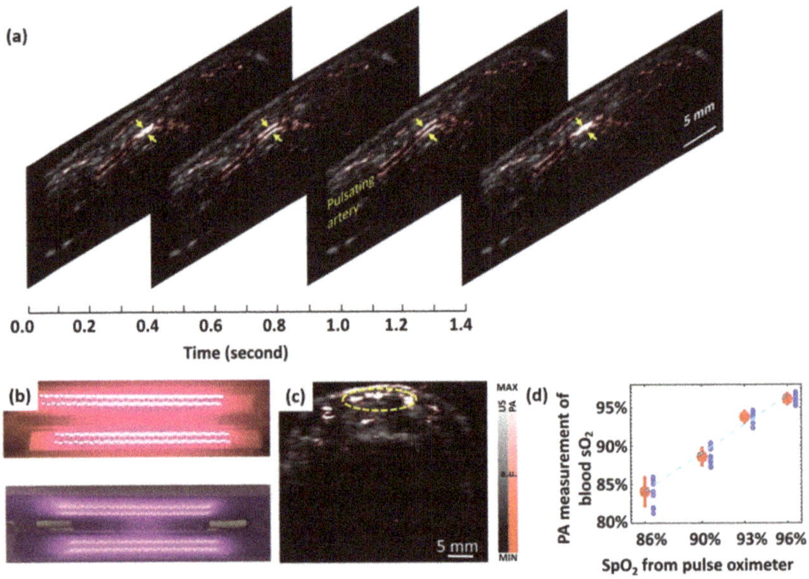

Figure 11. (**a**) Four frames from a PA imaging video (index finger, longitudinal section) presenting the pulsation of an artery marked by the arrows. (**b**) Photo of a pair of dual-wavelength LED bars that emits 690-nm and 850-nm light alternatively. (**c**) PA-US combined image showing a cross-section of the vasculature in a human finger. (**d**) Correlation between the LED-PAI-based blood sO2 in the finger and the SpO2 readouts from a pulse oximeter. Reprinted from Ref. [6].

Functional PA imaging of blood oxygenation saturation in blood vessels was performed on the index fingers of volunteers. Figure 11c is an example of a PA 2D image resolving blood vessels in an axial view of a finger, which is overlaid on a US image in grayscale. In order to quantify the functional information of PA imaging results, the oxygenation saturation levels of pixels in the ROI are averaged, as shown by the yellow dotted circle. The quantitative PA measurement of blood oxygenation saturation in the finger is then correlated with the reading of a pulse oximeter, which is the gold standard. As shown in Figure 11d, the blood oxygenation measured by LED-PAI and the reading of the pulse oximeter have a good correlation (R-squared ~ 0.98).

3.3.2. Simultaneous Imaging of Veins and Lymphatic Vessels

Using dual-wavelength LED arrays working at 820/940-nm wavelengths, Kuniyil Ajith Singh et al. demonstrated that LED-based PAI could differentiate veins and lymphatic vessels (after injecting ICG) in human volunteers [86]. Utilizing a simple image division algorithm (assuming that ICG will not absorb 940-nm light), they showed that LED-based PAI can clearly differentiate veins and lymphatic vessels in real time, as shown in Figure 12. This multispectral PA/US approach holds strong potential in guiding procedures like lymphaticovenous anastomosis, where it is crucial to differentiate venous blood and lymphatic vessels.

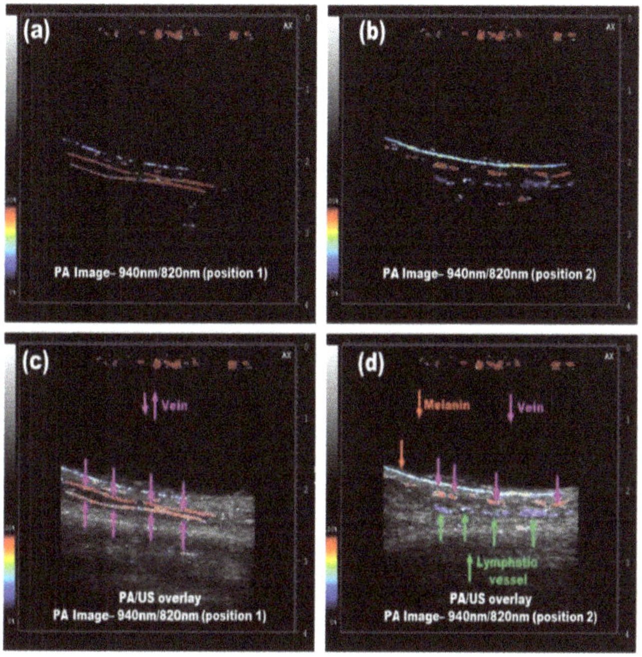

Figure 12. Real-time imaging of lymphatic vessels and veins simultaneously using LED-based PAI. (**a**) PA image: 940 nm/820 nm image generated when the probe was aligned to a vein (position 1); (**b**) PA image: 940 nm/820 nm image generated when the probe was aligned to a lymphatic vessel (position 2); (**c**) 940 nm/820 nm PA image overlaid on US image when the probe was in position 1; and (**d**) 940 nm/820 nm PA image overlaid on US image when the probe was aligned to position 2. Reprinted from Ref. [86].

3.3.3. Full View Tomography of Finger Joints

It is well known that 2D PA and US imaging using linear arrays suffers from limited-view artifacts. Additionally, due to the directivity limitations of US transducers, there is a high probability of loss of information. One way to overcome this limitation is to scan around the object of interest and generate full-view 3D tomographic images with higher spatial resolution and no limited view problems. However, most of the commercially available and lab-made 3D PA/US tomography systems utilize pulsed lasers, which are expensive, bulky and not suitable for a point-of-care setting. Joseph et al. recently demonstrated the possibility to generate full view LED-based 3D PA and US tomographic images of human finger joints (Figure 13) [87,88]. This proof-of-concept study gives a direct confirmation that inexpensive and portable LEDs along with commercially available linear array US probes can generate high-quality 3D tomographic images with rich information suitable for multiple point-of-care clinical applications. In another promising study, Agrawal et al. developed a multispectral molecular LED-based 3D PA tomography system and demonstrated the potential in unmixing and separating three optical absorbers (melanin, methylene blue and ICG) embedded inside a tissue mimicking phantom [89].

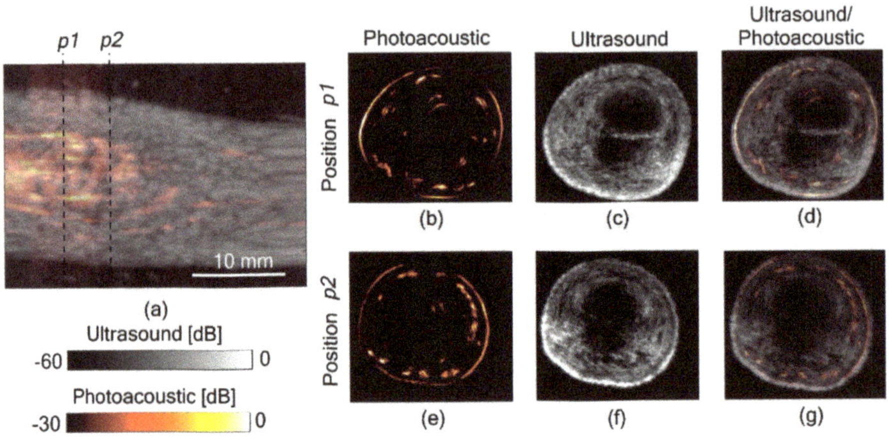

Figure 13. In-vivo finger joint tomographic imaging using multi-angle spatial compounding. (**a**) Overlaid LED-based PA and US MIP image showing finger joint from a linear scan. (**b**) PA, (**c**) US and (**d**) combined tomographic images of the finger joint (p1). (**e**) PA, (**f**) US and (**g**) combined tomographic images 5 mm in front of the joint (p2), respectively. Reprinted from [87].

3.4. Clinical Pilot Studies

3.4.1. Imaging of Inflammatory Arthritis

Since PAI can detect hyper-vascularization and hypoxia, two key early markers of rheumatoid arthritis, it has been explored extensively in multiple preclinical and early clinical pilot studies [90–94]. Relatively small joints are usually first affected by inflammatory arthritis, and PAI in combination with conventional US imaging holds strong potential in this area. Early results from animal models and human subjects clearly indicate that PAI is expected to be translated soon to clinics as an affordable point-of-care tool for early detection of inflammatory diseases [95–99]. Janggun et al. explored the feasibility of using an LED-based PA system to detect inflammation of soft tissues surrounding joints, and the ability to distinguish arthritic joints from normal ones by assessing enhanced blood signal in synovial tissue [31].

Figure 14 shows representative US Doppler and PA images of the three groups compared in this study, including clinically active arthritis joints, subclinical active arthritis joints, and normal healthy joints. Figure 14a shows a case of a clinically active case of arthritis, with hyperemia seen at the same location in both Doppler and PA images. Figure 14b shows a case of subclinical active arthritis, in which hyperemia is only visible in the PA image of the patient's metacarpophalangeal (MCP) joint, but not in the US Doppler image. The normal subjects also underwent US Doppler scans as the patient volunteers, followed by scanning of same location of joints using LED-based PA imaging. When compared with the results of arthritic joints, no significant hyperemia was found in the synovium area of normal subjects, which has also been confirmed by US Doppler imaging. Figure 14c shows the normal situation; no hyperemia was seen in the US Doppler or PA image. Finally, Figure 14d,e shows the statistical results indicating there is significant difference between different groups, confirmed using a two-tailed p-test. This clinical pilot study gives a direct confirmation that LED-based PAI offers higher sensitivity to angiogenic microvasculature than US doppler imaging, the current gold standard.

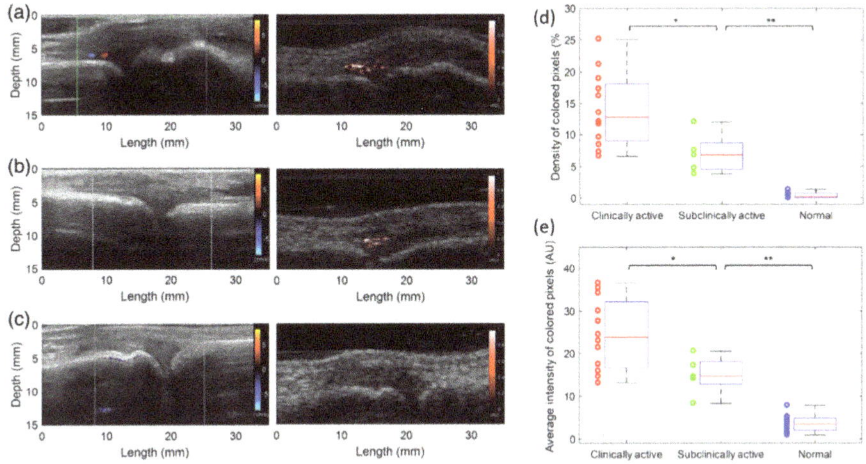

Figure 14. US Doppler (left) and PA (right) images acquired from LED-based system from three groups: (**a**) clinically active arthritis joints, (**b**) subclinical active arthritis joints, and (**c**) normal healthy joints). (**d**) The quantified results showing the density of colored pixels and (**e**) the average intensity of colored pixels in PA images of the three groups. * $p < 0.05$, ** $p < 0.005$. Reprinted from Ref. [31].

3.4.2. Diagnosis and Treatment Monitoring of Port Wine Stain

Port wine stain (PWS) is a benign capillary vascular malformation, with significant social and emotional impact. In a recent clinical pilot study, Cheng et al. demonstrated that LED-based PAI can be used as a point-of-care tool for clinical evaluation and treatment monitoring of PWS disease [100]. We foresee that LED-PAI will a have profound impact in this application, in which vascular contrast with depth information (~1 cm) is key for good diagnosis.

PWS is categorized as a benign capillary vascular malformation, which is difficult to cure. In general, PWS appears on the face, but it can affect other areas of the body too. The affected skin surface may thicken slightly and develop an irregular, pebbled surface in adulthood. PWS's cosmetic appearance causes substantial mental stress for the patients. In the study by Cheng et al., 22 patients were enrolled and separated into two groups based on their age (group 1: 3–6 years, group 2: above 6 years). The representative PA/US images (PWS region and normal region with clear difference in vascular contrast) are shown in Figure 15, along with the masked photograph of patient. The significant difference between the two different age groups also corresponds well with the given knowledge of PWS disease. Based on this result, it is clear that PAI as an imaging tool holds good potential in evaluation of PWS. Through this clinical pilot study, they demonstrated for the first time that LED-based PAI can be used as a point-of-care tool for clinical evaluation and PDT-treatment monitoring of PWS disease.

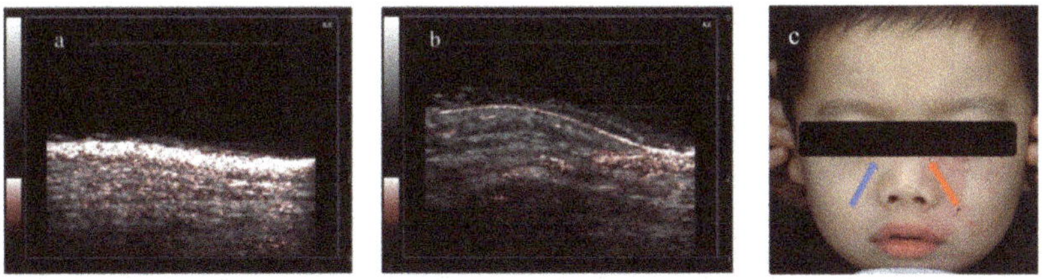

Figure 15. Typical PA/US overlay image of (**a**) PWS region (ROI) and (**b**) control region (HR). (**c**) Photograph of a PWS patient with imaging assisting marks. Reprinted with permission from Ref. [101].

4. Discussion

In this paper, we first reviewed the historical development of LED-based PAI, starting from the first report on single-point measurements to latest clinical pilot studies using high-power LED arrays. In the span of 10 years, there has been significant growth in this field, especially with the improvement of pulse energy (nJ to hundreds of μJ) and PRR (200 Hz to 16,000 Hz) of LEDs and also the advancements in low-noise data acquisition electronics. All these developments have resulted in commercialization of the technology, and it is worth mentioning that LED-PAI is now capable of functional imaging (oxygenation and blood flow imaging) of superficial and sub-surface tissue (more than 1 cm) at frame rates unachievable for laser-based systems (500 Hz). Even though LED-based PAI cannot be used for applications requiring larger imaging depth (for example, a full breast), it holds potential in several superficial imaging applications, specifically in rheumatology and dermatology.

Compared to solid-state lasers, the energy level of LED is two orders lower. However, the PRR of LED is much higher than that of a laser, which then can largely benefit the imaging quality by averaging more frames. Besides the emission energy level, another major difference between the pulsed LEDs and the solid-state lasers is the temporal pulse width. The pulse width of LED is tens of nanoseconds, whereas that of the solid-state lasers could be less than ten nanoseconds. The temporal pulse width imposes a limit on the spatial resolution of the imaging system. For example, the 70-ns pulse width of the 850-nm LED corresponds to a spatial resolution of 105 μm (=70 ns × 1500 μm/μs). This point, however, turns out to benefit the detection efficiency, especially when using bandlimited US probes for detection. The PA signal generated by a solid-state laser with a pulse width of 3.5 ns has a frequency component up to almost 300 MHz, in which anything above ~12 MHz will not be detected using a conventional mid-frequency range US probe. For a typical 5-MHz commercial US probe with 80% detection bandwidth, when using a 3.5-ns laser pulse, PA signal detection efficiency is 40 times less than the attainable efficiency when using an LED array generating 100-ns light pulses [102]. The resolution offered by a typical 5–10-MHz clinical US probe is 200 μm. Therefore, the pulse width of the LEDs can potentially be extended to 100 ns without affecting the spatial resolution (70 ns setting is used in all the studies reported in this paper).

In the second part of this paper, we reviewed some of the preclinical and clinical applications reported using LED-based PAI (in sequence of phantom, ex-vivo, small animal, and human in-vivo studies) as listed in Table 3. It is clear from the results that multispectral LED-based PA and US imaging holds strong potential in multiple applications, for example, guiding minimally invasive procedures, blood oxygen saturation imaging, diagnosis and staging of inflammatory arthritis, peripheral vascular assessment, guidance of surgical procedures like lymphaticovenous anastomosis, etc. In all the studies, an imaging depth of 0.5–1 cm was achieved at 10-Hz US and PA frame rates, which is good enough for multiple clinical applications. However, it is of paramount importance to improve the imaging depth for exploring more clinical applications, especially in the area of breast imaging and cardio-vascular medicine. To solve this issue and accelerate the clinical translation of LED-based PAI, several reconstruction and image processing techniques have been reported recently [103–105]. Use of clinically approved contrast agents (for example, ICG) may also be also useful to enhance the imaging depth.

The LED-based PA imaging system described here (AcousticX) is safe for both skin and eye exposure. Since LED emissions are incoherent, the ANSI safety limits for collimated laser beams do not apply. Instead, the international electrotechnical commission (IEC) 62471 is followed. According to IEC 62471, the exposure limit for skin is based on thermal injury due to the temperature rise in tissue. Assuming that the illumination on the same skin area lasts continuously for 5 s using two 850-nm LED bars working at a 4-KHz pulse repetition rate, the estimated exposure is 4.57×10^3 W·m^{-2}, which is below the thermal hazard limit for skin of 5.98×10^3 W·m^{-2}. For eye safety, two aspects need to be considered, which are retinal thermal hazard exposure limit (weak visual stimulus) and infrared radiation eye safety limit. Assuming a continuous illumination at the front of the eye for 5 s using two 850-nm LED bars working at a 4-KHz pulse repetition rate, the estimated exposures are

2.92×10^3 W·m^{-2}·sr^{-1} for retinal thermal exposure and 4.57×10^3 W·m^{-2} for infrared radiation exposure, both lower than the safety limits for eye of 1.34×10^5 W·m^{-2}·sr^{-1} and 5.38×10^3 W·m^{-2}, respectively.

Table 3. Summary of preclinical and clinical applications of LED-based PAI.

Target	Application		Depth (mm)	Contrast Agent	Wavelength (nm)
Medical needles, Vasculature	Guidance of minimally invasive procedures with peripheral tissue targets [32]	Phantom and ex vivo studies	38	N/A	850
Vasculature	Imaging of human placental vasculature [64]		7	N/A	850
Tumor	Imaging of intraocular tumors [6]		10	N/A	850
Vasculature	Non-invasive monitoring of angiogenesis [7]	Animal in vivo	10	N/A	850
Ulcer	Noninvasive imaging of pressure ulcers [76]		10	N/A	690
Oxygen saturation	Oxygen saturation imaging in Rheumatoid arthritis [77]		5	N/A	750/850
Molecular	Detection and monitoring of reactive oxygen and nitrogen species [29]		10	CyBA	850
Tumor/Contrast agents	Imaging of tumor using contrast enhancement [79]		10	NC	850
Cells/ Contrast agents	Imaging of molecular-labelled cells [27]		10	DiR	850
Vasculature	Imaging of peripheral microvasculature and function [6]	Healthy human in vivo	10	N/A	690/850
Vasculature	Simultaneous imaging of veins and lymphatic vessels [86]		10	ICG	940/820
Finger joints	Full view tomography of finger joints [87]		5	N/A	850
Finger joints	Imaging of inflammatory arthritis [31]	Patient in vivo	5	N/A	850
Skin	Imaging of port wine stain [101]		10	N/A	850

To date, the maximum imaging depth achieved by LED-based PA imaging in an in-vivo setting is 1 cm. Even though this is encouraging considering the low pulse energy, it is important to improve LED optical output for better usefulness in more clinical applications. To an extent, signal averaging helps to improve SNR without effecting frame rate. However, if the magnitudes of the acoustic signals generated by the weak illumination are significantly below the noise-equivalent pressure level of the US probe, averaging will not effectively improve SNR. In terms of translational potential, addition of PA imaging to a clinical US system will have relatively easier clinical acceptance. Considering the acoustic bandwidth limitation of conventional pulse-echo US probes, we believe that the lower pulse width (70 ns is the setting used in all the applications reported in this paper) of LEDs is not a bottleneck. We foresee work on two aspects to improve the imaging depth: (1) use of nanostack technology to squeeze more light out of LED elements, with multiple p-n junctions embedded in the epitaxial layer, increasing the region's light generation and thereby leading to a higher optical output, and (2) improving the driving electronics to increase the PRR further, consequently resulting in the possibility to average more frames and improve SNR without losing temporal resolution. LED might compete with LD with market share during clinical translation, but will not have much overlap with Nd: YAG OPO-based PAI system which focus on deep-tissue applications. However, LED-PAI may be a suitable option for point-of-care applications like guidance of peripheral vascular access procedures, rheumatoid arthritis screening, PWS diagnosis and treatment monitoring, etc.

Looking into the future, it is foreseen that advances in high-power LED technology, mainly driven by the lighting industry, and significant developments in machine/deep learning and signal processing algorithms will increase the use of LEDs in the context of PAI. LED arrays in different shapes could be also developed to find different applications, for example, developing ring-shape LEDs for breast 3D imaging and coupling light into optical fibers for minimally invasive and endoscopic procedures. Worth mentioning here, if sufficient focusing can be achieved, LEDs will be the ideal candidate for acoustic-resolution PA microscopy (PAM). There have also been some recent reports in this direction. In 2017, Dai et al. presented a PAM system based on miniature LEDs working at a 405-nm source, which showed the capability of in-vivo mapping of vasculature networks in biological tissue [106,107]. They used a high-power LED (power~1.2 W) working at 405 nm wavelength, and a pulse width of 200 ns. The repetition rate was extended to 40 KHz towards meeting the high demand of scanning speed in a PAM setting. They acquired a complete PAM image in vivo in about 1 h, which is an encouraging result for the first LED-based PAM proof-of-concept study, but this temporal resolution is not good enough for clinical studies. On the other hand, for optical resolution PAM, the use of LEDs is likely to be challenging, as it would be difficult to achieve the necessary micron-scale diffraction-limited spot sizes. It would be also interesting to develop new low-frequency US probes (2–3 MHz) with ultra-high bandwidth and sensitivity to achieve higher imaging depth without compromising spatial resolution.

5. Conclusions

The use of LEDs as an illumination source introduces some limitations. First, the LEDs cannot be spectrally tuned, which eliminates the possibility of PA spectroscopic applications in which multiple chromophores are involved. Second, the pulse width of the LED is low when compared to a laser (30–100 ns), which affects the stress confinement satisfaction and can impact the efficiency of acoustic wave generation. Third, LEDs have low optical output power—this can limit penetration depth at higher frame rates. However, LED-based PAI systems offers several advantages, including a significant reduction in cost, smaller footprint, no requirement of laser calibration and monitoring, and no need for optical goggles or light-tight shields. Thus, LED-based systems are not only suitable for point-of-care non-invasive applications complementing US imaging, but also ideal for personalized or wearable PA equipment, and we foresee that this technology could additionally have broad utility in a number of therapeutic drug monitoring applications.

With wide optical wavelength range, flexible pulse-width setting, small footprint, low cost, and energy efficiency, LED-based PAI holds strong potential in functional and molecular preclinical and clinical imaging. We foresee that the addition of LED-based PAI to conventional US imaging in a clinical scanner will have a huge impact in point-of-care diagnostic imaging and also accelerate the clinical translation of PAI.

Author Contributions: Conceptualization, M.K.A.S. and J.Y.; methodology, N.S.; software, N.S.; investigation, Y.Z., T.F. and Q.C.; resources, Q.C. and X.W.; data curation, Y.Z. and T.F.; writing—original draft preparation, Y.Z. and T.F.; writing—review and editing, Y.Z., T.F., M.K.A.S. and J.Y.; supervision, S.D. All authors have read and agreed to the published version of the manuscript.

Funding: This research was funded by the National Key Research and Development Program of China (No. 2017YFC0111402) and the Natural Science Funds of Jiangsu Province of China (No. BK20181256, No. BK 20170826). This work was funded by the National Natural Science Foundation of China (Grant No. 11704188); China Postdoctoral Science Foundation (Grant No. 2019M651564).

Conflicts of Interest: Mithun Kuniyil Ajith Singh and Naoto Sato are employed by CYBERDYNE, INC. The authors have no other financial interests or conflict of interest to disclose. We clarify that in some of the reported studies in this review article [6,32,64,77,86,87,101], CYBERDYNE, INC as a company was involved in technical developments or design of experiments either directly or through industry-academic collaborative research projects.

References

1. Wang, L.V.; Hu, S. Photoacoustic tomography: in vivo imaging from organelles to organs. *Science* **2012**, *335 6075*, 1458–1462. [CrossRef]
2. Wang, X.; Pang, Y.; Ku, G.; Xie, X.; Stoica, G.; Wang, L.V. Noninvasive laser-induced photoacoustic tomography for structural and functional in vivo imaging of the brain. *Nat. Biotechnol.* **2003**, *21*, 803–806. [CrossRef] [PubMed]
3. Beard, P. Biomedical photoacoustic imaging. *Interface Focus* **2011**, *1*, 602–631. [CrossRef]
4. Oladipupo, S.; Hu, S.; Kovalski, J.; Yao, J.; Santeford, A.; Sohn, R.E.; Shohet, R.; Maslov, K.; Wang, L.V.; Arbeit, J.M. VEGF is essential for hypoxia-inducible factor-mediated neovascularization but dispensable for endothelial sprouting. *Proc. Natl. Acad. Sci. USA* **2011**, *108*, 13264–13269. [CrossRef] [PubMed]

5. Oladipupo, S.S.; Hu, S.; Santeford, A.C.; Yao, J.; Kovalski, J.R.; Shohet, R.V.; Maslov, K.; Wang, L.V.; Arbeit, J.M. Conditional HIF-1 induction produces multistage neovascularization with stage-specific sensitivity to VEGFR inhibitors and myeloid cell independence. *Blood J. Am. Soc. Hematol.* **2011**, *11*, 4142–4153. [CrossRef] [PubMed]
6. Zhu, Y.; Xu, G.; Yuan, J.; Jo, J.; Gandikota, G.; Demirci, H.; Agano, T.; Sato, N.; Shigeta, Y.; Wang, X. Light emitting diodes based photoacoustic imaging and potential clinical applications. *Sci. Rep.* **2018**, *8*, 1–12. [CrossRef]
7. Zhu, Y.; Lu, X.; Dong, X.; Yuan, J.; Fabiilli, M.L.; Wang, X. LED-Based Photoacoustic Imaging for Monitoring Angiogenesis in Fibrin Scaffolds. *Tissue Eng. Part C Methods* **2019**, *25*, 523–531. [CrossRef]
8. Erpelding, T.N.; Kim, C.; Pramanik, M.; Jankovic, L.; Maslov, K.; Guo, Z.; Margenthaler, J.A.; Pashley, M.D.; Wang, L.V. Sentinel lymph nodes in the rat: noninvasive photoacoustic and US imaging with a clinical US system. *Radiology* **2010**, *256*, 102–110. [CrossRef]
9. Kim, C.; Cho, E.C.; Chen, J.; Song, K.H.; Au, L.; Favazza, C.; Zhang, Q.; Cobley, C.M.; Gao, F.; Xia, Y. In vivo molecular photoacoustic tomography of melanomas targeted by bioconjugated gold nanocages. *ACS Nano* **2010**, *4*, 4559–4564. [CrossRef]
10. Hu, S.; Wang, L. Neurovascular photoacoustic tomography. *Front. Neuroenerg.* 2010. [CrossRef]
11. Hu, S.; Yan, P.; Maslov, K.; Lee, J.-M.; Wang, L.V. Intravital imaging of amyloid plaques in a transgenic mouse model using optical-resolution photoacoustic microscopy. *Opt. Lett.* **2009**, *34*, 3899–3901. [CrossRef] [PubMed]
12. Hu, S.; Rao, B.; Maslov, K.; Wang, L.V. Label-free photoacoustic ophthalmic angiography. *Opt. Lett.* **2010**, *35*, 1–3. [CrossRef] [PubMed]
13. Jiao, S.; Jiang, M.; Hu, J.; Fawzi, A.; Zhou, Q.; Shung, K.K.; Puliafito, C.A.; Zhang, H.F. Photoacoustic ophthalmoscopy for in vivo retinal imaging. *Opt. Express* **2010**, *18*, 3967–3972. [CrossRef] [PubMed]
14. Zhang, E.Z.; Povazay, B.; Laufer, J.; Alex, A.; Hofer, B.; Pedley, B.; Glittenberg, C.; Treeby, B.; Cox, B.; Beard, P. Multimodal photoacoustic and optical coherence tomography scanner using an all optical detection scheme for 3D morphological skin imaging. *Biomed. Opt. Express* **2011**, *2*, 2202–2215. [CrossRef]
15. Favazza, C.P.; Wang, L.V.; Jassim, O.W.; Cornelius, L.A. In vivo photoacoustic microscopy of human cutaneous microvasculature and a nevus. *BIOMEDO* **2011**, *16*, 016015. [CrossRef]
16. Yang, J.-M.; Favazza, C.; Chen, R.; Yao, J.; Cai, X.; Maslov, K.; Zhou, Q.; Shung, K.K.; Wang, L.V. Toward dual-wavelength functional photoacoustic endoscopy: Laser and peripheral optical systems development. In Proceedings of the Photons Plus Ultrasound: Imaging and Sensing , San Francisco, CA, USA, 9 February 2012; p. 822316.
17. Yang, J.-M.; Maslov, K.; Yang, H.-C.; Zhou, Q.; Shung, K.K.; Wang, L.V. Photoacoustic endoscopy. *Opt. Lett.* **2009**, *34*, 1591–1593. [CrossRef]
18. Zhu, Y.; Johnson, L.A.; Huang, Z.; Rubin, J.M.; Yuan, J.; Lei, H.; Ni, J.; Wang, X.; Higgins, P.D.; Xu, G. Identifying intestinal fibrosis and inflammation by spectroscopic photoacoustic imaging: an animal study in vivo. *Biomed. Opt. Express* **2018**, *9*, 1590–1600. [CrossRef]
19. Zhu, Y.; Johnson, L.A.; Rubin, J.M.; Appelman, H.; Ni, L.; Yuan, J.; Wang, X.; Higgins, P.D.; Xu, G. Strain-photoacoustic imaging as a potential tool for characterizing intestinal fibrosis. *Gastroenterology* **2019**, *157*, 1196–1198. [CrossRef]
20. Zhu, Y.; Johnson, L.; Rubin, J.; Wang, X.; Higgins, P.; Xu, G. Characterization of intestinal fibrosis and inflammation with transcutaneous spectroscopic PA imaging (Conference Presentation). *Photons Plus Ultrasound Imaging Sens.* **2018**, 10494. [CrossRef]
21. Feng, T.; Perosky, J.E.; Kozloff, K.M.; Xu, G.; Cheng, Q.; Du, S.; Yuan, J.; Deng, C.X.; Wang, X. Characterization of bone microstructure using photoacoustic spectrum analysis. *Opt. Express* **2015**, *23*, 25217–25224. [CrossRef] [PubMed]
22. Kaiplavil, S.; Mandelis, A.; Wang, X.; Feng, T. Photothermal tomography for the functional and structural evaluation, and early mineral loss monitoring in bones. *Biomed. Opt. Express* **2014**, *5*, 2488–2502. [CrossRef] [PubMed]

23. Feng, T.; Kozloff, K.M.; Tian, C.; Perosky, J.E.; Hsiao, Y.-S.; Du, S.; Yuan, J.; Deng, C.X.; Wang, X. Bone assessment via thermal photo-acoustic measurements. *Opt. Lett.* **2015**, *40*, 1721–1724. [CrossRef] [PubMed]
24. Wang, X.; Feng, T.; Cao, M.; Perosky, J.E.; Kozloff, K.; Cheng, Q.; Yuan, J. Photoacoustic measurement of bone health: A study for clinical feasibility. In Proceedings of the 2016 IEEE International Ultrasonics Symposium (IUS), Tours, France, 18–21 September 2016.
25. Jansen, K.; Van Der Steen, A.F.; van Beusekom, H.M.; Oosterhuis, J.W.; van Soest, G. Intravascular photoacoustic imaging of human coronary atherosclerosis. *Opt. Lett.* **2011**, *36*, 597–599. [CrossRef]
26. Wang, B.; Yantsen, E.; Larson, T.; Karpiouk, A.B.; Sethuraman, S.; Su, J.L.; Sokolov, K.; Emelianov, S.Y. Plasmonic intravascular photoacoustic imaging for detection of macrophages in atherosclerotic plaques. *Nano Lett.* **2009**, *9*, 2212–2217. [CrossRef]
27. Hariri, A.; Lemaster, J.; Wang, J.; Jeevarathinam, A.S.; Chao, D.L.; Jokerst, J.V. The characterization of an economic and portable LED-based photoacoustic imaging system to facilitate molecular imaging. *Photoacoustics* **2018**, *9*, 10–20. [CrossRef]
28. Hariri, A.; Jeevarathinam, A.S.; Zhao, E.; Jokerst, J.V. Molecular imaging of oxidative stress sensing using LED-based photoacoustic imaging (Conference Presentation). *Photons Plus Ultrasound Imaging Sens.* **2019**, 10878. [CrossRef]
29. Hariri, A.; Zhao, E.; Jeevarathinam, A.S.; Lemaster, J.; Zhang, J.; Jokerst, J.V. Molecular imaging of oxidative stress using an LED-based photoacoustic imaging system. *Sci. Rep.* **2019**, *9*, 1–10. [CrossRef]
30. Xia, W.; Maneas, E.; Huynh, N.T.; Singh, M.K.A.; Brown, N.M.; Ourselin, S.; Gilbert-Kawai, E.; West, S.J.; Desjardins, A.E. Imaging of human peripheral blood vessels during cuff occlusion with a compact LED-based photoacoustic and ultrasound system. *Photons Plus Ultrasound Imaging Sens.* **2019**, 10878. [CrossRef]
31. Jo, J.; Xu, G.; Zhu, Y.; Burton, M.; Sarazin, J.; Schiopu, E.; Gandikota, G.; Wang, X. Detecting joint inflammation by an LED-based photoacoustic imaging system: a feasibility study. *BIOMEDO* **2018**, *23*, 110501. [CrossRef]
32. Xia, W.; Kuniyil Ajith Singh, M.; Maneas, E.; Sato, N.; Shigeta, Y.; Agano, T.; Ourselin, S.; J West, S.; E Desjardins, A. Handheld real-time LED-based photoacoustic and ultrasound imaging system for accurate visualization of clinical metal needles and superficial vasculature to guide minimally invasive procedures. *Sensors* **2018**, *18*, 1394. [CrossRef] [PubMed]
33. Allen, T.J.; Beard, P.C. High power visible light emitting diodes as pulsed excitation sources for biomedical photoacoustics. *Biomed. Opt. Express* **2016**, *7*, 1260–1270. [CrossRef] [PubMed]
34. Agano, T.; Sato, N.; Nakatsuka, H.; Kitagawa, K.; Hanaoka, T.; Morisono, K.; Shigeta, Y. Attempts to increase penetration of photoacoustic system using LED array light source. *Photons Plus Ultrasound Imaging Sens.* **2015**, 9323. [CrossRef]
35. Agano, T.; Sato, N.; Nakatsuka, H.; Kitagawa, K.; Hanaoka, T.; Morisono, K.; Shigeta, Y. Comparative experiments of photoacoustic system using laser light source and LED array light source. *Photons Plus Ultrasound Imaging Sens.* **2015**, 9323. [CrossRef]
36. Allen, T.J.; Beard, P.C. Pulsed near-infrared laser diode excitation system for biomedical photoacoustic imaging. *Opt. Lett.* **2006**, *31*, 3462–3464. [CrossRef]
37. Kolkman, R.G.; Steenbergen, W.; van Leeuwen, T.G. In vivo photoacoustic imaging of blood vessels with a pulsed laser diode. *Lasers Med. Sci.* **2006**, *21*, 134–139. [CrossRef]
38. Daoudi, K.; Van Den Berg, P.; Rabot, O.; Kohl, A.; Tisserand, S.; Brands, P.; Steenbergen, W. Handheld probe integrating laser diode and ultrasound transducer array for ultrasound/photoacoustic dual modality imaging. *Opt. Express* **2014**, *22*, 26365–26374. [CrossRef]
39. Kalva, S.K.; Upputuri, P.K.; Pramanik, M. High-speed, low-cost, pulsed-laser-diode-based second-generation desktop photoacoustic tomography system. *Opt. Lett.* **2019**, *44*, 81–84. [CrossRef]
40. Agano, T.; Singh, M.K.A.; Nagaoka, R.; Awazu, K. Effect of light pulse width on frequency characteristics of photoacoustic signal—An experimental study using a pulse-width tunable LED-based photoacoustic imaging system. *Int. J. Eng. Technol.* **2018**, *7*, 4300–4303.
41. Upputuri, P.K.; Pramanik, M. Pulsed laser diode based optoacoustic imaging of biological tissues. *Biomed. Phys. Eng. Express* **2015**, *1*, 045010. [CrossRef]
42. Wang, T.; Nandy, S.; Salehi, H.S.; Kumavor, P.D.; Zhu, Q. A low-cost photoacoustic microscopy system with a laser diode excitation. *Biomed. Opt. Express* **2014**, *5*, 3053–3058. [CrossRef]

43. Allen, T.J.; Beard, P.C. Light emitting diodes as an excitation source for biomedical photoacoustics. *Photons Plus Ultrasound Imaging Sens.* **2013**, 8581. [CrossRef]
44. Allen, T.J. High-Power Light Emitting Diodes; An Alternative Excitation Source for Photoacoustic Tomography. In *LED-Based Photoacoustic Imaging*; Springer: Berlin/Heidelberg, Germany, 2020; pp. 23–43.
45. Willert, C.; Stasicki, B.; Klinner, J.; Moessner, S. Pulsed operation of high-power light emitting diodes for imaging flow velocimetry. *Meas. Sci. Technol.* **2010**, *21*, 075402. [CrossRef]
46. Hansen, R.S. Using high-power light emitting diodes for photoacoustic imaging. *Ultrason. Imaging Tomogr. Ther.* **2011**, *7968*. [CrossRef]
47. Ikuta, T.; Shimizu, R. The dynamic response of magnetic domain walls to applied fields in yttrium orthoferrite, observed by a stroboscopic technique. *J. Phys. D Appl. Phys.* **1974**, *7*, 726. [CrossRef]
48. McFarlane, W. An inexpensive nanosecond light pulser for use in photomultiplier system testing. *Rev. Sci. Instrum.* **1974**, *45*, 286–289. [CrossRef]
49. Araki, T.; Misawa, H. Light emitting diode-based nanosecond ultraviolet light source for fluorescence lifetime measurements. *Rev. Sci. Instrum.* **1995**, *66*, 5469–5472. [CrossRef]
50. Araki, T.; Fujisawa, Y.; Hashimoto, M. An ultraviolet nanosecond light pulse generator using a light emitting diode for test of photodetectors. *Rev. Sci. Instrum.* **1997**, *68*, 1365–1368. [CrossRef]
51. Chaney, A.; Sundararajan, R. Simple MOSFET-based high-voltage nanosecond pulse circuit. *IEEE Trans. Plasma Sci.* **2004**, *32*, 1919–1924. [CrossRef]
52. Agano, T.; Sato, N. Photoacoustic Imaging System using LED light source. In Proceedings of the 2016 Conference on Lasers and Electro-Optics (CLEO), San Jose, CA, USA, 5–10 June 2016; pp. 1–2.
53. Narouze, S.N. *Atlas of Ultrasound-Guided Procedures in Interventional Pain Management*; Springer: Berlin/Heidelberg, Germany, 2018.
54. Wang, L.V.; Yao, J. A practical guide to photoacoustic tomography in the life sciences. *Nat. Methods* **2016**, *13*, 627. [CrossRef]
55. Kim, C.; Erpelding, T.N.; Maslov, K.I.; Jankovic, L.; Akers, W.J.; Song, L.; Achilefu, S.; Margenthaler, J.A.; Pashley, M.D.; Wang, L.V. Handheld array-based photoacoustic probe for guiding needle biopsy of sentinel lymph nodes. *BIOMEDO* **2010**, *15*, 046010. [CrossRef]
56. Su, J.L.; Karpiouk, A.B.; Wang, B.; Emelianov, S.Y. Photoacoustic imaging of clinical metal needles in tissue. *BIOMEDO* **2010**, *15*, 021309. [CrossRef] [PubMed]
57. Pratt, R.; Deprest, J.; Vercauteren, T.; Ourselin, S.; David, A.L. Computer-assisted surgical planning and intraoperative guidance in fetal surgery: a systematic review. *Prenat. Diagn.* **2015**, *35*, 1159–1166. [CrossRef] [PubMed]
58. Lewi, L.; Deprest, J.; Hecher, K. The vascular anastomoses in monochorionic twin pregnancies and their clinical consequences. *Am. J. Obstet. Gynecol.* **2013**, *208*, 19–30. [CrossRef] [PubMed]
59. Guiot, C.; Gaglioti, P.; Oberto, M.; Piccoli, E.; Rosato, R.; Todros, T. Is three-dimensional power Doppler ultrasound useful in the assessment of placental perfusion in normal and growth-restricted pregnancies? *Ultrasound Obstet. Gynecol.* **2008**, *31*, 171–176. [CrossRef] [PubMed]
60. Ntziachristos, V.; Razansky, D. Molecular imaging by means of multispectral optoacoustic tomography (MSOT). *Chem. Rev.* **2010**, *110*, 2783–2794. [CrossRef]
61. Maneas, E.; Xia, W.; Singh, M.K.A.; Sato, N.; Agano, T.; Ourselin, S.; West, S.J.; David, A.L.; Vercauteren, T.; Desjardins, A.E. Human placental vasculature imaging using an LED-based photoacoustic/ultrasound imaging system. In Proceedings of the Photons Plus Ultrasound: Imaging and Sensing 2018, San Francisco, CA, USA, 19 February 2018.
62. Xia, W.; Maneas, E.; Nikitichev, D.I.; Mosse, C.A.; Dos Santos, G.S.; Vercauteren, T.; David, A.L.; Deprest, J.; Ourselin, S.; Beard, P.C. Interventional photoacoustic imaging of the human placenta with ultrasonic tracking for minimally invasive fetal surgeries. In *International Conference on Medical Image Computing and Computer-Assisted Intervention*; Springer: Berlin/Heidelberg, Germany, 2015.
63. Xia, W.; Nikitichev, D.I.; Mari, J.M.; West, S.J.; Pratt, R.; David, A.L.; Ourselin, S.; Beard, P.C.; Desjardins, A.E. Performance characteristics of an interventional multispectral photoacoustic imaging system for guiding minimally invasive procedures. *BIOMEDO* **2015**, *20*, 086005. [CrossRef]

106. Dai, X.; Yang, H.; Jiang, H. In vivo photoacoustic imaging of vasculature with a low-cost miniature light emitting diode excitation. *Opt. Lett.* **2017**, *42*, 1456–1459. [CrossRef]
107. Dai, X.; Yang, H.; Jiang, H. Low-cost high-power light emitting diodes for photoacoustic imaging. In Proceedings of the Photons Plus Ultrasound: Imaging and Sensing 2017, San Francisco, CA, USA, 29 January–1 February 2017; p. 100644.

sensors

MDPI

Article

Optical-Resolution Photoacoustic Microscopy Using Transparent Ultrasound Transducer

Haoyang Chen [1], Sumit Agrawal [1], Ajay Dangi [1], Christopher Wible [1], Mohamed Osman [1], Lidya Abune [1], Huizhen Jia [1], Randall Rossi [2], Yong Wang [1] and Sri-Rajasekhar Kothapalli [1,3,4,*]

1 Department of Biomedical Engineering, The Pennsylvania State University, State College, PA 16802, USA; haoyangchen@psu.edu (H.C.); sua347@psu.edu (S.A.); axd571@psu.edu (A.D.); clw5710@psu.edu (C.W.); mqo5163@psu.edu (M.O.); lua28@psu.edu (L.A.); huj202@psu.edu (H.J.); yxw30@psu.edu (Y.W.)

2 Huck Institutes of the Life Sciences, The Pennsylvania State University, State College, PA 16802, USA; rmr29@psu.edu

3 Penn State Cancer Institute, The Pennsylvania State University, Hershey, PA 17033, USA

4 Graduate Program in Acoustics, The Pennsylvania State University, State College, PA 16802, USA

* Correspondence: srkothapalli@psu.edu

Received: 23 November 2019; Accepted: 8 December 2019; Published: 11 December 2019

check for updates

Abstract: The opacity of conventional ultrasound transducers can impede the miniaturization and workflow of current photoacoustic systems. In particular, optical-resolution photoacoustic microscopy (OR-PAM) requires the coaxial alignment of optical illumination and acoustic-detection paths through complex beam combiners and a thick coupling medium. To overcome these hurdles, we developed a novel OR-PAM method on the basis of our recently reported transparent lithium niobate ($LiNbO_3$) ultrasound transducer (Dangi et al., Optics Letters, 2019), which was centered at 13 MHz ultrasound frequency with 60% photoacoustic bandwidth. To test the feasibility of wearable OR-PAM, optical-only raster scanning of focused light through a transducer was performed while the transducer was fixed above the imaging subject. Imaging experiments on resolution targets and carbon fibers demonstrated a lateral resolution of 8.5 μm. Further, we demonstrated vasculature mapping using chicken embryos and melanoma depth profiling using tissue phantoms. In conclusion, the proposed OR-PAM system using a low-cost transparent $LiNbO_3$ window transducer has a promising future in wearable and high-throughput imaging applications, e.g., integration with conventional optical microscopy to enable a multimodal microscopy platform capable of ultrasound stimulation.

Keywords: photoacoustic microscopy; ultrasonic transducer; optical-resolution photoacoustic microscopy; transparent ultrasound transducer; ultrasound stimulation

1. Introduction

Optical-resolution photoacoustic microscopy (OR-PAM) has recently gained significant attention from the biomedical-imaging community as it provides labelfree optical contrast from physiologically relevant tissue chromophores that are located a few millimeters deep, with subcellular spatial resolution [1–5]. In OR-PAM, a tightly focused laser pulse illuminates the tissue and generates wideband acoustic waves from light-absorbing chromophores that are then detected by an acoustic transducer. Time-resolved photoacoustic waves, in combination with the two-dimensional raster scanning along the x-y plane (lateral dimension), generate three-dimensional data from which maximum amplitude projection (MAP) and volume-rendered photoacoustic images can be created.

Conventional OR-PAM setups use complex imaging geometries to coaxially align optical illumination and acoustic detection paths. In early OR-PAM setups [6], coaxial alignment was achieved using an acoustic-optic prism combiner consisting of one right-angle prism and one rhomboid prism pressed tightly to a thin layer of silicone oil. The laser light was focused by a system of optical lenses and then passed through the prism combiner before irradiating the tissue. A correction lens was attached to the prism combiner to refocus the light that was defocused through the combiner. Tissue-generated photoacoustic waves propagated through the rhomboid prism and were reflected by the silicone oil layer into the ultrasound detector attached to the prism. Since the entire imaging head, consisting of the above acoustic-optic prism combiner, the transducer, and the focused light, was moved to scan the subject, these systems exhibited slow acquisition speed, limited field of view (FOV), and significant acoustic loss.

The current generation of OR-PAMs reflect the light, instead of the acoustic waves, by sandwiching an aluminium foil in the acoustic-optic combiner. This allows dual axis optic only scanning using a two dimensional galvo mirror to improve the image acquisition speed and generate a wide FOV [7]. The entire imaging head, including the galvo mirror, is submerged in a large volume ($70 \times 40 \times 20$ mm^3) of a nonconducting liquid coupling medium that rests above the imaged subject. Such a bulky imaging head limits high throughput and wearable imaging applications because it constrains animal imaging performed under anesthesia and causes discomfort to living subjects. Moreover, acoustic loss here is still significant because acoustic waves travel through the large coupling medium and the prism combiner before being detected by the transducer.

Alternatively, some OR-PAMs include a ring-shaped single-element ultrasound transducer to eliminate the off-axis alignment problems of optical illumination and acoustic detection. The focused light is directly delivered through a hole at the center of the transducer, or coupled using a single-mode fiber integrated with a gradient-index (GRIN) lens. The imaging head is then two-dimensionally raster-scanned using mechanical stages to generate volumetric images [8–12]. Although the imaging head is miniaturized in these OR-PAM systems, the FOV, numerical aperture, and imaging speed (due to physical scanning of the imaging head) are still limited. Besides, they still require a-few-milimeter thick water coupling medium above the imaged subject due to long working distances.

The above drawbacks of conventional OR-PAM systems can be addressed if ultrasound detectors are transparent to light. To achieve this, all-optical ultrasound detection technologies, such as Fabry–Pérot etalons [13], microring resonators [14], and other photonic integrated circuits [15] were studied for PAM. Although these are transparent technologies offering high photoacoustic sensitivity, they require complex fiber integration with an additional laser source and other optical-detection instruments. More importantly, they lack ultrasound excitation capabilities for applications that require combined ultrasound sensing and ultrasound tissue stimulation [16,17]. Recently, transparent capacitive micromachined ultrasonic transducers (CMUTs) were developed [18,19]. However, CMUTs need specialized front-end application-specific integrated circuits (ASICs), and involve a complicated fabrication process inside a cleanroom.

In order to address all the above limitations, we recently reported a photoacoustic-imaging technique using an optically transparent bulk piezoelectric lithium niobate ($LiNbO_3$) ultrasound transducer [20]. $LiNbO_3$ offers several advantages over other piezoelectric materials:

1. It exhibits a good electromechanical coupling coefficient (53%);
2. It demonstrates a high Curie temperature (>1100 °C), making it easy to process through high-temperature sputtering without losing poling [21];
3. It shows promising results in high-frequency ultrasound and PAM applications [22–24].

On the basis of these advantages, our previous work [20] introduced two transparent photoacoustic-imaging schematics, one that directly integrated a multimode optical fiber with the transparent $LiNbO_3$ transducer, and the other that involved the optical-only scanning of the laser spot over a 10×10 mm^2 $LiNbO_3$ window transducer.

Extending the window-transducer approach, here we present a novel OR-PAM that allows the optical-only scanning of a tightly focused light beam through a 10×10 mm^2 single-element $LiNbO_3$ transparent-ultrasound-transducer (TUT) window and demonstrates its applicability to image biological samples. The spatial resolution and signal-to-noise ratio (SNR) of the OR-PAM system were characterized using imaging experiments on resolution test targets and carbon-fiber phantoms. The biological imaging capabilities of the OR-PAM were studied using *ex ovo* chick-embryo chorioallantoic-membrane (CAM) vasculature and imaging melanoma phantoms through a piece of mouse skin.

The proposed TUT-based OR-PAM approach simplifies the coaxial alignment of optic and acoustic paths without the need for additional optical components (such as acoustic-optic prism combiners and correction lenses) and a large acoustic-coupling medium. Our approach provides other advantages:

1. The TUT can be fixed onto the imaging object (such as the skull of a mouse) to facilitate wearable imaging without a thick coupling medium. In the future, this will likely help in imaging the brains of freely behaving or awake mice in combination with ultrasound stimulation;
2. Depending on the size of the TUT, it enables the high-speed scanning of large areas with single-channel data acquisition.
3. In the future, our TUT approach can also be integrated into conventional optical microscopes to realize a multimodal microscopy platform with ultrasound-stimulation capabilities.

The rest of this paper is organized as follows. Section 2 describes the process of TUT fabrication and a schematic representation of the OR-PAM setup. TUT characterization and validation studies, including carbon-fiber imaging, CAM vasculature, and melanoma depth profiling, are presented in Section 3. The advantages of the proposed TUT-based OR-PAM system, the limitations, and its future directions are discussed in Section 4.

2. Materials and Methods

2.1. Transducer-Fabrication Processes

As reported in our recent work [20], a 250 μm thick Y-cut 36° $LiNbO_3$ wafer was sputtered with 200 nm thick indium tin oxide (ITO) on both sides and then diced into square pieces of 10 × 10 mm^2, which resulted in ~80% optical transparency in the visible and near-infrared optical-wavelength regions. Square tubing (0.5 in. width, 0.032 in. wall thickness, and 10 mm height) was used as a conductive housing. The bottom electrode and brass tubing were connected using a conductive silver epoxy (E-solder 3022, Von Roll Isola Inc., New Haven, CT, USA) that outlined the 4 edges of the bottom electrode. The conductive silver epoxy had 1 mm thickness and 2 mm total width, with 1 mm covering the bottom electrode. This resulted in an FOV of 9 × 9 mm^2. The conductive epoxy also acted as an absorber for the surface-acoustic waves generated by the $LiNbO_3$ in response to pulsed-light incidence.

The top electrode was connected to an SMA to BNC connector using a microstranded wire. A nonconducting and transparent epoxy (Epotek-301, Epoxy Technologies Inc., Billerica, MA, USA) was poured until it filled the brass housing. This epoxy was used as the backing layer that would reduce the ringing effect by absorbing vibrational energy, and improve bandwidth [25]. Extra care was taken to ensure that no particles were trapped in the epoxy that may have diffracted the light or caused a shadowing effect.

The epoxy is known to shrink during the curing process that can lead to a curved surface inside the brass housing. This curvature can lead to light diffractions and aberrations. In order to ameliorate this effect, a microglass slide with 150 μm thickness was placed on top of the transducer to form a flat surface. A cross-sectional schematic view of the TUT is shown in Figure 1a. Figure 1b shows a photograph of the TUT on top of a Nittany Lion mascot.

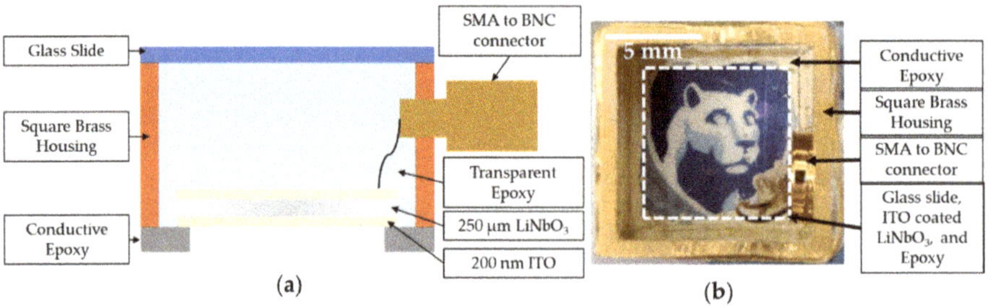

Figure 1. Transparent-ultrasound-transducer (TUT) design based on piezoelectric lithium niobate ($LiNbO_3$) material. (**a**) Schematic cross-sectional view of fabricated TUT, which had 10 mm height and 9×9 mm^2 field of view. ITO: Indium tin oxide; (**b**) Photograph of fabricated TUT clearly showing Nittany Lion mascot underneath.

2.2. Optical-Resolution Photoacoustic-Microscopy Experiment Setup

A schematic representing the top-down view of the OR-PAM setup is shown in Figure 2. The system employed a high-speed pulsed laser (GLPM-10, IPG Photonics; 532 nm wavelength; 1.4 ns pulse duration; tunable pulse-repetition rate in the range of 10–600 kHz; tunable pulse energy between 1.6 and 19 µJ). The 4 mm diameter laser beam passed through a beam sampler (BSF10-A, Thorlabs Inc., Newton, NJ, USA) that diverted 10% of its energy to a photodiode (DET10A, Thorlabs Inc.) used to synchronize with a high-speed (1 gigasample per second) 16 bit data-acquisition system (Razormax-16, Dynamic Signals LLC, Lockport, IL, USA) connected to a computer.

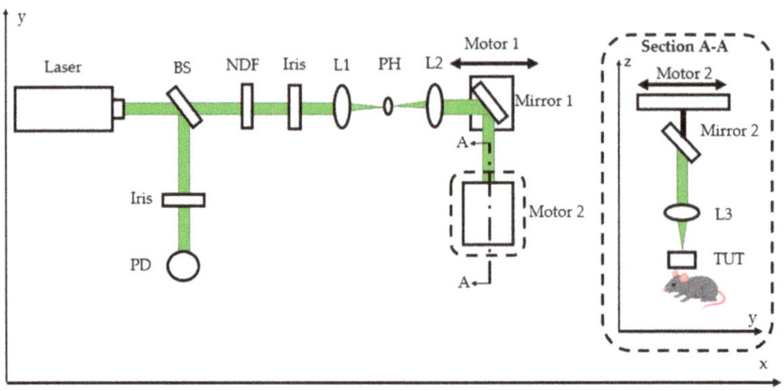

Figure 2. Top-down view schematic of optical-resolution photoacoustic-microscopy (OR-PAM) setup. Raster scanning achieved by Motor 1 moving Mirror 1 to perform x-axis scan. Motor 2 moved Mirror 2 and L3 to perform y-axis scan. BS: beam sampler; NDF: neutral density filter; PD: photodiode; PH: pinhole; L1, L2, L3: planoconvex lenses with 50, 75, and 50 mm focal lengths, respectively.

The remaining 90% of the beam energy passed through a neutral density filter (NDC-50C-4M, Thorlabs Inc., Newton, NJ, USA) and an iris before entering a spatial filter system. A 20 µm pinhole (P20D, Thorlabs Inc., Newton, NJ, USA) and 2 lenses, with focal lengths 50 (LA1131-A, Thorlabs Inc., Newton, NJ, USA) and 75 mm (LA1608, Thorlabs Inc., Newton, NJ, USA) respectively, were then used to filter and collimate the beam. In order to raster-scan the sample for imaging, 2 motorized stages (NRT-1000, Thorlabs Inc., Newton, NJ, USA) were used to guide the light along the x and y axes for scanning (Figure 2). A 45° mirror was mounted onto Motor-1, which moved along the x-axis. Motor-2 was mounted on a vertical stage (on the z-axis) and drove another 45° mirror and a 50 mm focal-length planoconvex lens (LA1131-A, Thorlabs Inc., Newton, NJ, USA) along the y-axis. The focused light passed through the TUT mounted just above the imaging sample. A thin layer (~1 mm) of deionized water was used as a coupling medium to receive photoacoustic waves generated from the tissue.

3. Results and Discussion

3.1. System and Transducer Characterization

First, the transducer was evaluated by analyzing its electrical impedance using a vector network analyzer (Agilent E5100A, Keysight Technologies, Inc., Santa Rosa, CA, USA). Impedance measurements are used to estimate the effective electromechanical coupling coefficient, k_{eff}, of the transducer, which represents its efficiency to convert between electrical and mechanical energy [26]. As seen in Figure 3a, the resonance and antiresonance frequencies were measured to be 12.05 and 14.22 MHz, respectively, with a resultant k_{eff} of ~0.53.

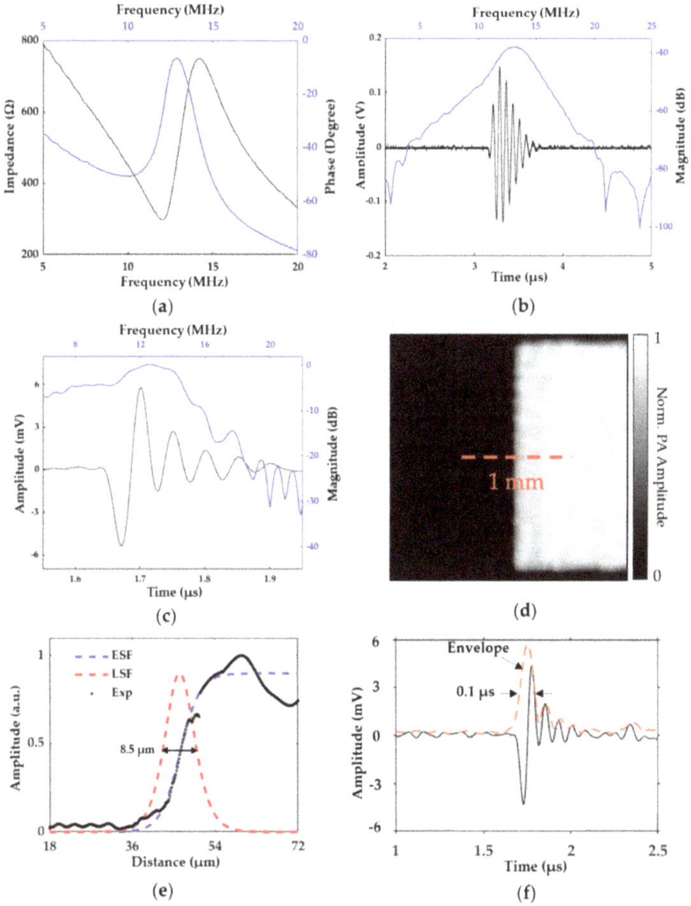

Figure 3. Results of transparent lithium niobate ultrasound-transducer-based OR-PAM system characterization. (**a**) Measured electrical impedance results; (**b**) Pulse-echo response; (**c**) Photoacoustic pulse response of USAF resolution test target; (**d**) Maximum-amplitude-projection (MAP) image of target via edge scanning. PA: photoacoustic; (**e**) Edge-response data and fitted line-spread-function (LSF) curve showed 8.5 μm lateral resolution. ESF: edge spread function; (**f**) Gaussian enveloped curve fitted profile showed 150 μm axial resolution.

Next, we performed pulse-echo and hydrophone measurements on the TUT using the methods described in our earlier work [20]. Pulse-echo measurement showed a center frequency of 13 MHz and a fractional bandwidth of 25%, as seen in Figure 3b. The hydrophone measurements showed a peak pressure of 85 kPa at 5.4 mm distance from the transducer surface. Our photoacoustic-pulse response was acquired via a USAF resolution test target (R3L3S1P, Thorlabs Inc., Newton, NJ, USA) and amplified by a preamplifier (Olympus 5073PR, Olympus NDT Inc., Waltham, MA, USA) with 39 dB gain. The laser energy measured by a pyroelectric energy meter (PE9-ES-C, Ophir-Spiricon, LLC, North Logan, UT, USA) after spatial filtering was found to be approximately 250 nJ. As seen in Figure 3c, our received photoacoustic signal was then averaged 100 times, and showed a center frequency of 13 MHz with a−6 dB fractional bandwidth of 60%. The SNR was calculated as the 20 $\log_{10}$ ratio of the PA signal amplitude to the standard deviation of noise, which was equal to 38 dB.

Next, the lateral resolution of the OR-PAM system was measured by linear scanning along the edge of a ~2 × 2 mm^2 square block of the USAF resolution-test target. During this test, the block was scanned with a 0.5 μm step size. Figure 3d shows the MAP image of the square edge, with a dashed line showing the 1 mm long scan length. The experiment data of the edge-spread function are shown in Figure 3e. The line-spread function was obtained by taking the first derivative of the fitted edge-spread function; its full-width half-maximum (FWHM) showed a lateral resolution of 8.5 μm. The axial resolution of the system was then estimated by taking the FWHM of a Gaussian envelope applied to a photoacoustic signal from the target. The FWHM was found to be 0.1 μs, which was equal to 150 μm in water, as seen in Figure 3f.

The axial resolution of the PAM system is inversely proportional to the bandwidth of the acoustic receiver and estimated to be 0.88 c/B [27], where B is the −6 dB bandwidth in MHz and c is the ultrasound velocity inside the tissue medium. Using this relation, the axial resolution of the proposed OR-PAM system was expected to be ~167 μm, which aligned well with the experimentally observed value of 150 μm. This axial resolution could further be improved by increasing the TUT's bandwidth using a stronger acoustic absorption material as the backing layer. This would also likely improve the SNR and spatial resolution of the OR-PAM system. Additionally, since no matching layer was used in the current TUT, the transmitted acoustic energy propagated through the tissue was ~17%, considering the acoustic impedance of the $LiNbO_3$ wafer and the tissue was 34 and 1.5 MRayls, respectively. If a transparent matching layer with proper acoustic impedance was added, such as a two-matching-layer design using glass slide and parylene coating, we could achieve transmission coefficient as high as ~45%, as the acoustic impedance mismatch between the piezoelectric material and the tissue would be reduced. This would increase our acoustic transmission and receiving sensitivities, and further result in an improved SNR.

3.2. Phantom- and Biological-Tissue-Imaging Experiments

Next, our OR-PAM system was validated by imaging a 12 μm diameter dense carbon-fiber network (that simulated capillary blood vessels) embedded in an agarose phantom gel. Step size was set at 2 μm to cover an area of 0.5 × 0.5 mm^2. The photoacoustic signal was then acquired via the high-speed data-acquisition system and averaged 100 times to generate the image. This MAP image of the carbon fiber can be seen in Figure 4a, where each fiber is clearly distinguishable with sufficient resolution and contrast.

Further, we demonstrated the feasibility of utilizing the OR-PAM for vasculature imaging using chicken embryos. Chicken embryos were used as an animal model to visualize different development phases [28], and OR-PAM could reveal their important vasculature information for clinical relevance [29].

For this study, Day 4 fertile chicken eggs (E4) were obtained from the Poultry Education and Research Center (PERC) at The Pennsylvania State University. These eggs were gently cracked, and the embryos carefully placed on weigh boats under sterile conditions. The embryos were then incubated at 38 °C with 3% CO_2 in a humidified incubator. For imaging, the CAM attached to each embryo was removed by cutting around its edges, and then each embryo was quickly and gently transferred to a Petri dish and rinsed with deionized water. Finally, each embryo was placed on top of an agarose-gel phantom bed for imaging. Figure 4b shows a photograph of the CAM: the scanning area is marked by a white box. Scan step size was set at 20 μm to cover a 2 × 2 mm^2 area, and imaging data were averaged 500 times to provide a sufficient SNR. The vasculature image from MAP, seen in Figure 4c, clearly shows the vascular-branch pattern marked in Figure 4b with adequate contrast and resolution.

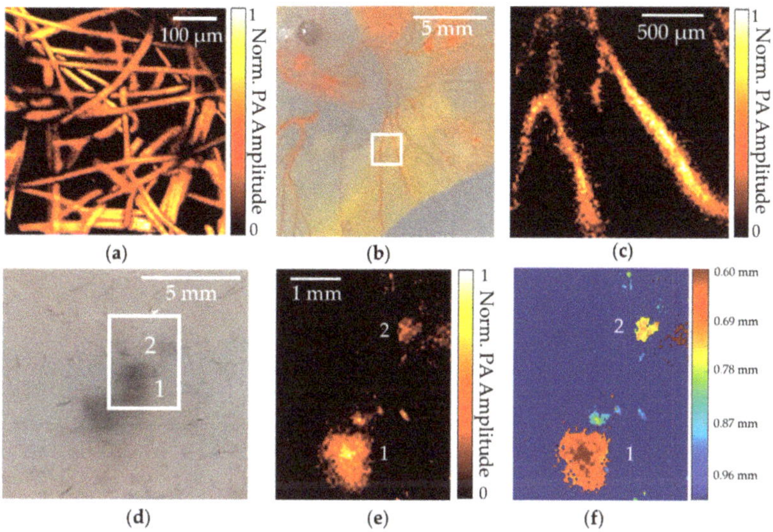

Figure 4. Imaging capabilities of transparent ultrasound-transducer-based OR-PAM system. (**a**) MAP image of carbon-fiber phantom with 0.5 × 0.5 mm^2 area. PA: photoacoustic; (**b**) Photograph of chick-embryo chorioallantoic membrane (CAM) with imaging area marked by blue box; (**c**) MAP image of CAM vasculature inside blue box of (**b**); (**d**) Photograph of melanoma phantom; (**e**) MAP image of melanin particles detected under mouse skin; (**f**) Color-coded depth profiling of melanoma phantom. Color bar represents depth relative to skin surface.

Next, the OR-PAM's application in melanoma imaging was demonstrated by scanning a melanoma phantom. The depth of melanoma invasion under the skin, also known as Breslow's depth, is one of the three most important prognostic factors in melanoma detection, and it reveals important details about how tumor cells invade [30]. To demonstrate the feasibility of TUT-based OR-PAM wearable imaging of melanoma patients without the need for thick gel coupling, we conducted the following melanoma-tissue experiment.

Approximately 2 mg of melanin particles (M8631, Sigma Aldrich, St. Louis, MO, USA; optical absorption coefficient ~1100 cm^{-1} at 532 nm [31]) was mixed with 100 mg of 1.5% agarose phantom gel and placed under a piece of mouse skin at different depths. The scan area was set as 3.5 × 4.5 mm^2 to cover two melanoma spots under the skin, as shown by the white box region in Figure 4d (animal protocols were approved by the Institutional Animal Care and Use Committee, Pennsylvania State University, University Park, State College, PA, USA).

Pulse energy after the spatial filter was set at ~600 nJ to yield an optical fluence of ~14.7 mJ/cm^2, which was below the American National Standards Institute (ANSI) safety skin maximum permissible exposure (MPE) limit of 20 mJ/cm^2 at 532 nm [32]. A-lines were averaged 300 times to generate the MAP image, as seen in Figure 4e. The image was then color-coded with a distance relative to the skin surface, which showed clear melanoma boundaries and depth information, as seen in Figure 4f. The feasibility of the high-contrast melanoma imaging demonstrated here could benefit clinical point-of-care depth detection and the monitoring of melanoma cells using wearable TUT-based OR-PAM.

The main limitations of our first-generation TUT-based OR-PAM system were scan speed (100 × 100 steps took 50 min) and SNR. Scan speed could be improved by using state-of-the-art scanning methods, such as galvo-mirror-based scanning of the optical beam like that employed in conventional OR-PAM systems. This could achieve a scanning speed of 1000 × 1000 steps in 100 s [7]. The SNR could be improved by further investigating novel high-acoustic absorption backing and proper impedance-matching layers. Using a better backing layer and the described two-matching-layer design, we could improve bandwidth by 60% [22,33] and transmission coefficient by ~160%.

Furthermore, fabricating the TUT with a higher-frequency $LiNbO_3$ wafer would further increase bandwidth [34] and lead to an improvement in the spatial resolution of the OR-PAM system. During the scanning, blood diffusion from certain areas of these *ex ovo* samples resulted in low-contrast vascular images: vascular contrast is expected to improve when imaging living subjects *in vivo*. Despite these limitations, the novel TUT-based OR-PAM approach presented here showed promising results that could lead to the miniaturization of OR-PAM for emerging wearable and high-throughput imaging applications. Additionally, the total cost of the proposed TUT is less than 50 USD, well below that of current OR-PAM setups that require additional optical components and instruments.

4. Conclusions

A novel OR-PAM system based on a transparent $LiNbO_3$ ultrasound transducer was developed, characterized, and validated using both inanimate and biological subjects. The transducer was ITO-coated with 80% optical transmission in the visible and near-infrared optical-wavelength regions, and had a center frequency of 13 MHz with a fractional photoacoustic bandwidth of 60%. The resultant transparency of the $LiNbO_3$ transducer facilitated a shared pathway for both light and acoustic-wave propagation.

Our approach also removed the need for additional optical components (such as acoustic-optic prisms) and large-coupling media used in conventional OR-PAM systems. Instead, the OR-PAM presented in this work has a much smaller, lighter imaging head—the TUT itself—with minimal acoustic coupling. Imaging experiments demonstrated an SNR of 38 dB, and a lateral and axial resolution of 8.5 and 150 µm, respectively.

The OR-PAM's feasibility of vascular imaging was demonstrated using Day 4 CAM and melanoma depth profiling using melanoma-tissue phantoms. In the future, the proposed TUT could be integrated in conventional optical-microscopy techniques to allow multimodal microscopy that is capable of both ultrasound stimulation and sensing. Eventually, the TUT technology could be further scaled to develop miniaturized photoacoustic endomicroscopy and microendoscopy devices for space-constrained point-of-care clinical applications, e.g., prostate and pancreas needle biopsies [35–37].

Author Contributions: Conceptualization, H.C., A.D., and S.-R.K.; software, H.C., S.A., and A.D.; investigation, S.-R.K.; biological resources, L.A., H.J., R.R., and Y.W.; data curation, H.C., A.D., and S.A.; writing—original-draft preparation, H.C. and S.A.; experiment design, H.C., S.A., A.D., and S.-R.K.; hardware, H.C., A.D., C.W., and M.O.; supervision, S.-R.K.; project administration, S.-R.K.; funding acquisition, Y.W. and S.-R.K.

Funding: This research was funded by NIH-NIBIB R00EB017729-04 (S.R.K.) and the Penn State Cancer Institute—Highmark seed grant (SRK), and partially supported by NIH-NIAMS R01 AR073364 (Y.W.).

Acknowledgments: We acknowledge Eugene Gerber for his kind support in the machining of parts.

Conflicts of Interest: The authors declare no conflict of interest.

References

1. Lan, B.; Liu, W.; Wang, Y.C.; Shi, J.; Li, Y.; Xu, S.; Sheng, H.; Zhou, Q.; Zou, J.; Hoffmann, U.; et al. High-speed widefield photoacoustic microscopy of small-animal hemodynamics. *Biomed. Opt. Express* **2018**, *9*, 4689–4701. [CrossRef] [PubMed]
2. Yao, J.; Wang, L.; Yang, J.M.; Maslov, K.I.; Wong, T.T.; Li, L.; Huang, C.H.; Zou, J.; Wang, L.V. High-speed label-free functional photoacoustic microscopy of mouse brain in action. *Nat. Methods* **2015**, *12*, 407. [CrossRef] [PubMed]
3. Wang, L.; Maslov, K.; Wang, L.V. Single-cell label-free photoacoustic flowoxigraphy *in vivo*. *Proc. Natl. Acad. Sci. USA* **2013**, *110*, 5759–5764. [CrossRef] [PubMed]

4. Zhang, W.; Li, Y.; Nguyen, V.P.; Huang, Z.; Liu, Z.; Wang, X.; Paulus, Y.M. High-resolution, *in vivo* multimodal photoacoustic microscopy, optical coherence tomography, and fluorescence microscopy imaging of rabbit retinal neovascularization. *Light Sci. Appl.* **2018**, *7*, 103. [CrossRef] [PubMed]
5. Cao, R.; Li, J.; Ning, B.; Sun, N.; Wang, T.; Zuo, Z.; Hu, S. Functional and oxygen-metabolic photoacoustic microscopy of the awake mouse brain. *Neuroimage* **2017**, *150*, 77–87. [CrossRef] [PubMed]
6. Maslov, K.; Zhang, H.F.; Hu, S.; Wang, L.V. Optical-resolution photoacoustic microscopy for *in vivo* imaging of single capillaries. *Opt. Lett.* **2008**, *33*, 929–931. [CrossRef]
7. Kim, J.Y.; Lee, C.; Park, K.; Han, S.; Kim, C. High-speed and high-SNR photoacoustic microscopy based on a galvanometer mirror in non-conducting liquid. *Sci. Rep.* **2016**, *6*, 34803. [CrossRef]
8. Wong, T.T.; Zhang, R.; Zhang, C.; Hsu, H.C.; Maslov, K.I.; Wang, L.; Shi, J.; Chen, R.; Shung, K.K.; Zhou, Q.; et al. Label-free automated three-dimensional imaging of whole organs by microtomy-assisted photoacoustic microscopy. *Nat. Commun.* **2017**, *8*, 1386. [CrossRef]
9. Chen, Q.; Guo, H.; Jin, T.; Qi, W.; Xie, H.; Xi, L. Ultracompact high-resolution photoacoustic microscopy. *Opt. Lett.* **2018**, *43*, 1615–1618. [CrossRef]
10. Rebling, J.; Estrada, H.; Gottschalk, S.; Sela, G.; Zwack, M.; Wissmeyer, G.; Ntziachristos, V.; Razansky, D. Dual-wavelength hybrid optoacoustic-ultrasound biomicroscopy for functional imaging of large-scale cerebral vascular networks. *J. Biophotonics* **2018**, *11*, e201800057. [CrossRef]
11. Dangi, A.; Agrawal, S.; Tiwari, S.; Jadhav, S.; Cheng, C.; Datta, G.R.; Trolier-McKinstry, S.; Pratap, R.; Kothapalli, S.R. Ring PMUT array based miniaturized photoacoustic endoscopy device. In Proceedings of the Photons Plus Ultrasound: Imaging and Sensing 2019, International Society for Optics and Photonics, San Francisco, CA, USA, 27 February 2019; Volume 10878, p. 1087811.
12. Dangi, A.; Agrawal, S.; Datta, G.R.; Srinivasan, V.; Kothapalli, S.R. Towards a Low-Cost and Portable Photoacoustic Microscope for Point-of-Care and Wearable Applications. *IEEE Sens. J.* **2019**. [CrossRef]
13. Ansari, R.; Zhang, E.Z.; Desjardins, A.E.; Beard, P.C. All-optical forward-viewing photoacoustic probe for high-resolution 3D endoscopy. *Light Sci. Appl.* **2018**, *7*, 75. [CrossRef] [PubMed]
14. Li, H.; Dong, B.; Zhang, X.; Shu, X.; Chen, X.; Hai, R.; Czaplewski, D.A.; Zhang, H.F.; Sun, C. Disposable ultrasound-sensing chronic cranial window by soft nanoimprinting lithography. *Nat. Commun.* **2019**, *10*, 1–9. [CrossRef] [PubMed]
15. Dong, B.; Sun, C.; Zhang, H.F. Optical detection of ultrasound in photoacoustic imaging. *IEEE Trans. Biomed. Eng.* **2016**, *64*, 4–15. [CrossRef] [PubMed]
16. Gougheri, H.S.; Dangi, A.; Kothapalli, S.R.; Kiani, M. A Comprehensive Study of Ultrasound Transducer Characteristics in Microscopic Ultrasound Neuromodulation. *IEEE Trans. Biomed. Circuits Syst.* **2019**, *13*, 835–847. [CrossRef]
17. Tufail, Y.; Matyushov, A.; Baldwin, N.; Tauchmann, M.L.; Georges, J.; Yoshihiro, A.; Tillery, S.I.H.; Tyler, W.J. Transcranial pulsed ultrasound stimulates intact brain circuits. *Neuron* **2010**, *66*, 681–694. [CrossRef]
18. Zhang, X.; Wu, X.; Adelegan, O.J.; Yamaner, F.Y.; Oralkan, Ö. Backward-mode photoacoustic imaging using illumination through a CMUT with improved transparency. *IEEE Trans. Ultrason. Ferroelectr. Freq. Control* **2017**, *65*, 85–94. [CrossRef]
19. Li, Z.; Ilkhechi, A.K.; Zemp, R. Transparent capacitive micromachined ultrasonic transducers (CMUTs) for photoacoustic applications. *Opt. Express* **2019**, *27*, 13204–13218. [CrossRef]
20. Dangi, A.; Agrawal, S.; Kothapalli, S.R. Lithium niobate-based transparent ultrasound transducers for photoacoustic imaging. *Opt. Lett.* **2019**, *44*, 5326–5329. [CrossRef]
21. Baba, A.; Searfass, C.T.; Tittmann, B.R. High temperature ultrasonic transducer up to 1000 C using lithium niobate single crystal. *Appl. Phys. Lett.* **2010**, *97*, 232901.
22. Fei, C.; Chiu, C.T.; Chen, X.; Chen, Z.; Ma, J.; Zhu, B.; Shung, K.K.; Zhou, Q. Ultrahigh frequency (100 MHz–300 MHz) ultrasonic transducers for optical resolution medical imagining. *Sci. Rep.* **2016**, *6*, 28360. [CrossRef] [PubMed]
23. Liu, W.; Shcherbakova, D.M.; Kurupassery, N.; Li, Y.; Zhou, Q.; Verkhusha, V.V.; Yao, J. Quad-mode functional and molecular photoacoustic microscopy. *Sci. Rep.* **2018**, *8*, 11123. [CrossRef] [PubMed]
24. Zhou, Q.; Cha, J.H.; Huang, Y.; Zhang, R.; Cao, W.; Shung, K.K. Alumina/epoxy nanocomposite matching layers for high-frequency ultrasound transducer application. *IEEE Trans. Ultrason. Ferroelectr. Freq. Control* **2009**, *56*, 213–219. [PubMed]

25. Shung, K.K.; Zippuro, M. Ultrasonic transducers and arrays. *IEEE Eng. Med. Biol. Mag.* **1996**, *15*, 20–30. [CrossRef]
26. Chen, Q.; Wang, Q.M. The effective electromechanical coupling coefficient of piezoelectric thin-film resonators. *Appl. Phys. Lett.* **2005**, *86*, 022904. [CrossRef]
27. Zhang, C.; Maslov, K.I.; Yao, J.; Wang, L.V. *In vivo* photoacoustic microscopy with 7.6-μm axial resolution using a commercial 125-MHz ultrasonic transducer. *J. Biomed. Opt.* **2012**, *17*, 116016. [CrossRef]
28. Dorrell, M.I.; Marcacci, M.; Bravo, S.; Kurz, T.; Tremblay, J.; Rusing, J.C. *Ex ovo* model for directly visualizing chick embryo development. *Am. Boil. Teach.* **2012**, *74*, 628–634.
29. Laufer, J.G.; Zhang, E.Z.; Treeby, B.E.; Cox, B.T.; Beard, P.C.; Johnson, P.; Pedley, B. *In vivo* preclinical photoacoustic imaging of tumor vasculature development and therapy. *J. Biomed. Opt.* **2012**, *44817*, 056016. [CrossRef]
30. Balch, C.M.; Soong, S.J.; Gershenwald, J.E.; Thompson, J.F.; Reintgen, D.S.; Cascinelli, N.; Urist, M.; McMasters, K.M.; Ross, M.I.; Kirkwood, J.M.; et al. Prognostic factors analysis of 17,600 melanoma of 14patients: Validation of the American Joint Committee on Cancer melanoma staging system. *J. Clin. Oncol.* **2001**, *19*, 3622–3634. [CrossRef]
31. Ansari, M.A.; Mohajerani, E. Mechanisms of laser-tissue interaction: Optical properties of tissue. *J. Lasers Med. Sci.* **2011**, *2*, 119.
32. American National Standards Institute. *American National Standard for Safe Use of Lasers*; Laser Institute of America: New York, NY, USA, 2007.
33. Szabo, T.L.; Lewin, P.A. Piezoelectric materials for imaging. *J. Ultrasound Med.* **2007**, *26*, 283–288. [CrossRef] [PubMed]
34. Shung, K.K.; Cannata, J.; Zhou, M.Q.; Lee, J. High Frequency Ultrasound: A New Frontier for Ultrasound. In Proceedings of the 2009 Annual International Conference of the IEEE Engineering in Medicine and Biology Society, Hilton Minneapolis, Minnesota, 2–6 September 2009; pp. 1953–1955.
35. Kothapalli, S.R.; Sonn, G.A.; Choe, J.W.; Nikoozadeh, A.; Bhuyan, A.; Park, K.K.; Cristman, P.; Fan, R.; Moini, A.; Lee, B.C.; et al. Simultaneous transrectal ultrasound and photoacoustic human prostate imaging. *Sci. Trans. Med.* **2019**, *11*, 507. [CrossRef] [PubMed]
36. Huang, S.; Qin, Y.; Chen, Y.; Pan, J.; Xu, C.; Wu, D.; Chao, W.Y.; Wei, J.T.; Tomlins, S.A.; Wang, X.; et al. Interstitial assessment of aggressive prostate cancer by physio-chemical photoacoustics: An *ex vivo* study with intact human prostates. *Med. Phys.* **2018**, *45*, 9. [CrossRef] [PubMed]
37. Tummers, W.S.; Miller, S.E.; Teraphongphom, N.T.; Gomez, A.; Steinberg, I.; Huland, D.M.; Hong, S.; Kothapalli, S.R.; Hasan, A.; Ertsey, R.; et al. Intraoperative pancreatic cancer detection using tumor-specific multimodality molecular imaging. *Ann. Surg. Oncol.* **2018**, *25*, 1880–1888. [CrossRef] [PubMed]

sensors

MDPI

Article

Optimizing Irradiation Geometry in LED-Based Photoacoustic Imaging with 3D Printed Flexible and Modular Light Delivery System

Maju Kuriakose [1], Christopher D. Nguyen [1], Mithun Kuniyil Ajith Singh [2] and Srivalleesha Mallidi [1,*]

[1] Department of Biomedical Engineering, Tufts University, Medford, MA 02155, USA; maju.kuriakose@tufts.edu (M.K.); Christopher.Nguyen@tufts.edu (C.D.N.)

[2] Research and Business Development Division, CYBERDYNE INC, 3013 AK Rotterdam, The Netherlands; mithun_ajith@cyberdyne.jp

* Correspondence: Srivalleesha.Mallidi@tufts.edu

Received: 21 May 2020; Accepted: 29 June 2020; Published: 6 July 2020

Abstract: Photoacoustic (PA) imaging–a technique combining the ability of optical imaging to probe functional properties of the tissue and deep structural imaging ability of ultrasound–has gained significant popularity in the past two decades for its utility in several biomedical applications. More recently, light-emitting diodes (LED) are being explored as an alternative to bulky and expensive laser systems used in PA imaging for their portability and low-cost. Due to the large beam divergence of LEDs compared to traditional laser beams, it is imperative to quantify the angular dependence of LED-based illumination and optimize its performance for imaging superficial or deep-seated lesions. A custom-built modular 3-D printed hinge system and tissue-mimicking phantoms with various absorption and scattering properties were used in this study to quantify the angular dependence of LED-based illumination. We also experimentally calculated the source divergence of the pulsed-LED arrays to be 58° ± 8°. Our results from point sources (pencil lead phantom) in non-scattering medium obey the cotangential relationship between the angle of irradiation and maximum PA intensity obtained at various imaging depths, as expected. Strong dependence on the angle of illumination at superficial depths (−5°/mm at 10 mm) was observed that becomes weaker at intermediate depths (−2.5°/mm at 20 mm) and negligible at deeper locations (−1.1°/mm at 30 mm). The results from the tissue-mimicking phantom in scattering media indicate that angles between 30–75° could be used for imaging lesions at various depths (12 mm–28 mm) where lower LED illumination angles (closer to being parallel to the imaging plane) are preferable for deep tissue imaging and superficial lesion imaging is possible with higher LED illumination angles (closer to being perpendicular to the imaging plane). Our results can serve as a priori knowledge for the future LED-based PA system designs employed for both preclinical and clinical applications.

Keywords: LED; photoacoustic imaging; ultrasound; 3-D printed photoacoustic probe holder; light delivery optimization; LED divergence

1. Introduction

Photoacoustic (PA) imaging has gained significant popularity for imaging functional and molecular information in both preclinical and clinical settings [1–5]. The technique involves sending light pulses (a few nanosecond pulse-width) into imaging planes that get absorbed by endogenous (e.g., hemoglobin) or exogenous (e.g., Indocyanine Green) tissue chromophores and generate acoustic waves, which can be detected by conventional ultrasound (US) transducers [2,6,7]. Based on the endogenous contrast provided by hemoglobin, PA imaging has shown promise in vascular functional imaging of human neonatal brains [8,9], malignant lesions [10–14], and monitoring therapies such as photodynamic therapy [1,15,16], etc. As PA imaging uniquely possesses the best properties of optical imaging (high spatial resolution, functional properties, and imaging speed) and US imaging (structural properties and penetration depth reaching tens of cm), its relevance and popularity are continuously increasing in clinical settings [2,4,17–19].

In PA phenomena, the acoustic pressure ($\mathbf{P_0}$) generated is proportional to the optical absorption coefficient (μ_a, m^{-1}) of the light absorber and locally available light fluence or radiant exposure (f_0, Jm^{-2}). This can be represented by [2,6,20,21]:

$$\mathbf{P}_0 = u_a f_0 \tag{1}$$

where, is the dimensionless, material thermal property dependent Gruneisen coefficient.

Light attenuates as it travels down through a material or tissue due to scattering and absorption. Moreover, for a limited aperture illumination, angle of illumination also plays an important role in defining local fluence (Figure 1). As a result, f_0 and thus PA signal intensity, $\mathbf{P_0}$, changes as a function of depth (distance from transducer or excitation source) and as a function of the illumination angle. Therefore, the optimization of light delivery is crucial for efficient PA imaging and obtaining high signal-to-noise (SNR) ratio at deeper penetration depths [22–26]. Specifically, for reflection mode PA imaging (transducer and light source on the same side of the sample), several studies demonstrated the dependence of irradiation angle and fiber (source)-to-transducer positioning on PA signal at various depths experimentally or through simulations [23,27–31]. Either unilateral or bilateral positioning of fiber bundles aligned with the nanosecond pulsed laser have been employed in these studies. For example, Haisch et al. utilized a mechanical setup that allowed unilateral illumination (one-side of the transducer) with 20–80 degrees range of motion [32]. The fiber bundle aperture size used in that study was shorter than the ultrasound transducer, which may hinder full aperture illumination of near field absorbers and can presumably suitable for intermediate to deep tissue imaging and is more suitable to image smaller lesions on the skin. In another study by Sivasubramanian et al., two fiber bundles were placed on either side of the transducer at a fixed angle. Change in the illumination angle would require changing the holder setup [33]. More recently, Sangha et al. designed a motorized system to change the bilateral light illumination in the 0°–60° range and concluded that the illumination geometry optimization is important to achieve high SNR at different depths [34]. All these studies point out that change in illumination geometry effects PA SNR at different depths and strongly indicate the need for a flexible handheld system that can deliver light at different angles depending on the depth of the lesions.

The light delivery optimization studies mentioned above were performed with spatially low diverging coherent laser sources. Though conventional lasers (e.g., Q-switched optical parameter oscillator (OPO)) can deliver the required pulse energy at various NIR wavelengths, their bulkiness, minimal-portability, and difficulty in operation prevent them from effortless usage in a clinical setting. Interestingly, nanosecond pulsed light-emitting diodes (LED) show promise in being an alternative to lasers, while offering cost-effectiveness and ease of operation has been recently proven to be successful in several studies [35–39]. Despite the low power of LEDs (about 3 orders of magnitude lower than the conventional Q-switched laser sources), their high pulse repetition rate (PRR) (maximum reported up to 16 kHz opposed) gives the opportunity to average several frames in real-time to achieve an SNR on par with conventional laser-based PA imaging (PAI). In addition, the large spatial divergence of LED arrays (~60°), could aid in irradiating larger sample area and potentially also provide a quasi-uniform illumination over several millimeters without a light diffuser. Given these attributes, LED-based PA systems demonstrate strong potential for clinical translation. As the use of LED's in PA imaging is still in its infancy, it is important to characterize and optimize the light delivery strategies for better SNR at various depths.

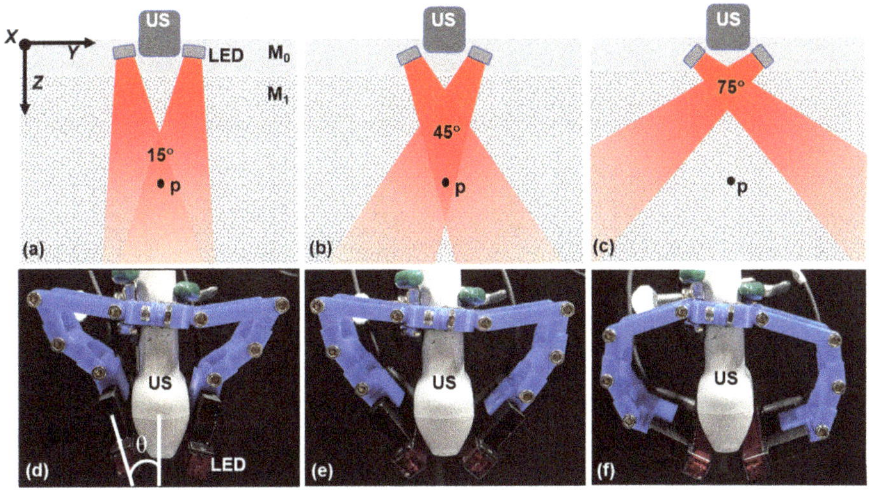

Figure 1. (**a–c**) Schematics of light emitting diode (LED) illumination cross-sectional view of photoacoustic (PA) setup at representative angles: 15°, 45°, and 75°, respectively, orthogonal to the imaging plane *XZ*, that contains a hypothetical absorber, p. M_0 is the medium that facilitates acoustic coupling between the transducer and the phantom material M_1; (**d–f**) Photographs of LED source pivoted at representative angles, θ = 15°, 45°, and 75°, respectively, using 3D printed modular hinge system.

LED array-based PAI studies so far have used a fixed orientation either in the reflection mode [40–42] or transmission mode [38,43]. In this study, we designed a flexible modular light delivery system for reflection mode PAI that is capable of orienting light from the LED arrays at various angles in the range of 0°–90° (Figure 1). Utilizing the flexible light delivery system, we evaluated PA image contrast in various tissue-mimicking phantoms (point targets in non-scattering and scattering liquid media, absorbing lesion under non-scattering liquid and scattering tissue such as the chicken breast) for the PA signal dependency as a function of irradiation angle and depth. We believe that our findings have an important impact on optimizing the design of LED-based PA probes and accelerate its clinical translation towards imaging both deeper and shallower lesions.

2. Materials and Methods

2.1. Photoacoustic System and Modular Arrangement for Varying Illumination Direction

2.1.1. AcousticX

An LED-based photoacoustic system (AcousticX from Cyberdyne Inc., Tsukuba, Japan) with linear US transducer (7 MHz central frequency, 128 elements, 0.315 mm pitch, and 38.4 mm aperture size, elevation focus of 15 mm) and two 850 nm LED arrays (30 to 150 ns pulse width, 4 kHz maximum repetition rate, 200 µJ pulse energy for each array, 5 mm × 40 mm aperture size, 60° divergence) on both sides of the US detector was used for the experiments [44]. PA and US raw data were sampled at 40 MHz and 20 MHz, respectively, and data was reconstructed in real-time using an inbuilt Fourier-domain reconstruction algorithm of the system. For offline analysis, both PA and US data were reconstructed using a previously reported frequency domain beamforming algorithm [45]. Radiant exposure per pulse at the LED array surface is about 100 µJ/cm^2 (200 µJ/pulse in an array area of 2 cm^2). Given the LED source divergence of 60° and ~10.5 mm distance between the US transducer and the phantom surface to accommodate LED arrays for different angles, maximum radiant exposure at the phantom surface was estimated to be about 29 μJ cm^{-2} (for an area of about 6.93 cm^2) from a single array at 0° LED illumination angle.

2.1.2. Flexible LED Holder: Modular Design for Adjusting Irradiation Direction

Two identical modular hinge systems were designed, 3D printed and used to pivot LED arrays at different illumination directions with respect to the imaging plane. The modular LED holder consisted of three parts that were designed on Autodesk and printed using polylactic acid (PLA) on a MakerBot system. All pieces were joined together with 8–32 socket head screws. (Figure 2). Each of the modules consisted of four hinges that were attached to one another. These hinges can be adjusted or pivoted to create the required angle of illumination. Modules were attached to the US transducer on its one end and the LED arrays were gripped through the heat sink of the arrays on the pivoting end of the module, as shown in the figure. The inter LED array distance (between their adjacent edges) was about 1 cm to accommodate the US transducer. Experiments were done for 0°, 15°, 30°, 45°, 60°, 75°, and 90° angles using this modular arrangement. The LED array angles were adjusted with respect to the central axis of the US transducer using a custom-made protractor as shown in Figure 1d. During the experiments, both the US transducer and sample position were unaltered and only the LED sources were adjusted, to avoid PA intensity variations due to sample motion with respect to the US transducer.

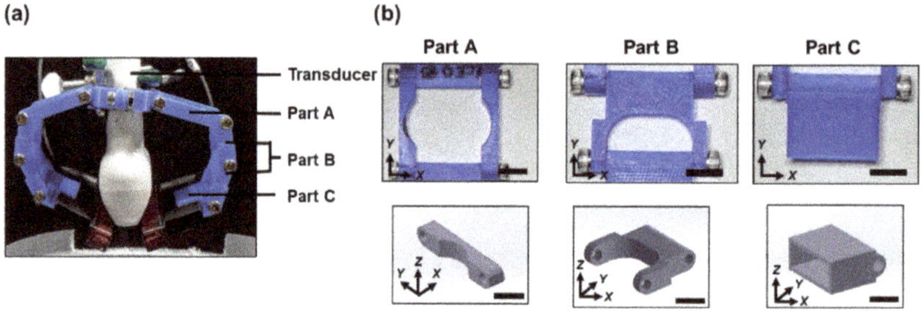

Figure 2. (**a**) Modular hinge system holding the LED arrays and attached to the US transducer. (**b**) The photographs of the individual pieces are shown in the top panel. The 3D renderings of the hinge pieces are shown on the bottom panel. Part A fits around the transducer and extends the horizontal reach of the holder to allow the LEDs to be placed at angles approaching 90°. Part B in conjunction with Part A allows precise horizontal and vertical height adjustment. The holder consists of two-part B pieces, and schematic of only one piece is shown in the panel. Part C holds the LEDs using the heat sinks and provides flexibility for any final adjustments on the LED illumination angle. Scale bar = 10 mm.

2.2. *Phantoms*

2.2.1. Graphite Pencil Lead Phantoms

A matrix of pencil leads (Graphite 2B 0.5 mm manufactured by June Gold, Bountiful, UT USA), arranged in 4 rows × 5 columns with a spacing of about 5 mm (columns) and 6 mm (rows), was constructed using two 3D printed plastic holders as shown in Figure 3a. The phantom construction with pencil lead is immersed in a container with water or 1% Intralipid (Sigma Aldrich Inc., Atlanta, GA, USA) solution (Figure 3b). The scattering coefficient of 1% intralipid solution is 1.8 mm^{-1} [46,47] close to the values reported for tumor tissue (1–2 mm^{-1}) [48–50]. All experiments were conducted at room temperature (22 °C).

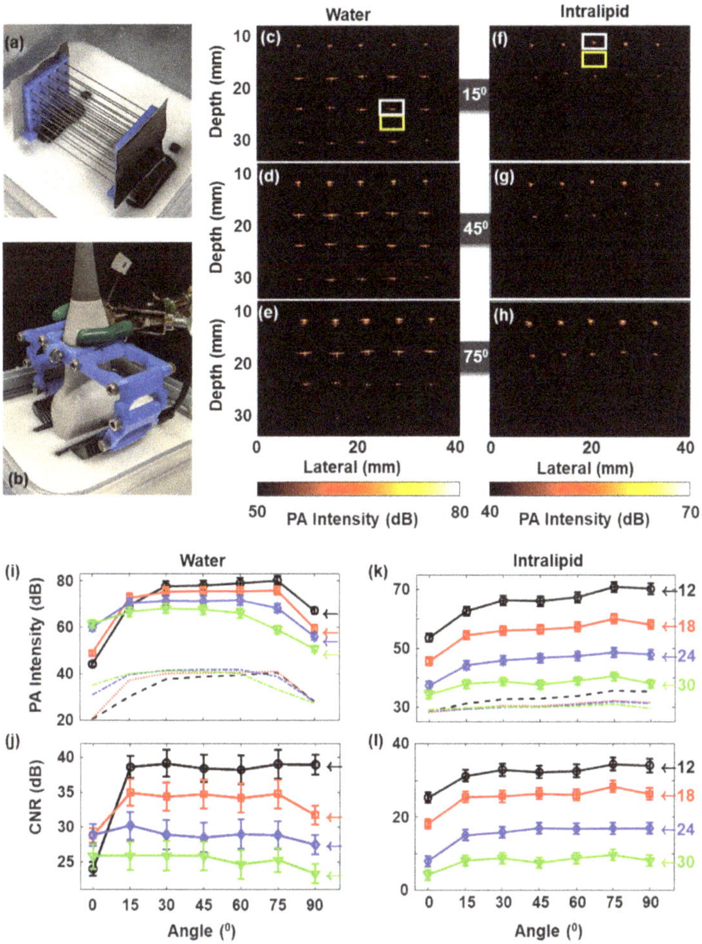

Figure 3. (**a**) Photograph of Pencil lead matrix; (**b**) photograph of an experimental arrangement using intralipid medium; (**c–h**) PA image acquired at representative angles 15° (**c**,**f**), 45° (**d**,**g**), and 75° (**e**,**h**) in water (**c–e**) and 1% intralipid (**f–h**); (**i**,**k**) Mean PA signal intensities and their standard deviations (of 5 lateral positions at each depth) plotted as a function of LED angles in water (**j**) and intralipid (**k**); and (**j**,**l**) corresponding contrast to noise ratios (CNRs) obtained as a function of LED angles in water (**j**) and intralipid (**l**). Different depths from the transducer are indicated by 12 mm (line with black circles), 18 mm (line with red squares), 24 mm (line with blue diamonds), and 30 mm (line with green downward triangles). The noise background levels corresponding to 12, 18, 24, and 30 mm are represented by black, red, blue, and green dash-dotted lines, respectively, in (**i**) in water and (**j**) in intralipid. The backgrounds were obtained right below from each signal regions, e.g., as indicated by the yellow rectangles in (**c**,**f**). Images for all the angles can be found in Figure S2 (for water) and Figure S3 (for intralipid).

2.2.2. Tissue Mimicking Phantom Containing Lesion with High Optical Absorbance

Tissue mimicking phantoms were prepared using agar powder (Sigma-Aldrich Inc., Atlanta, GA, USA). Titanium (IV) oxide, anatase powder (99.8%, Sigma-Aldrich Inc., Atlanta, GA, USA) was added to provide acoustic contrast and enhance the optical scattering properties of the agar. The preparation was done by slowly adding 1% wt./vol. of agar powder and 1% wt./vol. of TiO_2 powder into continually stirred deionized water at ambient conditions to avoid clumps. The final solution was then heated above 80° C, above the melting temperature of agar, and exposed to a vacuum level of about 0.1 atm for 5 min to degas the solution and cooled it down to room temperature to obtain the final phantom. A cylindrical light-absorbing lesion with acoustic scatterers was prepared in a similar aforementioned method. Additionally, 0.5% wt./vol. graphite powder (<20 μm, synthetic graphite, Sigma Aldrich Inc., Atlanta, GA, USA) and 0.5% wt./vol. TiO_2 powder were added. The concentration of the absorbing and scattering particles was chosen to mimic tumor tissue with an absorption coefficient (~0.2 cm^{-1}) and reduced scattering coefficient (~10 cm^{-1}) as previously reported in the literature [46–51].

2.3. *Signal Analysis*

2.3.1. PA Intensity & Contrast to Noise Ratio (CNR) Calculation

The PA signal intensity of each pencil lead in the phantom was calculated by taking the maximum pixel value from the region of interest (ROI) around the target (white rectangle in Figure 3c). All the five laterally positioned PA intensities were then averaged to find mean and standard deviation (σ) of PA intensities corresponding to each depth location and illumination angles. Background (Bg) was calculated from an ROI below each point target (yellow rectangle in Figure 3c). It is important to note that the noise/background values were calculated from regions close to the signal ROIs. We chose the ROIs within close proximity of the signal ROI and not from regions at the corner or with only electronic noise, including reconstruction related artifacts that may be present when changing the illumination angle. To plot PA intensity changes as a function of angle in the tissue-mimicking phantom, PA signal intensities from the lesion was obtained by choosing the median PA intensities above the Bg level. Here also, similar to the pencil phantom case, Bg and σ were chosen from a region close to the lesion ROI. PA intensity was plotted in decibel (dB) with the formula:

$$PA\ intensity\ in\ dB = 10log_{10}(PA\ intensity) \tag{2}$$

and the CNR was calculated using the formula:

$$CNR\ in\ dB = 10log_{10}\left(\frac{PA\ intensity - Bg}{\sigma}\right) \tag{3}$$

2.3.2. Divergence of the LED Source

Divergence of the LED source is an angular measure of the increase in irradiation area (and corresponding diameter or radius) with distance from the source. Laser light sources are known to have very low divergence while LED sources have high divergence. Sources with high divergence have lower radiant exposure per unit area on the target at a given depth than sources with lower divergence. These changes in radiant exposure can influence PA signal and hence it is critical to evaluate the divergence of the source and choose appropriate illumination angle to obtain maximum CNR. Assuming the LED is a line source aligned in the *X*-axis (parallel to the US transducer) while the emitted wavefronts take quasi-cylindrical shape in the imaging volume, an approximate divergence of the source in the *YZ* plane (Figure 4), in degrees can be computed using:

$$Divergence\ angle = 2 \times \left\{\tan^{-1}\left(\frac{target\ depth}{LED\ to\ Detector\ distance}\right)\right\} \tag{4}$$

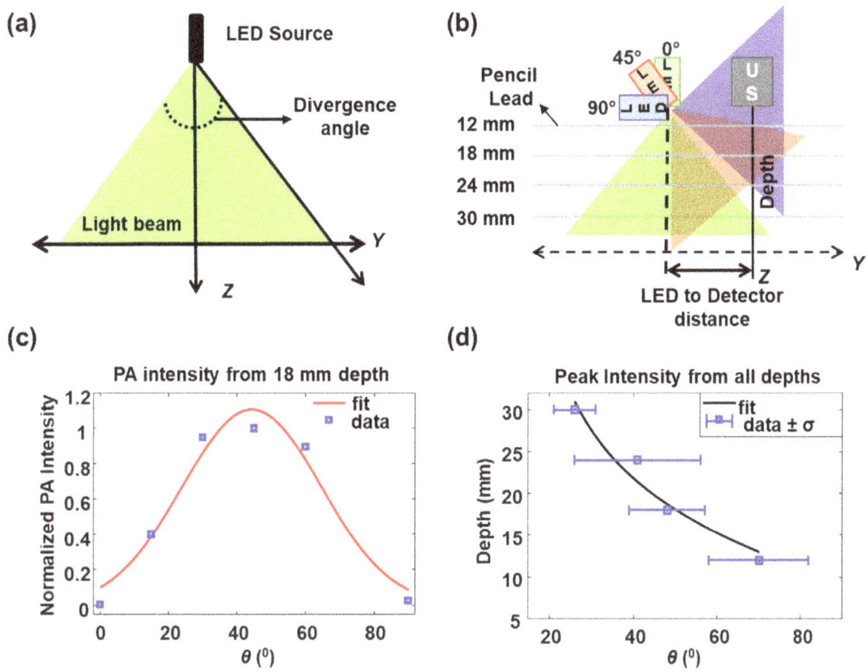

Figure 4. (**a**) Illustration of the source wavefront profile, while assuming the LED array as a line source with Gaussian profile having divergence described by Equation (4); (**b**) Source profile after pivoting LED arrays at three different angles in the imaging plane–green at 0°, red at 45°, and blue at 90°; (**c**) Normalized PA intensity obtained from 18 mm target vs. LED illumination angle, θ, is indicated in blue squares. The red solid curve shows the Gaussian fit using Equation (5), with an R-squared value of 0.95; (**d**) Blue squares with error bar show the peak PA intensity with standard deviation for each depth plotted as a function of θ for pencil lead phantom data in water. The black solid curve displays the best fit using a cotangent function (R^2 = 0.95).

The multiplying factor 2 is used to include both sides of the illumination plane. Target depth and LED to detector distance refer to the absorber (pencil lead) position in depth below the detector (US transducer) and its lateral distance to the source (LED array), respectively. In our experiment, the target and LED to detector distance were fixed while the source was pivoted, as shown in Figure 4b. We assume that the LED source exhibits a Gaussian spatial intensity profile and thus a Gaussian function is used to fit PA intensity vs. LED- illumination angle data to find the divergence of the LED arrays, given by

$$f(x) = a\, e^{-\left(\frac{x-b}{c}\right)^2} \tag{5}$$

where x is the source angle, a is the peak PA intensity and that corresponds to angle b, and c is the half of the angle span at which PA intensity shows half maximum (50%) or −6 dB roll-off. The divergence can then be calculated by multiplying c by two, resulting in the full width of the angle at half maximum of the PA intensity (FWHM). The model can be fitted using a least-squares minimization method to obtain best-fit parameters. Moreover, the position of the peak intensity for a chosen depth can be described by

$$d = m + n \cdot \cot(\theta) \tag{6}$$

where θ is the LED illumination angle (Figure 1), m is the offset of imaging plane in depth, n is the separation between LED and detector, and d is the imaging depth.

3. Results

Experiments were conducted in two different phantom environments: liquid and chicken tissue. The first phantom consisted of pencil lead (point source) as absorbing targets (Figure 3) in water or 1% intralipid media. The second tissue-mimicking phantom consisted of an absorbing cylindrical lesion made of graphite powder (Figure 5), which was placed on top of a chicken breast tissue arranged obliquely to the imaging plane, while the top layer was interchanged between water or chicken breast tissue. The experiments were designed to probe the phantom at all depths simultaneously for each angle of choice, thereby reducing experimental uncertainties.

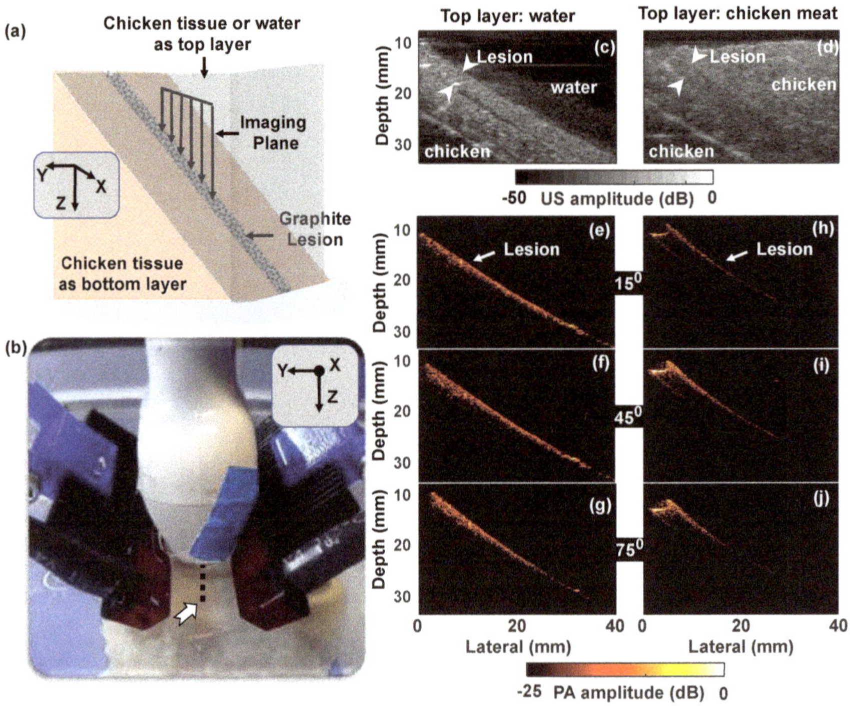

Figure 5. (**a**) Schematic of the tissue-mimicking phantom: Absorbing lesion (cylindrical in shape, 2 mm in diameter) placed on chicken tissue and arranged obliquely in the *XZ* plane. (**b**) Photograph of the experiment. The dotted line indicated by arrowhead shows the projection of lesion to the *x*-axis; (**c**,**d**) US image of the sample showing water top layer (**c**) or chicken layer (**d**) lesion and bottom chicken tissue in both cases; (**e**–**j**) PA intensity images captured using water as the top layer (**e**–**g**) or chicken tissue as the top layer (**h**–**j**), by choosing LED array directions: 15° (**e**,**h**), 45° (**f**,**i**), and 75° (**g**,**j**) with respect to the imaging plane.

3.1. Pencil Lead Phantom Experiments

3.1.1. Pencil Lead in Scattering and Non-Scattering Media Shows Weak Dependency on the LED Illumination Angle

Initially, PA intensities were monitored as a function of the illumination angle using a 4 × 5 pencil-lead matrix (as detailed previously) aligned orthogonal to the imaging axis to form point targets at defined locations. LED directions were varied from 0° to 90° in steps of 15°, as shown in Figure 1. Figure 3a,b shows the photographs of the lead matrix and an experimental arrangement in intralipid, respectively. PA intensity images (in dB scale) obtained at three representative angles: 15°, 45°, and 75°

in water and 1% intralipid medium are shown in Figure 3c–e, Figure 3f–h, respectively. PA images from all the angles between 0° and 90° can be found in Figure S2 (for water) and Figure S3 (for intralipid). Angle dependent PA intensities and CNR were calculated for four depths (approximately at 12 mm, 18 mm, 24 mm, and 30 mm) as described in Section 2.3 and plotted in Figure 3i, Figure 3k, Figure 3j, Figure 3l, respectively. An approximate gap of 10.5 mm from the US transducer to the sample was kept avoiding near field reconstruction errors and also to allow room for LED adjustments. For non-scattering media (water), the PA signal intensity increased as LED illumination angle increased from parallel orientation (0°), and then reached a maximum value at intermediate angles and finally decreased in its strength as the angle approached 90° (LED illumination perpendicular to the transducer). It is obvious from the plots that the angles corresponding to the maximum PA intensity value were in an inverse relation to the target depth, by showing a maximum value at a steeper angle for the 12 mm target and a maximum value at a shallower angle corresponding for the 30 mm target.

Figure 3j demonstrates CNR obtained in water as a function of the angle of irradiation. It reveals that the PA contrast did not change significantly after reaching a maximum level for an extended angle range of 15° to 75°. This was due to an increase in the background (due to various artifacts) along with the target signal increase that in turn reduced the angle dependency. PA intensities using intralipid (Figure 3k,l) showed a similar trend as water, especially for angles between 15° to 75°. In the cases of 0° and 90°, even though the trend was similar to that of the water phantom, the changes in PA intensity were less for intralipid phantom as expected. This reduction in intensity is due to reduced fluence due to optical scattering that reduces the incoming light directionality (and thus the angle dependency) as opposed to the case of non-scattering water medium. Comparing the depth-dependent CNR within the quasi angle-independent regime (15°–75°), the total drop was larger in scattering media (~30 dB) than in the water phantom (~15 dB). This can be due to a larger attenuation promoted by increased light scattering.

3.1.2. LED Source Divergence and Optimum Illumination Angle from Pencil Lead Targets in Water

Knowing the target location in depth and its lateral separation from the illumination plane, the source divergence (Figure 4) orthogonal to the illumination plane (Y) can be computed. In our experiments, the center of the light source (approximated here as a line source) was located at 10.5 ± 2 mm away from the imaging plane (imaging axis of the US transducer). Utilizing Equation (5), the divergence of the LED source was calculated to be 58 ± 8° at FWHM from the Gaussian fit of PA intensities from pencil lead target at 18 mm depth (Figure 4c). The experimentally derived divergence value is in good agreement with the manufacturer's data (60°, from Cyberdyne Inc., Tsukuba, Japan). The peak PA intensity (coefficient b in Equation (5)) at 18 mm depth was observed at 44° ± 3°. We further analyzed the data in Figure 3i to infer the angle at which maximum PA intensity could be obtained as a function of depth (Figure 4d) using the coefficient b in Equation (5). The black solid curve in Figure 4d shows the cotangent fit (Equation (6)) to the data shown in blue squares with a goodness of fit (R-squared) value equal to 0.95. Fitting was done by choosing m (depth offset) and n (separation of LED to the detector) as free fit parameters. Best fits (and 95% confidence interval bounds) obtained for, m is 9 mm (−0.6, 18.6 mm) and n is 10.7 mm (3, 18 mm). A large offset value might be due to the experimental error coming from the spatial width of the LED array, which takes up ~10.5 mm below the US transducer, in the imaging plane, at its steepest angle (90°). The value of n matched well to our experimentally set approximate value of 10.5 mm. From the fit, maximum PA signal intensity value at 12 mm can be obtained with a LED illumination angle of 74° while 27° can be used to image lesions at 30 mm depth. It should be noted that there exists a strong dependency of the illumination angle at superficial regions (slope of −5°/mm at 10 mm), which weakens as it goes to deeper locations (−2.5°/mm at 20 mm and −1.1°/mm at 30 mm) due to the nature of the cotangent function. The diameter of the projected beam onto the imaging plane, Z, at a given θ is estimated to be about 11.2 mm (θ = 84.5°) and 11.8 mm (θ = 27°) at 10 mm and 30 mm depths, respectively. Interestingly, less than 4% variation in the beam diameter in the imaging range is observed with different LED illumination angle. So, the large source divergence of about 11–12 mm overcomes the strong angle dependencies and can be used to lower the number of illumination angles for imaging lesions at various depths as is the case with Laser illumination. The LED illumination angle effects are even less notable in scattering medium, where deep tissue imaging can be achieved with smaller LED illumination angles (closer to being parallel to the imaging plane) and superficial lesion imaging is possible with larger LED illumination angles (closer to being perpendicular to the imaging plane).

3.2. Effect of Surrounding Media and Illumination Direction on PA Signal from Tumor Mimicking Lesion

A second set of experiments were conducted using tumor mimicking light-absorbing lesion placed obliquely in the imaging plane on top of chicken breast tissue (backing layer) while using water (Figure 5a) or scattering chicken tissue as the top layer (Figure 5b). This phantom study aimed to simulate superficial or deep-seated tumor lesions filled with blood vessels or contrast agents. US images corresponding to experiments using a top water layer (Figure 5c) and top chicken layer (Figure 5d) show the lesion placement and surrounding layers for comparison with corresponding PA images. The PA images using water as the top layer or chicken tissue as the top layer at LED source angles 15°, 45°, and 75° are shown in Figure 5e–g, Figure 5h–j, respectively. An apparent lateral shift in lesion positions between the images generated with water and chicken top layers was due to a change in imaging transducer placement about 2 mm in the horizontal axis, between the experiments (Figure 5h–j), which had negligible or no effects in our depth-dependent analysis.

Experiments with the top water layer show PA signals from the lesion for the entire imaging depth (Figure 5e–g), which can be associated with the negligible light scattering in water, opposed to chicken tissue (Figure 5h–j). Angle dependent PA intensity variations were also visible in both cases, where deeper tissue illumination was achieved at lower incident angles while increased PA intensity in the detector vicinity was observed for higher illumination angles. These results are very similar to those observed in Figure 3 using pencil lead phantom in water and intralipid. For a detailed analysis of depth-dependent PA intensity variation, Figure 6 was presented with analyses from three depth locations at 12 mm, 20 mm, and 28 mm, in which PA signals were plotted as a function of illumination angle, for the tissue-mimicking phantom with the top water and chicken layers. The white parallelograms in Figure 6a indicate the selected ROI from where the median PA intensity was calculated and the yellow parallelogram ROIs were considered for the background. Calculated PA intensities and CNRs as a function of angle for different depths are shown, respectively, in Figure 6b,c while using water as the top layer, and Figure 6d,f for chicken breast as the top layer. In the case of water as the top layer, the illumination angle parallel to the imaging plane, i.e., 0°, produced highest PA intensity (~23 dB) at the bottom (28 mm) than at the top regions; at 12 mm, the intensity dropped to ~17 dB as demonstrated in Figure 6b. On the other hand, experiment with the chicken top layer (Figure 6d) showed almost equal but low PA intensity (~17 dB) for all depths at 0°. It is interesting to note that the PA signal from 12 mm was quasi-constant for all angles from 30° and above. For the intermediate depth (20 mm), the intensities showed an increase up to 15°, then stayed almost unchanging until 60° and showed a slight decrease in the mean value with further increase in angle from 75° and above. It should also be noted that the signal from 28 mm depth while using chicken top layer was very close to the background level (CNR is less than 5 dB), which shows the maximum penetration depth achieved within our experimental limits.

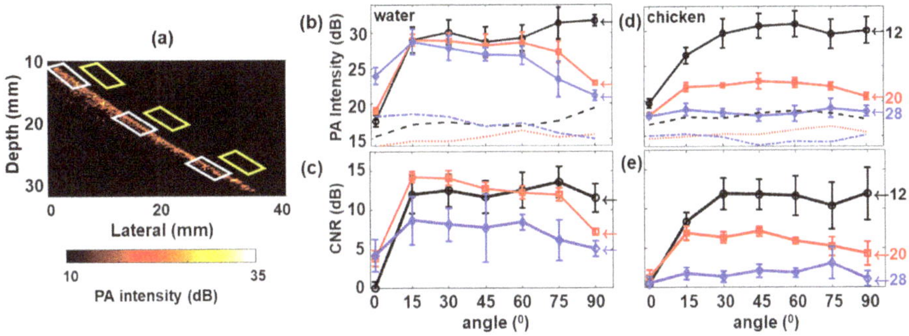

Figure 6. (**a**) PA image of a sample containing absorbing lesion (2 mm diameter) with the top water layer. White parallelograms indicate the regions selected for PA intensity and yellow for the background; (**b**,**d**) PA signal intensity plotted as a function of LED angles for selected depths in mm as labeled by numbers with the corresponding color, right next to each plot. Black circle with line corresponds to 12 mm, red squares with line corresponds to 20 mm and blue diamond with line indicates 28 mm. The black dotted lines, red dash-dotted lines, and blue dashed lines represent the background levels for 12, 20, and 28 mm depths (from the transducer) respectively; a gap of 10.5 mm exists between the transducer and the phantom surface. (**c**) CNR of the lesion under water; (**e**) CNR of the lesion under chicken breast. The error bars represent the standard deviation.

4. Discussion

The dependence of image contrast on irradiation parameters such as the angle of irradiation relative to the transducer, wavelength of the light irradiation, and distance between the transducer and the light source is undisputed. In this paper, we studied the dependence of signal intensity and contrast in PA images generated by an LED light source at various illumination angles. The 3D modular hinge design gave extreme flexibility to adjust the illumination angle as well as the transducer to LED-array distance. In the current work, the distance between the LED light source and transducer was relatively constant with the LED array positioned very close to the transducer (Figure 1). Studies addressing different separation distances between the transducer and the LED array can potentially give complementary information to our results. With respect to the 3D modular hinge system itself (Figure 2), several design criteria could be improved. Currently, the footprint of the modular system is large (11.5 cm at its widest dimension). Though the complete probe is lightweight due to lack of any heavy motors or metallic pieces, it is still larger than the other 3D printed fixed angle holders, e.g., by Sivasubramanian et al. [33]. The hinge system can be further modified with computer-controlled micro-hinges, can be adapted to any fiber bundles irrespective of their shape or size and can also be integrated with any laser or pulsed diode-based systems.

In all our results, it is evident that the background signal is also increasing along with the PA signal of interest (Figure S1), resulting in less impact on CNR when the illumination angle is increased beyond a certain limit. We believe that the background signal may be affected by various artifacts like reflection artifacts, out-of-plane clutter, and side lobes. For example, it is well known that more reflection artifacts are caused by high PA signal from tissue/phantom surface reflecting off acoustically dense structures when the illumination angle is steep and the fluency is high just beneath the US probe. These reverberation type of reflection artifacts are visible in our results too (Figure 3 and Figure S1) and would have impacted the CNR calculation. At lower illumination angle setups, light can scatter outside the imaging plane, get absorbed by different features, and generate out-of-plane artifacts and this also may have an impact in the CNR [52]. We strongly believe that it is also important to consider these common artifacts in the CNR analysis hence we chose a region very close to the target of interest for the calculations, instead of choosing a blank image (image generated with no light) or electronic noise. Furthermore, we used 850 nm irradiation, a wavelength at which there is relatively high penetration depth in tissue. The utility of other illumination wavelengths will impact the PA signal intensity and CNR based on the absorption properties of lesions at those wavelengths. Our future studies will involve evaluating these observations especially in an in vivo situation with both subcutaneous (superficial lesions) and orthotopic tumors (deep lesions).

A recent study by Agarwal et al. demonstrated that single-shot laser-based PAI and LED-based PAI achieved the same SNR from lesions 2–3 cm deep in chicken tissue. Though high frame averaging (2560 frames) was performed in LED-based PAI, they achieved real-time imaging due to the high pulse repetition frequency of LED sources [53]. The utility of LED-based PAI systems has been extensively reviewed by Zhu et al. [37] and one of the key factors currently hindering the clinical translation of LED-based photoacoustic imaging systems regardless of demonstrated promise in multiple preclinical and clinical imaging applications is the optical output power of the LED arrays [37,54]. It is of paramount importance to improve the optical output power of LEDs to enhance its usage in a wide range of deep-tissue imaging applications, and thus accelerate the clinical translation. The pulse repetition rate of LEDs is several-fold higher than pulsed LASER systems and it is feasible to average multiple image frames to improve SNR without compromising on real-time imaging capability. However, averaging N frames can improve SNR by only $\sqrt{N}$, and thus this approach has its limitations. Recently several developments in beamforming methodologies are made to improve the CNR and SNR of the LED-based PA systems [55,56]. We believe that the improvement of optical pulse energy and the development of novel image reconstruction and enhancement algorithms will be critical to accelerate the clinical translation of LED-based PAI [54].

5. Conclusions

Our results show that the optical excitation using an LED source behaves differently than the laser excitation due to a large source divergence of LED arrays, which we calculated to be 58° ± 8°. This in turn reduces the source direction dependency of PA signal at different depths. Our analysis in the non-scattering medium shows a strong dependence of illumination angle vs. depth at near field regions (−5°/mm at ~10 mm) and weak dependence at deeper locations (−2.5°/mm at 20 mm; −1.1°/mm at 30 mm). On the other hand, results from tissue-mimicking phantom in scattering media showed significantly weaker angle dependence of PA signal intensity than the phantom in water. So, utilizing an LED-based system (or a source with similar divergence) for either deep tissue lesions or superficial lesions would be less cumbersome in terms of source alignment by offering the freedom to choose a wide range of irradiation angles without losing CNR in the images. In contrast, spatially coherent sources would require stringent alignment strategies for illuminating lesions at various depths [57,58]. The modular light delivery system and results presented in this study can serve as a priori knowledge for future LED-based PA system designs and aid in further catapulting its utility in both preclinical and clinical applications.

Supplementary Materials: The following are available online at http://www.mdpi.com/1424-8220/20/13/3789/s1, Figure S1: (**a**) A photograph of Pencil lead array placed in water bath in the presence of PA transducer on top; (**b**) Schematic of the pencil lead array; (**c–h**) PA image acquired at representative angles 15° (**c**,**f**), 45° (**d**,**g**), and 75° (**e**,**h**) in water (**c–e**) and 1% intralipid (**f–h**); Differ to Figure 2 in the main article, where the reconstructed images are shown in a dynamic range of 30 dB, these images are shown in 40 dB to show the presence of background. Figure S2: Pencil lead in water at different angles of illumination mentioned as in each image title. PA intensity values are given in the bottom color bar. Figure S3: Pencil lead in intralipid at different angles of illumination mentioned as in each image title. PA intensity values are given in the bottom color bar.

Author Contributions: Conceptualization, M.K., M.K.A.S., and S.M.; methodology, M.K., C.D.N., and S.M.; software, M.K.; validation, M.K. and C.D.N.; formal analysis, M.K.; investigation, M.K.; resources, S.M.; data curation, M.K. and C.D.N.; writing—original draft preparation, M.K. and S.M.; writing—review and editing, M.K., M.K.A.S., C.D.N., and S.M.; visualization, M.K. and S.M.; supervision, S.M.; project administration, S.M.; funding acquisition, S.M. All authors have read and agreed to the published version of the manuscript.

Funding: This research was funded by the School of Engineering, Tufts University. Partial salary support of M.K and S.M. by the National Institute of Health RO1CA231606 grant is gratefully acknowledged.

Acknowledgments: The authors would like to thank Michael D. Kennedy, Department of Biomedical Engineering, Tufts University, Medford, MA, USA, for 3D printing the modular holder for LED arrays.

Conflicts of Interest: M.K.A.S. is employed by CYBERDYNE INC. The authors have no financial interests or conflicts of interest to disclose.

References

1. Hester, S.C.; Kuriakose, M.; Nguyen, C.D.; Mallidi, S. Role of Ultrasound and Photoacoustic Imaging in Photodynamic Therapy for Cancer. *Photochem. Photobiol.* **2020**, *96*, 260–279. [CrossRef]
2. Beard, P. Biomedical photoacoustic imaging. *Interface Focus* **2011**, *1*, 602–631. [CrossRef]
3. Brown, E.; Brunker, J.; Bohndiek, S.E. Photoacoustic imaging as a tool to probe the tumour microenvironment. *Dis. Model Mech.* **2019**, *12*, dmm039636. [CrossRef]
4. Steinberg, I.; Huland, D.M.; Vermesh, O.; Frostig, H.E.; Tummers, W.S.; Gambhir, S.S. Photoacoustic clinical imaging. *Photoacoustics* **2019**, *14*, 77–98. [CrossRef]
5. Zackrisson, S.; van de Ven, S.; Gambhir, S.S. Light in and sound out: Emerging translational strategies for photoacoustic imaging. *Cancer Res.* **2014**, *74*, 979–1004. [CrossRef]
6. Liu, G. Theory of the photoacoustic effect in condensed matter. *Appl. Opt.* **1982**, *21*, 955–960. [CrossRef] [PubMed]
7. Wang, L.V. Tutorial on Photoacoustic Microscopy and Computed Tomography. *IEEE J. Sel. Top. Quantum Electron.* **2008**, *14*, 171–179. [CrossRef]
8. Craig, B.; Sussman, C.R.; Zhang, Q.; Jiang, H.; Zheng, T.; Steindler, D.; Young, L.; Weiss, M.D. Photoacoustic tomography can detect cerebral hemodynamic alterations in a neonatal rodent model of hypoxia-ischemia. *Acta Neurobiol. Exp.* **2012**, *72*, 253–263.
9. Wang, X.; Chamberland, D.L.; Xi, G. Noninvasive reflection mode photoacoustic imaging through infant skull toward imaging of neonatal brains. *J. Neurosci. Methods* **2008**, *168*, 412–421. [CrossRef]
10. Lin, L.; Hu, P.; Shi, J.; Appleton, C.M.; Maslov, K.; Li, L.; Zhang, R.; Wang, L.V. Single-breath-hold photoacoustic computed tomography of the breast. *Nat. Commun.* **2018**, *9*, 2352. [CrossRef]
11. Dogra, V.S.; Chinni, B.K.; Valluru, K.S.; Moalem, J.; Giampoli, E.J.; Evans, K.; Rao, N.A. Preliminary results of ex vivo multispectral photoacoustic imaging in the management of thyroid cancer. *AJR Am. J. Roentgenol.* **2014**, *202*, W552–W558. [CrossRef] [PubMed]
12. Yang, M.; Zhao, L.; He, X.; Su, N.; Zhao, C.; Tang, H.; Hong, T.; Li, W.; Yang, F.; Lin, L.; et al. Photoacoustic/ultrasound dual imaging of human thyroid cancers: An initial clinical study. *Biomed. Opt. Express* **2017**, *8*, 3449–3457. [CrossRef] [PubMed]
13. Mallidi, S.; Luke, G.P.; Emelianov, S. Photoacoustic imaging in cancer detection, diagnosis, and treatment guidance. *Trends Biotechnol.* **2011**, *29*, 213–221. [CrossRef] [PubMed]
14. Valluru, K.S.; Wilson, K.E.; Willmann, J.K. Photoacoustic Imaging in Oncology: Translational Preclinical and Early Clinical Experience. *Radiology* **2016**, *280*, 332–349. [CrossRef]
15. Mallidi, S.; Watanabe, K.; Timerman, D.; Schoenfeld, D.; Hasan, T. Prediction of tumor recurrence and therapy monitoring using ultrasound-guided photoacoustic imaging. *Theranostics* **2015**, *5*, 289–301. [CrossRef]
16. Moore, C.; Jokerst, J.V. Strategies for Image-Guided Therapy, Surgery, and Drug Delivery Using Photoacoustic Imaging. *Theranostics* **2019**, *9*, 1550–1571. [CrossRef]
17. Kim, J.; Park, S.; Jung, Y.; Chang, S.; Park, J.; Zhang, Y.; Lovell, J.F.; Kim, C. Programmable Real-time Clinical Photoacoustic and Ultrasound Imaging System. *Sci. Rep.* **2016**, *6*, 35137. [CrossRef] [PubMed]
18. Schellenberg, M.W.; Hunt, H.K. Hand-held optoacoustic imaging: A review. *Photoacoustics* **2018**, *11*, 14–27. [CrossRef]
19. Upputuri, P.K.; Pramanik, M. Photoacoustic imaging in the second near-infrared window: A review. *J. Biomed. Opt.* **2019**, *24*, 1–20. [CrossRef]
20. Xu, M.; Wang, L.V. Photoacoustic imaging in biomedicine. *Rev. Sci. Instrum.* **2006**, *77*, 041101. [CrossRef]
21. Oraevsky, A.A.; Karabutov, A.A. Optoacoustic Tomography. In *Biomedical Photonics Handbook*; Vo-Dinh, T., Ed.; CRC: Boca Raton, FL, USA, 2003; Chapter 34.
22. Sivasubramanian, K.; Pramanik, M. High frame rate photoacoustic imaging at 7000 frames per second using clinical ultrasound system. *Biomed. Opt. Express* **2016**, *7*, 312–323. [CrossRef]

23. Sowers, T.; Yoon, H.; Emelianov, S. Investigation of light delivery geometries for photoacoustic applications using Monte Carlo simulations with multiple wavelengths, tissue types, and species characteristics. *J. Biomed. Opt.* **2020**, *25*, 1–16. [CrossRef]
24. Wang, Y.; Lim, R.S.A.; Zhang, H.; Nyayapathi, N.; Oh, K.W.; Xia, J. Optimizing the light delivery of linear-array-based photoacoustic systems by double acoustic reflectors. *Sci. Rep.* **2018**, *8*, 13004. [CrossRef]
25. Wang, Z.; Ha, S.; Kim, K. A new design of light illumination scheme for deep tissue photoacoustic imaging. *Opt. Express* **2012**, *20*, 22649–22659. [CrossRef]
26. Yang, G.; Amidi, E.; Nandy, S.; Mostafa, A.; Zhu, Q. Optimized light delivery probe using ball lenses for co-registered photoacoustic and ultrasound endo-cavity subsurface imaging. *Photoacoustics* **2019**, *13*, 66–75. [CrossRef] [PubMed]
27. Carles, G.; Zammit, P.; Harvey, A.R. Holistic Monte-Carlo optical modelling of biological imaging. *Sci. Rep.* **2019**, *9*, 15832. [CrossRef] [PubMed]
28. Flock, S.T.; Patterson, M.S.; Wilson, B.C.; Wyman, D.R. Monte Carlo modeling of light propagation in highly scattering tissue–I: Model predictions and comparison with diffusion theory. *IEEE Trans. Biomed. Eng.* **1989**, *36*, 1162–1168. [CrossRef]
29. Held, K.G.; Jaeger, M.; Ricka, J.; Frenz, M.; Akarcay, H.G. Multiple irradiation sensing of the optical effective attenuation coefficient for spectral correction in handheld OA imaging. *Photoacoustics* **2016**, *4*, 70–80. [CrossRef] [PubMed]
30. Watte, R.; Aernouts, B.; Van Beers, R.; Herremans, E.; Ho, Q.T.; Verboven, P.; Nicolai, B.; Saeys, W. Modeling the propagation of light in realistic tissue structures with MMC-fpf: A meshed Monte Carlo method with free phase function. *Opt. Express* **2015**, *23*, 17467–17486. [CrossRef] [PubMed]
31. Held, G.; Preisser, S.; Akarcay, H.G.; Peeters, S.; Frenz, M.; Jaeger, M. Effect of irradiation distance on image contrast in epi-optoacoustic imaging of human volunteers. *Biomed. Opt. Express* **2014**, *5*, 3765–3780. [CrossRef]
32. Haisch, C.; Eilert-Zell, K.; Vogel, M.M.; Menzenbach, P.; Niessner, R. Combined optoacoustic/ultrasound system for tomographic absorption measurements: Possibilities and limitations. *Anal. Bioanal. Chem.* **2010**, *397*, 1503–1510. [CrossRef]
33. Sivasubramanian, K.; Periyasamy, V.; Wen, K.K.; Pramanik, M. Optimizing light delivery through fiber bundle in photoacoustic imaging with clinical ultrasound system: Monte Carlo simulation and experimental validation. *J. Biomed. Opt.* **2017**, *22*, 41008. [CrossRef]
34. Sangha, G.S.; Hale, N.J.; Goergen, C.J. Adjustable photoacoustic tomography probe improves light delivery and image quality. *Photoacoustics* **2018**, *12*, 6–13. [CrossRef]
35. Jo, J.; Xu, G.; Zhu, Y.; Burton, M.; Sarazin, J.; Schiopu, E.; Gandikota, G.; Wang, X. Detecting joint inflammation by an LED-based photoacoustic imaging system: A feasibility study. *J. Biomed. Opt.* **2018**, *23*, 1–4. [CrossRef] [PubMed]
36. Leskinen, J.; Pulkkinen, A.; Tick, J.; Tarvainen, T.; Ntziachristos, V.; Zemp, R. Photoacoustic tomography setup using LED illumination. In Proceedings of the Opto-Acoustic Methods and Applications in Biophotonics IV, Munich, Germany, 24–25 June 2019.
37. Zhu, Y.; Feng, T.; Cheng, Q.; Wang, X.; Du, S.; Sato, N.; Kuniyil Ajith Singh, M.; Yuan, J. Towards Clinical Translation of LED-Based Photoacoustic Imaging: A Review. *Sensors* **2020**, *20*, 2484. [CrossRef]
38. Agrawal, S.; Fadden, C.; Dangi, A.; Yang, X.; Albahrani, H.; Frings, N.; Heidari Zadi, S.; Kothapalli, S.R. Light-Emitting-Diode-Based Multispectral Photoacoustic Computed Tomography System. *Sensors* **2019**, *19*, 4861. [CrossRef]
39. Xia, W.; Kuniyil Ajith Singh, M.; Maneas, E.; Sato, N.; Shigeta, Y.; Agano, T.; Ourselin, S.; J West, S.; E Desjardins, A. Handheld real-time LED-based photoacoustic and ultrasound imaging system for accurate visualization of clinical metal needles and superficial vasculature to guide minimally invasive procedures. *Sensors* **2018**, *18*, 1394. [CrossRef]
40. Zhu, Y.; Xu, G.; Yuan, J.; Jo, J.; Gandikota, G.; Demirci, H.; Agano, T.; Sato, N.; Shigeta, Y.; Wang, X. Light Emitting Diodes based Photoacoustic Imaging and Potential Clinical Applications. *Sci. Rep.* **2018**, *8*, 9885. [CrossRef]
41. Jo, J.; Xu, G.; Cao, M.; Marquardt, A.; Francis, S.; Gandikota, G.; Wang, X. A Functional Study of Human Inflammatory Arthritis Using Photoacoustic Imaging. *Sci. Rep.* **2017**, *7*, 15026. [CrossRef]

42. Hariri, A.; Lemaster, J.; Wang, J.; Jeevarathinam, A.S.; Chao, D.L.; Jokerst, J.V. The characterization of an economic and portable LED-based photoacoustic imaging system to facilitate molecular imaging. *Photoacoustics* **2018**, *9*, 10–20. [CrossRef]
43. Joseph Francis, K.; Boink, Y.E.; Dantuma, M.; Ajith Singh, M.K.; Manohar, S.; Steenbergen, W. Tomographic imaging with an ultrasound and LED-based photoacoustic system. *Biomed. Opt. Express* **2020**, *11*, 2152–2165. [CrossRef]
44. Toshitaka Agano, M.K.A.S.; Nagaoka, R.; Awazu, K. Effect of light pulse width on frequency characteristics of photoacoustic signal—An experimental study using a pulse-width tunable LED-based photoacoustic imaging system. *Int. J. Eng. Technol.* **2018**, *7*, 4300–4303.
45. Jaeger, M.; Schüpbach, S.; Gertsch, A.; Kitz, M.; Frenz, M. Fourier reconstruction in optoacoustic imaging using truncated regularized inversek-space interpolation. *Inverse Probl.* **2007**, *23*, S51–S63. [CrossRef]
46. Driver, I.; Feather, J.W.; King, P.R.; Dawson, J.B. The optical properties of aqueous suspensions of Intralipid, a fat emulsion. *Phys. Med. Biol.* **1989**, *34*, 1927. [CrossRef]
47. Pogue, B.W.; Patterson, M.S. Review of tissue simulating phantoms for optical spectroscopy, imaging and dosimetry. *J. Biomed. Opt.* **2006**, *11*, 041102. [CrossRef] [PubMed]
48. Van Leeuwen-van Zaane, F.; Gamm, U.A.; van Driel, P.B.A.A.; Snoeks, T.J.A.; de Bruijn, H.S.; van der Ploeg-van den Heuvel, A.; Mol, I.M.; Löwik, C.W.G.M.; Sterenborg, H.J.C.M.; Amelink, A.; et al. In vivo quantification of the scattering properties of tissue using multi-diameter single fiber reflectance spectroscopy. *Biomed. Opt. Express* **2013**, *4*, 696–708. [CrossRef] [PubMed]
49. Elena Vladimirovna, S.; Brian, J.; John, N.; Anna, N.Y. Optical properties of normal and cancerous human skin in the visible and near-infrared spectral range. *J. Biomed. Opt.* **2006**, *11*, 1–9.
50. Vogt, W.C.; Jia, C.; Wear, K.A.; Garra, B.S.; Joshua Pfefer, T. Biologically relevant photoacoustic imaging phantoms with tunable optical and acoustic properties. *J. Biomed. Opt.* **2016**, *21*, 101405. [CrossRef]
51. Bohndiek, S.E.; Bodapati, S.; Van De Sompel, D.; Kothapalli, S.R.; Gambhir, S.S. Development and Application of Stable Phantoms for the Evaluation of Photoacoustic Imaging Instruments. *PLoS ONE* **2013**, *8*, e75533. [CrossRef]
52. Nguyen, H.N.Y.; Steenbergen, W. Three-dimensional view of out-of-plane artifacts in photoacoustic imaging using a laser-integrated linear-transducer-array probe. *Photoacoustics* **2020**, *19*, 100176. [CrossRef]
53. Sumit, A.; Mithun Kuniyil Ajith, S.; Xinyi, Y.; Hussain, A.; Ajay, D.; Sri-Rajasekhar, K. Photoacoustic imaging capabilities of light emitting diodes (LED) and laser sources: A comparison study. In Proceedings of the SPIE 11240, Photons Plus Ultrasound: Imaging and Sensing 2020, San Francisco, CA, USA, 17 February 2020.
54. Kuniyil Ajith Singh, M.; Sato, N.; Ichihashi, F.; Sankai, Y. Clinical Translation of Photoacoustic Imaging—Opportunities and Challenges from an Industry Perspective. In *LED-Based Photoacoustic Imaging: From Bench to Bedside*; Kuniyil Ajith Singh, M., Ed.; Springer: Singapore, 2020; pp. 379–393. [CrossRef]
55. Miri Rostami, S.R.; Mozaffarzadeh, M.; Ghaffari-Miab, M.; Hariri, A.; Jokerst, J. GPU-accelerated Double-stage Delay-multiply-and-sum Algorithm for Fast Photoacoustic Tomography Using LED Excitation and Linear Arrays. *Ultrason. Imaging* **2019**, *41*, 301–316. [CrossRef] [PubMed]
56. Farnia, P.; Najafzadeh, E.; Hariri, A.; Lavasani, S.N.; Makkiabadi, B.; Ahmadian, A.; Jokerst, J.V. Dictionary learning technique enhances signal in LED-based photoacoustic imaging. *Biomed. Opt. Express* **2020**, *11*, 2533–2547. [CrossRef]
57. Alijabbari, N.; Alshahrani, S.S.; Pattyn, A.; Mehrmohammadi, M. Photoacoustic Tomography with a Ring Ultrasound Transducer: A Comparison of Different Illumination Strategies. *Appl. Sci.* **2019**, *9*, 94. [CrossRef] [PubMed]
58. Luo, H.; Yang, G.; Zhu, Q. Fiber endface illumination diffuser for endo-cavity photoacoustic imaging. *Opt. Lett.* **2020**, *45*, 632–635. [CrossRef] [PubMed]

sensors

MDPI

Article

Photoacoustic Imaging of Human Vasculature Using LED versus Laser Illumination: A Comparison Study on Tissue Phantoms and In Vivo Humans

Sumit Agrawal [1,†], Mithun Kuniyil Ajith Singh [2,†], Kerrick Johnstonbaugh [1], David C. Han [3,4], Colette R. Pameijer [4] and Sri-Rajasekhar Kothapalli [1,5,6,*]

1 Department of Biomedical Engineering, Pennsylvania State University, University Park, State College, PA 16802, USA; sua347@psu.edu (S.A.); kjohnstonbaugh97@gmail.com (K.J.)
2 Research & Business Development Division, CYBERDYNE INC, Cambridge Innovation Center, 3013 AK Rotterdam, The Netherlands; mithun_ajith@cyberdyne.jp
3 Department of Surgery, Penn State Heart and Vascular Institute, Hershey, PA 16802, USA; dch15@psu.edu
4 Penn State Hershey College of Medicine and Milton S. Hershey Medical Center, Hershey, PA 17033, USA; cpameijer@pennstatehealth.psu.edu
5 Penn State Cancer Institute, Pennsylvania State University, Hershey, PA 17033, USA
6 Graduate Program in Acoustics, Pennsylvania State University, University Park, State College, PA 16802, USA
* Correspondence: srkothapalli@psu.edu
† These authors contributed equally to this work.

Abstract: Vascular diseases are becoming an epidemic with an increasing aging population and increases in obesity and type II diabetes. Point-of-care (POC) diagnosis and monitoring of vascular diseases is an unmet medical need. Photoacoustic imaging (PAI) provides label-free multiparametric information of deep vasculature based on strong absorption of light photons by hemoglobin molecules. However, conventional PAI systems use bulky nanosecond lasers which hinders POC applications. Recently, light-emitting diodes (LEDs) have emerged as cost-effective and portable optical sources for the PAI of living subjects. However, state-of-art LED arrays carry significantly lower optical energy (<0.5 mJ/pulse) and high pulse repetition frequencies (PRFs) (4 KHz) compared to the high-power laser sources (100 mJ/pulse) with low PRFs of 10 Hz. Given these tradeoffs between portability, cost, optical energy and frame rate, this work systematically studies the deep tissue PAI performance of LED and laser illuminations to help select a suitable source for a given biomedical application. To draw a fair comparison, we developed a fiberoptic array that delivers laser illumination similar to the LED array and uses the same ultrasound transducer and data acquisition platform for PAI with these two illuminations. Several controlled studies on tissue phantoms demonstrated that portable LED arrays with high frame averaging show higher signal-to-noise ratios (SNRs) of up to 30 mm depth, and the high-energy laser source was found to be more effective for imaging depths greater than 30 mm at similar frame rates. Label-free in vivo imaging of human hand vasculature studies further confirmed that the vascular contrast from LED-PAI is similar to laser-PAI for up to 2 cm depths. Therefore, LED-PAI systems have strong potential to be a mobile health care technology for diagnosing vascular diseases such as peripheral arterial disease and stroke in POC and resource poor settings.

Keywords: deep tissue imaging; hemangioma; laser; light-emitting diodes (LED); mobile health; peripheral arterial disease; photoacoustic imaging; stroke; vascular malformations

Citation: Agrawal, S.; Kuniyil Ajith Singh, M.; Johnstonbaugh, K.; C. Han, D.; R. Pameijer, C.; Kothapalli, S.-R. Photoacoustic Imaging of Human Vasculature Using LED versus Laser Illumination: A Comparison Study on Tissue Phantoms and In Vivo Humans. *Sensors* **2021**, *21*, 424. https://doi.org/10.3390/s21020424

Received: 16 December 2020
Accepted: 6 January 2021
Published: 9 January 2021

1. Introduction

Vascular diseases are the leading cause of death worldwide. Some common vascular diseases include cardiovascular disease, stroke and peripheral artery disease (PAD) [1–3]. Many of these vascular diseases need point-of-care (POC) diagnosis and monitoring using nonionizing, noninvasive and cost-effective approaches. Although Doppler ultrasound meets all these requirements, it only maps blood flow, which is operator dependent and influenced by motion artifacts, resulting in limited sensitivity and specificity to detect the disease in its early stage [4]. A POC technique that provides direct label-free molecular and functional information of vasculature is needed to reliably detect and monitor vascular diseases [5,6].

Photoacoustic imaging (PAI) is a hybrid imaging modality that provides rich optical spectroscopic contrast at ultrasonic penetration depths and resolutions [7]. Over the past two decades, PAI has emerged as a promising tool for label-free imaging of individual blood vessels [8,9], detection of angiogenesis [10] and has also helped in extracting several physiologically relevant parameters such as blood oxygen saturation [11,12] and changes in the blood volume [13], which are vital for monitoring disease progression [14,15]. The basic mechanism of PAI includes a nanosecond pulsed optical excitation that illuminates the biological subject. The light absorbing molecules inside the tissue undergo thermoelastic expansion and generate broadband acoustic waves which are subsequently detected using conventional ultrasound (US) detectors [7]. Biomedical PAI mainly capitalizes on the intrinsic absorbers present in human tissue [8], such as oxy- and deoxyhemoglobin [10,11], melanin [16], lipids [17,18], water [19], RNA and DNA [20], as each of these exhibits a characteristic absorption spectrum. However, if the spectral contrast from these intrinsic chromophores is not sufficient to reveal the disease, a wide range of extrinsic contrast agents [21–25] can be functionalized to target different diseased biomarkers to increase molecular sensitivity and specificity.

While ultrasound and certain optical technologies are available in the size of a mobile phone [26,27], PAI systems still have to reach that level of portability. Conventional PAI systems employ bulky class-IV laser sources (100 mJ/pulse) and data acquisition systems to increase the peak imaging performance [28–31]. Such lasers not only increase the system cost and footprint but also carry a high risk of class-IV exposure. However, to translate PAI technology to POC clinical applications and to resource-limited settings, a significant reduction in both cost and size is required. To address this challenge, several cost-effective alternatives for both the optical excitation [32–35] and the ultrasound detection [36–39] components have been explored, including for wearable applications [40,41].

Recently, low-power laser diodes [42,43] and light-emitting-diodes (LEDs) [44] have been proposed as alternatives to laser sources. Specifically, pertaining to their noncoherent nature, LEDs carry a huge potential to be the safe and cost-effective alternative illumination sources for PAI [43,44]. However, compared to the class-IV lasers, LED arrays carry much less optical energy (i.e., order of 100's μJ) per pulse, and thus their PAI capabilities as a function of imaging depth need to be studied in detail to employ a suitable optical source for a given POC application [45,46].

To enhance the performance of LEDs, an arrayed arrangement of LED elements was developed [47,48], thereby increasing the pulse energies from a few μJ to hundreds of μJ. In addition to this, higher pulse repetition frequency (PRF) rates (i.e., ~4 KHz) of the LEDs allowed a sufficient PA frame averaging which led to significant signal-to-noise ratio (SNR) improvements for deep tissue targets [46]. These LED array B-mode PAUS [47,48] and tomographic imaging [49,50] setups have been demonstrated using several preclinical small animals [51,52] and in vivo human imaging studies [53–55].

To date, there is no study that quantitatively compares the PAI performance of LEDs and laser illumination head-to-head. Given these tradeoffs between portability, cost, optical energy and frame rate, this work systematically studies the deep tissue PAI performance of LED and laser excitation to help select a suitable source for a given biomedical application. First, a setup for sequentially performing PAI with these two optical illuminations has been developed. Controlled studies on different tissue phantoms have been performed for detailed evaluation of the imaging performance. Further, in vivo human hand vasculature imaging in 2-D and 3-D was performed using these two optical sources. The rest of the paper is organized as follows. Section 2 describes our proposed setup for comparing the two sources. Comparison studies including tissue-mimicking phantoms and the in vivo human imaging are presented in Section 3. Section 4 provides a detailed discussion of the results.

2. Materials and Methods

In this section, a detailed description of the experimental setup for studying the PAI capabilities of high-power laser and LED array sources is presented. First, the commercial LED array-based B-mode PA and ultrasound (US) imaging system (AcousticX, Cyberdyne Inc., Ibaraki, Japan), referred to in this paper as LED-PAUS, is presented in Section 2.1 and then the modifications performed to use the same AcousticX data acquisition system for laser-illumination-based PA, referred to as laser-PAUS, without interrupting parameters of the imaging system, are presented in Section 2.2.

2.1. LED Array-Based US/PA (LED-PAUS) Imaging System Description

Conventionally, a high-power and bulky laser source is employed for most B-mode PA and US systems [28,29]. The commercial LED-PAUS system, as shown in Figure 1, consists of a host controller (Figure 1a), data acquisition hardware (Figure 1b) and an interactive graphical user interface shown on the display in Figure 1d. Here, a linear ultrasound probe is sandwiched between two LED arrays to capture interleaved B-mode PA and US images, as shown in Figure 1e. Each LED array consists of four rows of 36 LED elements (1×1 mm size). The LED arrays used in this study consist of 850 nm LED elements and each array provides an output energy of 200 μJ/pulse. The excitation pulse widths for each of these LED arrays can be controlled with the software in the range of 30 to 150 ns. For all the experiments in this study, a pulse width of 70 ns was used that offered optimum energy at 850 nm [56]. The optical illumination profile achieved with the two 850 nm LED arrays operated with pulse widths of 70 ns is shown in Figure 1g. The shape of the beam falling on skin is approximately a rectangle with an area of 9 cm^2 (5 by 1.8 cm), leading to an incident optical fluence of 0.044 mJ/cm^2, considering 400 μJ total energy per pulse with the two LED arrays. The pulse repetition frequency (PRF) of these LEDs can also be controlled in the range of 1 to 4 KHz allowing multiple averaging options leading to different frame rates. To assess the effect of changing the frame rate over the PA image quality, several different combinations of PRF and frame averaging are used in this study, as discussed in the subsequent sections. The US probe used in this study is a 128-element linear US array with a pitch of 0.3 mm, center frequency of 7 MHz, elevational focus of 15 mm and a measured 6 dB bandwidth of 75%. The system provides PA and US acquisition sampling rates of 40 and 20 MHz, respectively.

2.2. Experimental Setup for Comparing LED-PAUS and Laser-PAUS

Figure 1 presents the overall experimental setup developed to compare the performance of LED-PAUS and laser-PAUS imaging. The LED-PAUS system, described in Section 2.1, was adapted for laser-PAUS imaging. In laser-PAUS, a portable optical parametric oscillator (OPO) laser source (Phocus Mobile, Opotek, Inc., Carlsbad, CA, USA), tunable in the range of 690–950 nm, shown in Figure 1c, provided the laser illumination. The laser has a fixed pulse width of 5–7 ns, a fixed PRF of 10 Hz and an output energy of 140 mJ per pulse at 750 nm. For this study, 850 nm wavelength laser illumination was used with the optical energy tuned down to 40 mJ/pulse. Light output from the laser was coupled to a 2 m long custom designed optical fiber bundle (Fiberoptic System Inc., Simi Valley, CA, USA). The fiber had a fused end with a diameter of 6.5 mm that entered into the tunable output port of the laser, as shown in Figure 1c. The distal end of this fiber was split into twenty smaller (1.45 mm inner diameter) fibers with numerical apertures of 0.55, each sharing equal optical energy. Ten out of these twenty fibers were inserted into each of the two custom designed 3-D printed fiber holders attached each side of the US probe, as shown in Figure 1f. This design allowed laser illumination similar to the case of LED arrays, in terms of the illumination angle and the overall geometry around the

US probe. Further, to achieve a uniform illumination profile on the tissue surface, two glass diffusers (N-BK7 Ground Glass Diffuser 1500 Grit, Thorlabs Inc., Newton, NJ, USA) were attached at the output end of the fiber holders, as shown in Figure 1f. The resulting laser-illumination profile at 850 nm is shown in Figure 1h. The shape of the beam falling on skin is approximately a rectangle with an area of 6 cm^2 (5 by 1.2 cm), leading to an incident optical fluence of 6.66 mJ/cm^2, considering 40 mJ total energy per pulse with the laser illumination. In our experimental setup, switching from LED arrays to a laser fiber setup and vice versa was convenient and did not disturb the US probe as well as the imaging subject. Figure 1i,j show the heads of LED-PAUS and laser-PAUS systems for acquiring respective B-mode PA and US images of human hand vasculature.

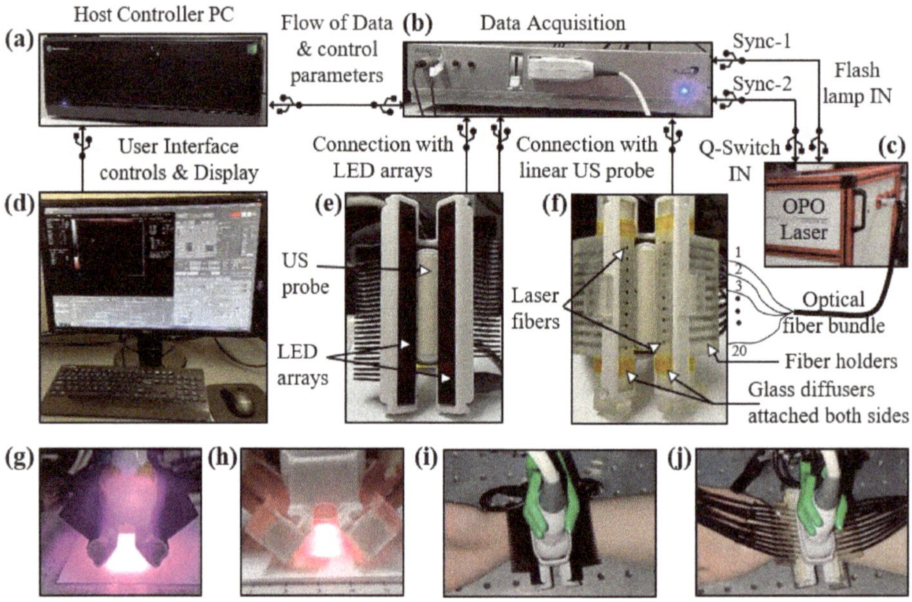

Figure 1. Description of the experimental setup designed for comparing light-emitting diode (LED)-based and high-power laser-based photoacoustic (PA) and ultrasound (US) imaging. The setup consists of the following key components: Commercial B-mode LED-PAUS system (AcousticX, Cyberdyne Inc., Ibaraki, Japan) with (**a**) a host controller PC and (**b**) data acquisition hardware. (**c**) A portable high-power laser (Phocus Mobile, Opotek Inc., Carlsbad, CA, USA) with its output coupled to the input end of an optical fiber bundle. The fiber bundle is split into twenty smaller fibers at the distal end. (**d**) Computer display: displays B-mode US (grayscale), PA (red scale), and coregistered US + PA (overlaid red PA on gray US). The interface also enables switching between LED and laser operation. (**e**) Arrangement of two 850 nm LED arrays around the US probe. (**f**) Arrangement of twenty laser fibers inserted into the two fiber holders around the US probe. Two glass diffusers attached at the fiber output ends to provide uniform laser illumination on the tissue surface. Optical illumination profile achieved with (**g**) two LED array sources and (**h**) laser source. (**i**,**j**) Pictures of a human wrist under imaging with the LED and laser arrangements, respectively.

3. Validation Experiments and Results

In this section, an extensive evaluation of the imaging performance of low-power LED-PAUS and high-power laser-PAUS systems is presented with the help of rigorous SNR and resolution studies on several tissue phantoms, in vivo imaging of the human wrist and in vivo 3-D vasculature mapping of the human forearm.

3.1. Photoacoustic Imaging Comparison of LED Arrays and Laser Source Using a Scattering Phantom

In this subsection, a controlled study evaluating the deep tissue PAI capabilities of low-power LED arrays and a high-power laser source is presented. An acrylic tank with four holes was fabricated and a pencil lead with a diameter of 0.5 mm was inserted into each hole. The tank was filled with an intralipid solution to mimic an optical reduced scattering coefficient (μ_s') of 20 cm^{-1}. Considering an approximate optical absorption coefficient (μ_a) of 0.05 to 0.1 cm^{-1}, the effective attenuation coefficient (μ_{eff}) of the intralipid phantom was in the range of 1.73 to 2.45 cm^{-1} ($\mu_{eff} = \sqrt{3\mu_a(\mu_a + \mu_s')}$). Figure 2a shows the schematic of the experimental setup with four pencil lead targets diagonally arranged along the depth of imaging inside the intralipid solution. The measured depth of these four pencil leads from the surface of the US transducer were 15, 23, 28, and 34 mm, respectively.

The laser-illumination setup discussed in Section 2.2 was first employed to capture the PA images. The AcousticX data acquisition system software was synchronized to the laser acquisition mode. Two external triggers from the system were used for driving the laser. One trigger from the synch-1 port of the system went into the flash-lamp input port of the OPO laser and the other trigger from the synch-2 port fed the Q-Switch IN port of the laser, as shown in Figure 1b,c. After switching ON the 2-D B-mode RF data acquisition, the laser flash-lamp was first turned ON followed by (after 10s delay) the laser, from a PC connected to the laser. Once the laser was ON and synchronized with the data acquisition, PA data could be captured at a 10 Hz frame rate (limited by the PRF of the laser). Figure 2b shows a PA image captured with the laser illumination at 850 nm wavelength and 40 mJ/pulse optical energy resulting in <20 mJ/cm^2 optical fluence. As shown, all four pencil lead targets generated PA signals due to their higher optical absorptions compared to the intralipid medium.

To acquire the PA images with LED arrays, the laser arrangement was removed and the two 850 nm wavelength LED arrays were attached to the US transducer as shown in Figure 1e. The software of the AcousticX system was set to the LED acquisition mode. The PRF for the LED arrays was selected as 4 KHz. Figure 2c shows a PA image captured using the LED array setup with a frame averaging of 128. The achieved frame rate in this configuration was 30 Hz (128 averaging at 4 KHz PRF). Further, to study the effect of frame averaging over the LED-PA imaging performance, the averaging was increased from 128 to 256, 384, 640, 1280 and 2560 frames, leading to frame rates of 15, 10, 6, 3, and 1.5 Hz, respectively. The corresponding PA images are shown in Figure 2d–h.

In order to perform an effective comparison of the captured PA images with the laser and the LED array illuminations, the raw PA data were extracted and analyzed in a local computer using MATLAB software. First, frequency domain reconstruction [57] was performed to beamform the PA images from the raw data. All beamformed PA images were then log-compressed, maintaining the same 70 dB scale, as presented in Figure 2b–h, for effective comparison. Further, to quantitatively compare these PA images, the SNR study was performed over all four pencil lead targets. For calculating the SNR, peak PA signal at the target locations (over a circular region surrounding the targets) and the mean noise adjacent to each target (over a similar circular region at same depth as targets) were calculated over the linear beamformed PA images. The calculated values of peak signal, mean noise and SNR for the four targets over the PA images corresponding to the varying frame averaging of laser acquisition (1, 2, 4, 10, 128 frame averages leading to 10 Hz, 5 Hz, 2.5 Hz, 1 Hz, and 78 mHz frame rates when computed offline) and the LED array acquisition at varying frame rates are presented in Table 1. The calculations at 78 mHz with the laser system were specifically performed in order to compare the SNR of two systems at the same frame averaging—i.e., 128 frames (LED 30 Hz). As shown, the peak PA signal for the laser acquisition is about two log orders higher than that for the LED arrays. However, the mean noise in the case of laser illumination is even higher (up to three log orders), leading to a lower SNR, especially when compared at shallow imaging depths.

Plots in Figure 2i–l present the SNR trends observed with varying frame rates of LED array acquisitions for the four targets. The corresponding value of SNR for the laser acquisition at 10 Hz is also marked in each of these plots. For the targets lower than 30 mm depth, the LED-based PAI at 10 Hz frame rate continued to show a higher SNR. Target-4 at 34 mm depth was detected at a higher SNR with the laser illumination as compared to the LED arrays at the same frame rate. However, increasing the PA frame averaging further for the LED arrays acquisition led to a lower frame rate (<10 Hz), and helped in boosting the SNR value for higher depth targets. Further, when maintaining same frame averaging (128 frames) for the laser (78 mHz frame rate) and the LED arrays (30 Hz frame rate), significantly high SNRs were observed for all four targets with the laser, sacrificing the real-time imaging.

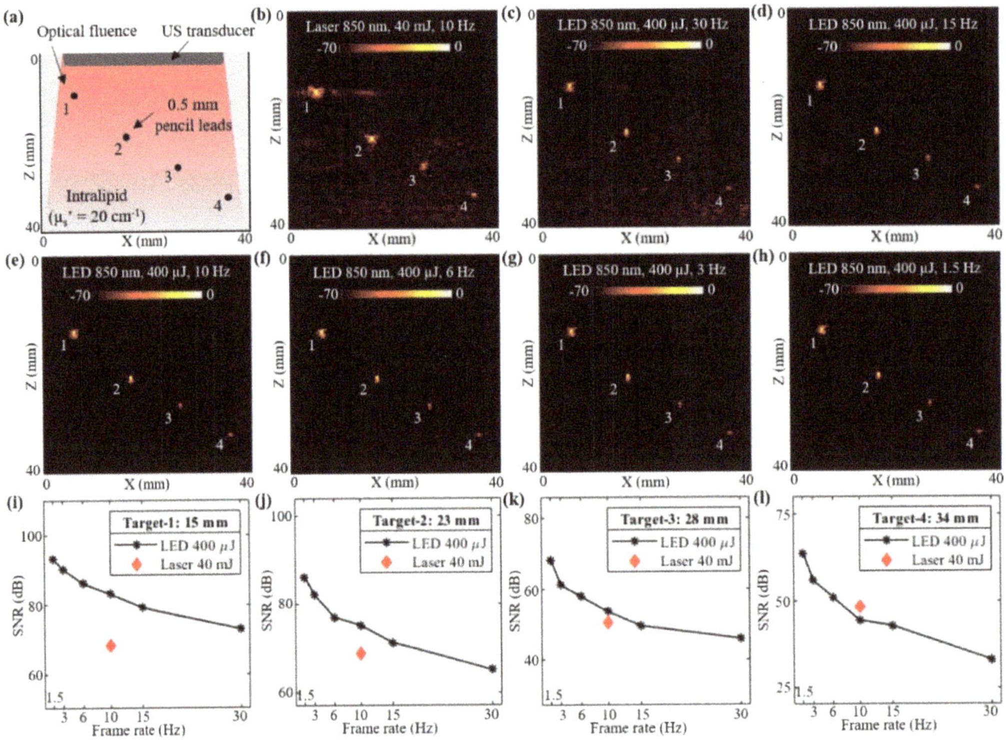

Figure 2. Performance evaluation of LED array-based and laser-illumination-based photoacoustic (PA) imaging in intralipid scattering phantom. (**a**) Shows the schematic of a scattering phantom with four 0.5 mm diameter pencil leads placed at 15, 23, 28 and 34 mm depth from the ultrasound (US) transducer inside intralipid medium mimicking an optical reduced scattering coefficient of 20 cm^{-1}. (**b**) Shows the PA imaging results with 850 nm laser illumination at 10 Hz and 40 mJ optical energy with <20 mJ/cm^2 optical fluence on the phantom surface. (**c–h**) Show the PA imaging results with LED array-based illumination at an 850 nm wavelength, a total of 400 μJ output energy from the two LED arrays and frame rates of 30, 15, 10, 6, 3 and 1.5 Hz, respectively. (**i–l**) Show the plots of signal-to-noise ratio (SNR) with respect to the acquisition frame rates comparing LED array and laser-illumination-based PA imaging performance for the pencil lead targets at four depths.

Table 1. Peak PA signal, average background noise and signal-to-noise ratio (SNR) values for four pencil lead targets located 15, 23, 28 and 34 mm deep inside scattering phantom imaged with LED array and laser illuminations at varying frame rates.

Configuration	Target-1: 15 mm			Target-2: 23 mm			Target-3: 28 mm			Target-4: 34 mm		
	Signal (a.u.)	Noise (a.u.)	SNR (dB)	Signal (a.u.)	Noise (a.u.)	SNR (dB)	Signal (a.u.)	Noise (a.u.)	SNR (dB)	Signal (a.u.)	Noise (a.u.)	SNR (dB)
Laser 10 Hz	2.62E14	1.04E11	68.01	4.09E14	1.49E11	68.77	7.96E13	2.38E11	50.48	4.19E13	1.62E11	48.24
Laser 5 Hz	2.60E14	5.01E10	74.31	4.25E14	7.39E10	75.18	8.58E13	1.43E11	55.55	4.38E13	8.81E10	53.93
Laser 2.5 Hz	2.58E14	2.94E10	78.85	4.07E14	4.26E10	79.59	8.17E13	8.26E10	59.90	3.8E13	4.01E10	59.55
Laser 1 Hz	2.58E14	1.44E10	85.05	4.05E14	2.25E10	85.10	7.89E13	6.64E10	61.50	3.79E13	1.66E10	67.18
Laser 78 mHz	2.47E14	5.66E09	92.79	3.39E14	7.95E09	92.58	5.94E13	3.86E10	63.80	2.89E13	3.35E09	78.50
LED 1.5 Hz	3.62E12	8.06E07	93.06	1.95E12	9.59E07	86.17	2.06E11	8.16E07	68.04	8.59E10	5.70E07	63.55
LED 3.0 Hz	3.57E12	1.11E08	90.14	1.94E12	1.50E08	82.33	1.84E11	1.62E08	61.13	7.80E10	1.27E08	55.79
LED 6.0 Hz	3.69E12	1.84E08	86.01	1.89E12	2.65E08	77.04	2.14E11	2.72E08	57.90	8.39E10	2.39E08	50.92
LED 10 Hz	3.66E12	2.57E08	83.06	2.06E12	3.55E08	75.26	1.95E11	4.08E08	53.58	7.64E10	4.70E08	44.22
LED 15 Hz	3.56E12	3.98E08	79.03	1.86E12	5.06E08	71.31	2.07E11	6.98E08	49.45	8.36E10	6.18E08	42.62
LED 30 Hz	3.70E12	8.49E08	72.79	1.95E12	1.08E09	65.12	2.38E11	1.21E09	45.84	7.22E10	1.63E09	32.92

In order to validate the optical properties of our phantom, we calculated the μ_{eff} value from the experimental peak PA signal values for LED 10 Hz data. Using the Beer Lambert's principle, i.e., $I(z) = I_o e^{-\mu_{eff} z}$, where *I(z)* is the PA intensity at depth (z) cm and I_o is the intensity at zero depth, and taking the ratio of two equations at two different depths (z_1 at 1.5 cm and z_2 at 3.4 cm) will cancel out the I_o and lead to the μ_{eff} 2.034 cm^{-1}. This closely matches to the previously reported values in the range of 1.734 to 2.456 cm^{-1}, mentioned above in the phantom description.

3.2. Photoacoustic Imaging Comparision of LED Arrays and Laser Illuminations over Chicken Tissue Phantom

To compare the deep tissue PAI capabilities of the low-power LED arrays and the high-power laser illuminations, a multilayer chicken tissue phantom was designed. Figure 3a shows the schematic of the chicken tissue phantom with five layers of chicken breast tissue stacked inside a water tank, with an estimated optical absorption and reduced scattering coefficients of 0.1 to 0.2 cm^{-1} and 1.0 to 5 cm^{-1}, respectively, with an 850 nm wavelength [58], leading to an effective attenuation coefficient (μ_{eff}) in the range of 0.575 to 1.766 cm^{-1} ($\mu_{eff} = \sqrt{3\mu_a(\mu_a + \mu_s')}$). Four pencil leads with diameters of 0.5 mm were placed in between the chicken tissue layers as shown. The measured depths of these pencil lead targets from the top layer of the chicken tissue were 11, 18, 24, and 31 mm, respectively.

With the above-described phantom, a laser-illumination setup was used to acquire the US and PA images at 10 Hz, maintaining the same 40 mJ output energy (<20 mJ/cm^2 optical fluence on the phantom surface) and laser with an 850 nm wavelength. The captured raw data from the AcousticX software were reconstructed in MATLAB to further perform the quantitative comparison. Figure 3b shows the beamformed B-mode US image of the chicken tissue phantom, at a log scale of 70 dB, clearly highlighting the chicken tissue structure. The US image also shows the distance of the four pencil lead targets from the top layer of chicken tissue. Similarly, the PA raw data captured from the software were reconstructed, log-compressed and was overlaid on the US image to generate a co-registered US + PA image of the phantom. Figure 3c shows the coregistered US and PA image at a 60 dB scale in order to keep the noise floor at the threshold. The beamformed PA image with 70 dB log scale is shown in Figure 3d.

Without disturbing the phantom, the PA images were subsequently acquired with the LED arrays setup. The laser arrangement was removed and the two 850 nm LED arrays were attached to the US probe. With the 4 KHz PRF of the LED arrays, the PA images were captured at varying frame averaging settings, similar to the settings used for Section 3.1, leading to the frame rates of 30, 15, 10, 6, 3 and 1.5 Hz. The captured raw data from the AcousticX software were again extracted in the MATLAB software and reconstructed to generate B-mode PA images for comparing with the laser PA images. All PA images were compressed at the 70 dB log scale for effective comparison. Four representative PA images at 30, 15, 10, and 6 Hz, respectively, are shown in Figure 3e–h.

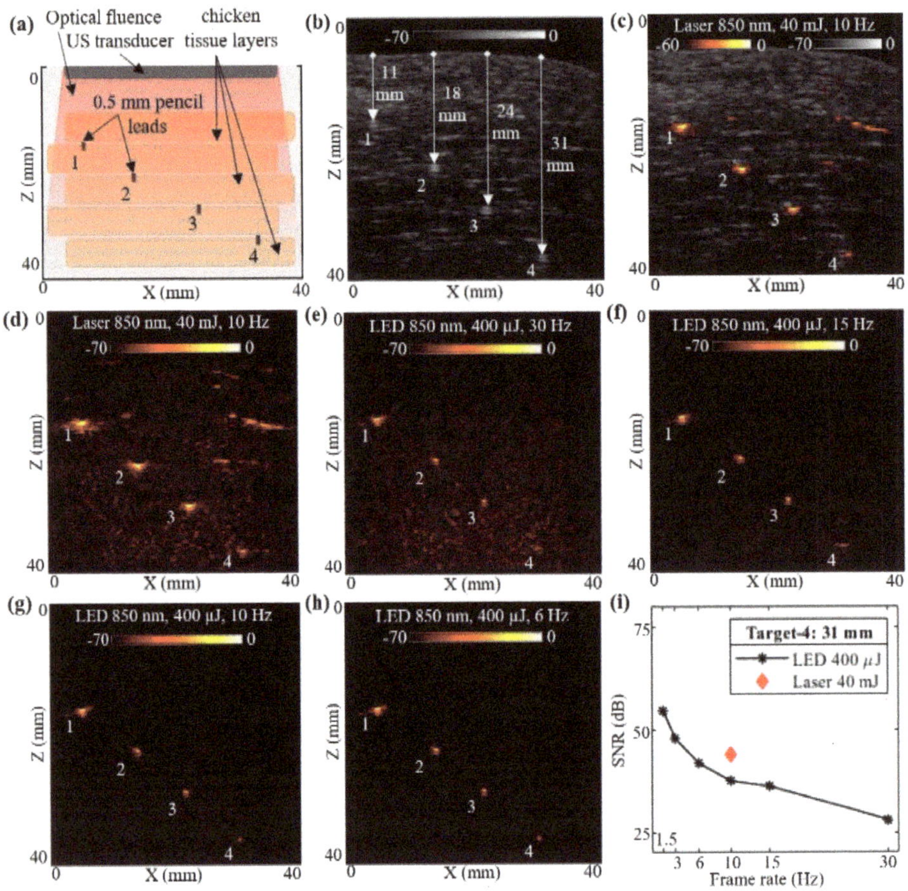

Figure 3. Performance evaluation of LED array-based and laser-based photoacoustic (PA) imaging in chicken tissue phantom. (**a**) Shows a schematic of chicken tissue phantom with five layers of chicken breast tissue stacked inside water tank. The positions of four 0.5 mm diameter pencil lead targets placed in between the chicken tissue layers are also shown. (**b–d**) Show the B-mode ultrasound (US), coregistered US + PA and PA image, respectively, captured with laser-based illumination at 850 nm wavelength and 10 Hz high pulse repetition frequency (PRF). (**e–h**) Show the PA imaging results obtained by imaging the chicken tissue phantom using LED array-based illumination, at 850 nm illumination and frame rates of 30, 15, 10, and 6 Hz, respectively. (**i**) Shows the plot of signal-to-noise ratio (SNR) with respect to the frame rate comparing the LED array and laser-illumination-based PA imaging performance for the deepest pencil lead target (target-4).

Further, for a quantitative comparison of high-power laser versus low power LED array-based acquisitions over the chicken tissue phantom, the SNR study was performed over the linear beamformed images in MATLAB. For this experiment, we studied the SNR of the shallowest target (target-1 located 11 mm) and the deepest target (target-4 located 31 mm) deep inside chicken breast tissue. For calculating the SNR, the peak signal at target locations and the mean noise adjacent to these targets were calculated. The values of peak PA signal, mean adjacent noise and the SNR for laser-based and LED array-based acquisitions at varying frame rates are listed in Table 2. The peak signal as well as mean noise for laser acquisition at 10 Hz is about two to three log orders of magnitude higher compared to the LED arrays. However, the SNR for LEDs is still comparable to the lasers. Figure 3i also demonstrates the trend of SNR for target-4, comparing the laser and the LED array illuminations at varying frame rates. For a 31 mm deep target, LED arrays provide close to 37.51 dB SNRs, whereas laser illumination provides about 43.75 dB SNR at a 10 Hz frame rate. When increasing the frame averaging, hence a reduction in the frame rate, LED arrays show significant improvement of SNR for the same target, up to 54.47 dB at 1.5 Hz, which is higher than the SNR of laser illumination at 10 Hz. These results demonstrate the capabilities of LED arrays to image deeper inside realistic tissue medium and motivated us to further study how they compare with high-power laser sources for imaging in vivo human vasculature.

Table 2. Peak PA signal, average background noise and signal-to-noise ratio (SNR) values for pencil lead targets located 11 and 31 mm deep inside chicken breast tissue imaged with LED array and laser illuminations at varying frame rates.

Configuration	Target-1: 11 mm			Target-4: 31 mm		
	Signal (a.u.)	Noise (a.u.)	SNR (dB)	Signal (a.u.)	Noise (a.u.)	SNR (dB)
Laser 10 Hz	1.4E14	1.2E11	61.63	2.1E12	1.4E11	43.75
LED 1.5 Hz	2.1E12	6.8E07	89.71	3.5E10	6.6E07	54.47
LED 3.0 Hz	2.1E12	9.2E08	87.35	3.4E10	1.4E08	47.76
LED 6.0 Hz	2.1E12	1.6E08	82.47	2.8E10	2.2E08	41.89
LED 10 Hz	2.0E12	2.8E08	77.13	3.2E10	4.3E08	37.51
LED 15 Hz	2.1E12	3.7E08	75.03	3.6E10	5.7E08	36.08
LED 30 Hz	2.1E12	7.4E08	69.06	2.9E10	1.2E09	27.98

Similar to Section 3.1, we validated the optical properties of this chicken tissue phantom by calculating the μ_{eff} value from the experimental peak PA signal values for LED 10 Hz data. The experimentally calculated μ_{eff} (2.067 cm^{-1}) is slightly higher than the previously calculated range mentioned above in the phantom description. The small discrepancy could be due to chicken tissue heterogeneity—prolonged imaging of chicken tissue in water medium that increases the optical attenuation.

3.3. *Photoacoustic Imaging Comparison of LED Arrays and Laser Sources: Resolution Study*

In this subsection, the spatial resolutions of the two optical illumination setups, the LED-PAUS and the laser-PAUS, are characterized. A 30 µm carbon fiber was placed in a bath containing water mixed with intralipid to obtain a scattering medium of 3 cm^{-1}. First, the two 850 nm LED arrays were attached to the US probe and a B-mode PA image of the phantom was acquired. The raw data captured were extracted in the MATLAB software and reconstructed to generate a B-mode image. The log-compressed B-mode PA image at a 30 dB scale is shown in Figure 4a. Figure 4b presents a sample zoomed time trace of an A-line across the target region for the PA data acquired with LED array illumination. To calculate the spatial resolution for this carbon fiber target, the line-spread functions of the PA amplitudes are plotted in the lateral and axial directions, respectively, as shown in Figure 4c,d. The obtained lateral and axial resolutions with a full-width-half-maximum (FWHM) approach are 350 and 210 µm, respectively.

To acquire the PA data with laser-illumination, the LED arrays were removed from the US probe and the laser fiber setup was attached without disturbing the phantom. The captured PA data were extracted and beamformed. Figure 4e shows the B-mode PA image, at a 30 dB scale, obtained for the same carbon fiber phantom using laser-illumination at 850 nm with 40 mJ output energy and a 10 Hz frame rate. A sample zoomed time trace of an A-line across the target region for the PA data acquired with the laser-illumination is shown in Figure 4f. The line-spread functions of the PA amplitudes in the lateral and axial directions with the laser illumination are shown in Figure 4g,h. The obtained lateral and axial resolutions using an FWHM approach for the laser illumination are 355 and 203 µm, respectively.

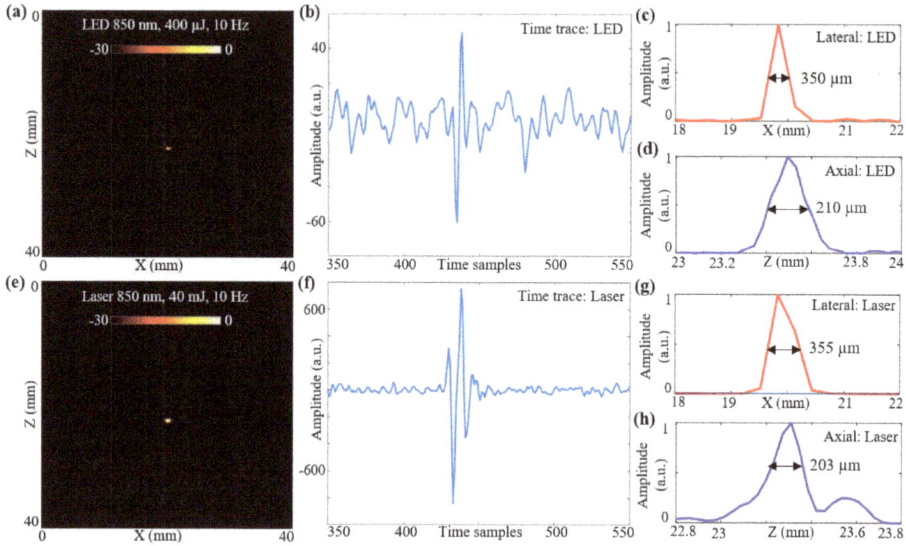

Figure 4. Resolution study for LED array-based and laser-illumination-based photoacoustic (PA) imaging. (**a**) B-mode PA image obtained for a 30 μm carbon fiber placed in an intralipid-based phantom using LED illumination at 850 nm with a frame rate of 10 Hz. (**b**) Shows zoomed time trace of an A-line across the target region for the PA data acquired with LED illumination. (**c**,**d**) Show the line-spread functions of the PA amplitudes for the carbon fiber target plotted in the lateral and axial directions, respectively, when imaged with LED setup. The obtained lateral and axial resolutions with the full-width-half-maximum (FWHM) approach are 350 and 210 μm, respectively. (**e**) B-mode PA image obtained for same carbon fiber phantom using laser illumination at 850 nm and 10 Hz frame rate. (**f**) Shows the time trace for the PA data acquired with laser illumination. (**g**,**h**) Show the line-spread functions of the PA amplitudes in lateral and axial directions with the laser illumination. The obtained lateral and axial resolutions using the FWHM approach are 355 and 203 μm, respectively.

3.4. In Vivo Photoacoustic Imaging Comparison of LED Arrays and Laser Illumination over In Vivo Human Wrist

In this subsection, the photoacoustic vascular imaging capabilities of the LED array and the laser illumination were compared by in vivo imaging of a healthy human volunteer's wrist vasculature. The volunteer was a healthy 25-year-old European male, and the experiment was conducted by following the internal imaging protocol of CYBERDYNE, INC (Rotterdam, The Netherlands) for healthy-volunteer imaging studies. For this study, the volunteer's right hand was positioned inside a large water tank, as shown in Figure 5a. The probe was positioned such that one of the major blood vessels, which supplies blood to the forearm and hand, is within the field-of-view (FOV).

The hand was first imaged with the laser-PAUS setup by attaching the laser fiber holders to the US probe, as shown in Figure 5a. The laser was operated at 850 nm wavelength, 10 Hz PRF, and delivered output optical energy of 40 mJ. This allowed ANSI safety limits of <20 mJ/cm^2 optical fluence on the hand surface [59]. During the real-time data acquisition, the probe was aligned such that the major blood vessel could be seen running parallel to the skin surface in the PA images. The captured US and PA raw data using the AcousticX software were later extracted in the MATLAB software and were reconstructed to generate the beamformed images. The beamformed, log-compressed B-mode US, PA and the coregistered US + PA images are shown in Figure 5b–d. The US image showed the anatomical features along the depth of human wrist, whereas the PA image highlighted the major blood vasculature. There was fairly strong correspondence between the locations of the blood vessel in the PA image and the appearance of anechoic regions in the US image.

Based on the anatomy of the vasculature in human wrist, the PA signals ~5 mm below the skin surface may have corresponded to the radial artery that travels across the front of the elbow, deep under the muscle until it comes to the wrist where it comes close to the skin surface. This is also marked with a white arrow in the PA image in Figure 5c.

To compare these laser-illumination results of the human wrist with the LED arrays, the laser-fiber attachments were gently removed without disturbing the location of the US probe. The two 850 nm LED arrays were then attached to the US probe, as shown in Figure 5e. With a PRF of 4 KHz and frame averaging of 384, leading to a frame rate of 10 Hz, the US and PA frames were captured using the LED array setup. The US and PA raw data were then reconstructed in MATLAB. Figure 5f–h show the beamformed log-compressed B-mode US, PA and coregistered US + PA images for the human wrist. As in the case of laser illumination, the LED array-based PA images also imaged the same vasculature below the skin surface. The radial artery present ~5 mm below the skin was clearly visible with the LED array-based acquisition as well.

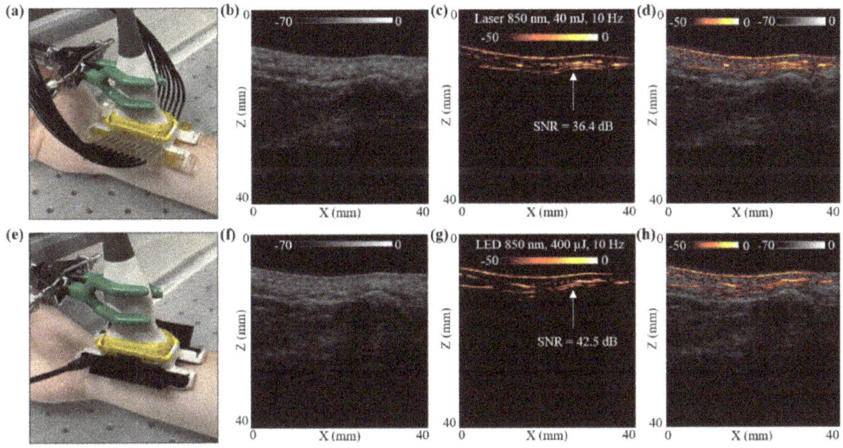

Figure 5. In vivo comparison of LED array-based and laser-based PA vasculature imaging over the right-hand wrist of a healthy 25-year-old male human volunteer. (**a**) Shows the experimental setup with the right-hand wrist placed inside a big water bath for the laser-based PA imaging. (**b–d**) Show the obtained US, PA and coregistered US + PA images for the setup shown in (**a**). (**e**) Shows the setup with LED arrays. (**f–h**) Show the obtained US, PA, and coregistered US + PA images for the setup shown in (**e**).

To further compare the two setups quantitatively, an SNR comparison study was performed for the radial artery, as marked with white arrows in Figure 5c,g. To calculate the SNR, the peak PA signal at the artery and the mean noise adjacent to the artery region was calculated over the linear beamformed PA images resulting from the laser-illumination and the LED array-based acquisitions. Table 3 presents the values of peak PA signal, mean noise and the SNR. The SNR values are also marked in the Figure 5c,g. Both the peak signal and mean noise with the laser are up to three log orders of magnitude higher compared to the LEDs. However, the SNR value for the laser-illumination-based PA image was about 6 dB lower than the SNR with the LED array acquisition. This follows the trend observed for the controlled tissue phantom studies discussed in Section 3.1, where the shallow targets (<30 mm) inside an intralipid medium were detected with higher SNRs using LED arrays compared to the laser illumination, maintaining the same frame rate.

Table 3. Peak PA signal, average background noise and signal-to-noise ratio (SNR) values for blood vessel target located ~5 mm below the skin surface of right-hand wrist of a healthy 25-year-old male human volunteer imaged with LED array and laser illumination at 10 Hz frame rate.

Configuration	Target: Blood Vessel ~5 mm below Skin Surface of Human Wrist		
	Signal (a.u.)	Noise (a.u.)	SNR (dB)
Laser 10 Hz	3.7E13	5.7E11	36.37
LED 10 Hz	5.3E10	4.0E08	42.49

3.5. In Vivo Photoacoustic Imaging Comparison of LED Arrays and Laser Illumination: In Vivo Human Forearm

This subsection presents in vivo 3-D mapping of the vasculature inside a human volunteer's forearm, using the LED array-based and the laser-based illuminations. For this study, the volunteer was a healthy 25-year-old European male, and the experiment was conducted by following the internal imaging protocol of CYBERDYNE, INC (Rotterdam, The Netherlands) for healthy-volunteer imaging experiments. The forearm of the volunteer was submerged in water. The US probe was fixed on a linear translation stage, translating in the Y-direction, as shown in Figure 6a.

For the laser acquisition, the laser-illumination setup was attached to the US probe. After switching ON the laser software, the RF data acquisition on the AcousticX software were first turned ON. Note that the 3-D data acquisition feature of the AcousticX software does not work for the laser mode. Therefore, to capture the data corresponding to the Y-direction scan, a manual stage motion feature was used while acquiring the data in the 2-D RF acquisition mode. The linear stage was translated for 60 mm in the Y-direction while capturing the RF data. The laser was tuned to 850 nm at 40 mJ output energy (<20 mJ/cm^2 optical fluence on the hand surface) and 10 Hz PRF. The captured PA raw data were later extracted and reconstructed in MATLAB. Initial PA frames captured prior to the stage motion were discarded for the 3-D volumetric reconstruction. Figure 6b presents the maximum-intensity-projection (MIP) of the 3-D PA volume for full depth, highlighting the major vasculature present in the human forearm. To better visualize the vasculature deeper than ~5 mm, a deep tissue MIP of the PA volume is also shown in Figure 6c. Considering the initial 15 mm stand-off of the US probe from the human skin surface, the deep tissue MIP fell within Z = 20 to 40 mm, where Z is the depth dimension. Since deeper blood vessels show weak PA intensities, the MIP image in Figure 6c was scaled to 30 dB as opposed to the full depth MIP in Figure 6b, which was scaled to 50 dB.

After the laser acquisition, the stage was manually moved back to the home position and the laser setup was detached from the US probe without disturbing the position of the forearm with respect to the probe. To acquire the LED array-based PA 3-D scan data, two 850 nm LED arrays were attached to the US probe. In this case, the AcousticX software directly allows 3-D PA data acquisition using the automatic 3-D scan feature. The same Y-direction translation stage was automatically scanned for 60 mm and the raw data captured were extracted in MATLAB. The beamformed PA frames were stitched in the Y-dimension to reconstruct a 3-D PA volume. Figure 6d,e present the full depth MIP and the deep tissue MIP of the 3-D PA volume, respectively.

Qualitatively, Figure 6b,e look very similar, indicating that both laser and LED array-based illumination are able to map the major vasculature present inside human forearm. Minor differences in the SNR for deep vessels can be noticed when comparing the deep tissue MIPs presented in Figure 6c,f. For example, a deeper blood vessel running parallel to the Y-axis at around X = 20 mm (in Figure 6c,f) was detected with a better SNR with the laser source as compared to the LED arrays. Similarly, the deep tissue MIP from laser acquisition show other minor PA signals (at coordinates X,Y = 30 mm, 50 mm), that are relatively weaker in the deep tissue MIP from the LED acquisitions, emphasizing the need for a high-power laser when imaging very deep (>2 cm) vasculature.

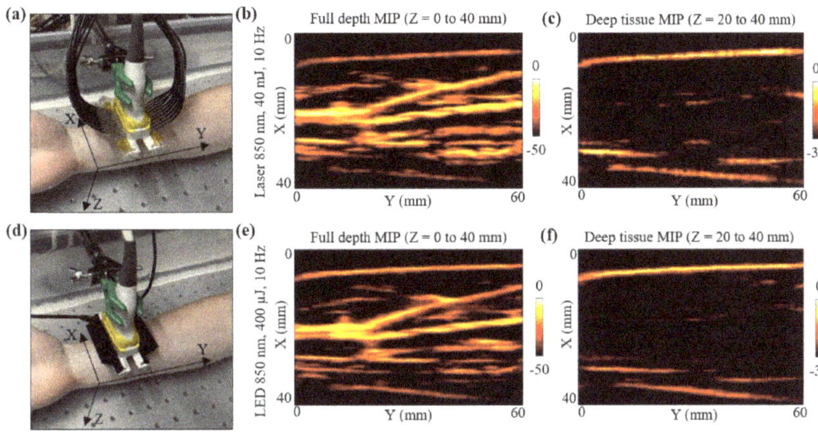

Figure 6. In vivo comparison of LED array-based and laser-based PA vasculature imaging over forearm of a healthy 25-year-old male human volunteer. (**a**) Shows the experimental setup with the right forearm placed inside big water bath for the laser-based PA imaging. (**b**,**c**) Show the obtained full depth and deep tissue maximum-intensity-projections (MIPs) of the reconstructed 3-D PA volume for the setup shown in (**a**). (**d**) Shows the setup with LED arrays. (**e**,**f**) Show the obtained full depth and deep tissue MIPs of the reconstructed 3-D PA volume for the setup shown in (**d**).

4. Discussion

To successfully translate PAI to the POC applications and to resource-limited settings, the cost and the overall size of the PAI systems need to be significantly cut down. LED arrays are one of the highly explored optical sources in the PA literature that are portable and significantly lower in cost and footprint. This study was focused on comparing the capabilities of these LED arrays (approximately USD 15K, including driver electronics) with a state-of-the-art high-power OPO laser (approximately USD 100K) for PAI. A commercial LED-PAUS system, AcousticX, was first adapted to perform sequential acquisitions of LED-PAUS and laser-PAUS. This ensured that the PA and US signals were detected by the same ultrasound transducer array and processed by the same data acquisition system. To further draw a fair comparison, the experimental setup shown in Figure 1 allowed for (1) similar optical illumination on the tissue surface in terms of the geometry, angle and aperture of illumination, and (2) a convenient, uninterrupted sequential acquisition of PA images with the two sources. To achieve uniform laser illumination on the tissue surface, two glass diffusers were attached to the output end of the laser fiber holders. For all the experimental studies presented in this work, the output optical energy for the laser and the LED arrays were maintained at 40 mJ and 400 µJ/pulse, respectively, with an 850 nm wavelength of illumination. A constant PRF of 10 Hz for the laser and 4 KHz for the LED arrays were used for all studies. With no frame averaging, the obtained PA frame rate for laser was 10 Hz. For the LED arrays, the PA frame averaging was varied from 128 to 256, 384, 640, 1280, and 2560, achieving the frame rates of 30, 15, 10, 6, 3, and 1.5 Hz, respectively.

The first comparison study presented in Section 3.1 consisted of controlled experiments on an intralipid phantom shown in Figure 2. The four 0.5 mm pencil lead targets arranged diagonally along the depth of the scattering phantom were imaged using both laser and LED arrays. Detailed analysis of the SNR presented in Table 1 and plotted in Figure 2i–l led to the following conclusions on the quantitative performance of the two sources. (1) At similar frame rates (10 Hz), LED-PAUS imaging showed higher SNR compared to laser-PAUS imaging for targets up to an imaging depth of 30 mm. (2) Laser-PAUS imaging provided a better SNR for the deeper targets (>30 mm) as compared to the LED-PAUS at 10 Hz. (3) With increased frame averaging and hence a reduced overall frame rate (<10 Hz), the deep tissue (>30 mm) performance of LED arrays can be improved. For example, the target-4 located 34 mm deep was imaged with a 48.24 dB SNR using laser illumination and with 44.22 dB SNR using LED arrays, both at 10 Hz. However, with increase in PA averaging (lower frame rates), LED arrays were able to provide an improved SNR of up to 50.92, 55.79, and 63.55 dB, when imaged at 6, 3, and 1.5 Hz frame rates, respectively. Further, comparison of the magnitudes of peak PA signals from the target and mean noise surrounding the target for the two optical illuminations led to the following observations. While higher optical energy from the laser source generates a PA signal two to three log orders of magnitude higher compared to the lower-power LED arrays, the average noise floor from the background regions was also observed to be up to three to four log orders higher for the laser sources compared to LED arrays. In conclusion, the low-power LED arrays were effective due to lower noise floor that comes from the higher frame averaging possible with high PRF (4 KHz) of LED arrays.

Our next comparison study, presented in Section 3.2, involved a multilayer chicken tissue phantom with four 0.5 mm pencil lead targets embedded in between the five layers of chicken breast tissue. Figure 3 presented the qualitative comparison of laser-PAUS and LED-PAUS, whereas Table 2 highlighted the quantitative comparison of PAI performance for the deepest pencil lead target—i.e., target-4 at 31 mm. This study results further confirmed our observations with the intralipid scattering phantom presented in Section 3.1. For example: (1) both peak PA signal and mean noise from the laser illumination were up to three log orders of magnitude higher compared to the LED array illumination. (2) For the target-4 at 31 mm, the laser showed a ~6 dB higher SNR compared to the LED arrays at a 10 Hz PA frame rate. (3) With the increase in the average, and hence reduced, frame rate (<10 Hz), LED arrays show higher SNR values than laser illumination, obtaining SNRs of 54.47 and 47.76 dB at 1.5 and 3 Hz, respectively, for the same target-4 as compared to an SNR of 43.75 dB with the laser.

The study presented in Section 3.3 compared the lateral and axial resolutions of the two setups, LED-PAUS and laser-PAUS, by imaging carbon fiber with a 30 μm diameter inside an intralipid-based phantom. Although the pulse widths for the laser (5–7 ns) and the LED arrays (70 ns) in this study were not similar, the spatial resolution were shown to be in the same ranges—i.e., ~350 μm lateral and ~205 μm axial. This confirmed the fact that the spatial resolution here is mainly limited by the acoustic detection.

In the final study, in Section 3.4 we presented an in vivo LED-PAUS and laser-PAUS imaging of the wrist of a healthy human volunteer. Figure 5 presents the qualitative and quantitative comparison results for a radial artery seen ~5 mm below the wrist skin surface. Analysis of the peak PA signal, mean noise and SNR for this radial artery is presented in Table 3. As observed with the studies presented in previous sections, for a shallow depth target, the LED-PAUS showed better SNR than the laser-PAUS at the same 10 Hz frame rates. The ~5 mm deep radial artery was imaged with a 36.37 dB SNR with the laser compared to a 42.49 dB SNR with LED arrays. This study substantiates that the LED-PAUS imaging is an attractive choice for several preclinical and clinical applications.

Further, in Section 3.5, to test the capabilities of deep tissue in vivo vascular imaging, we scanned the right-hand forearm of a healthy human volunteer. The 3-D vasculature map, presented in Figure 6, demonstrated that the LED arrays could image all the vessels that could be seen with a laser source, especially at shallow depths of up to 1.5 cm. For deeper (>30 mm) vessels, laser sources provided better signals. However, the amount of noise observed with the laser at those depths was also significantly high and thus limited the imaging performance when viewing deeper vessels. With an increased frame averaging with the LED arrays, the SNR of deeper vessels can potentially be improved.

The conclusive observations made by studying the performance of the two optical sources bring us to the following remarks. (1) LED arrays exhibit a strong potential for translating PAI systems to the resource-limited settings. (2) Higher frame averaging enhances the LED-PAUS imaging performance, especially for deeper targets (>30 mm), making it suitable for deep tissue imaging; however, this sacrifices the frame rate. To further improve the performance of LED-PAUS imaging, the state-of-the-art machine learning approaches can be employed that can help in (1) further boosting the frame rates by avoiding the need for higher frame averaging [60,61], and (2) improving the SNR for deep tissue targets [61,62].

5. Conclusions

Photoacoustic imaging capabilities of low-cost and low-power LED arrays were compared head-to-head with a high-power laser source using both tissue-mimicking phantoms and in vivo human subjects. The experimental observations on different tissue mimicking phantom studies demonstrated (1) high PRFs of LED arrays can be leveraged for averaging PA frames to achieve high SNR up to an imaging depth of 30 mm with 10 Hz frame rate, (2) high-power laser sources show higher SNRs for deeper targets (>30 mm) compared to the LED arrays at 10 Hz and (3) with increased frame averaging (<10 Hz), the SNR of a target at 34 mm depth with LED-PAUS closely matched that of laser-PAUS. The in vivo human hand vasculature imaging studies presented in this work also demonstrated similar observations. (1) The ~5 mm deep radial artery was imaged with a higher SNR using LED arrays in comparison to the laser source, and (2) for deeper vessels in the subject's forearm, the laser at the 10 Hz frame rate had a slightly higher SNR compared to the LED arrays at 10 Hz. In summary, due to the low power of LED arrays, a higher frame averaging is required to image deep tissue targets. LED-PAUS holds strong potential in point-of-care diagnosis of vascular diseases.

Author Contributions: Conceptualization: S.A., M.K.A.S. and S.-R.K.; Methodology: S.A., and K.J.; Software: S.A.; Hardware design: S.A.; Validation: S.A., M.K.A.S. and K.J.; Formal analysis: S.A. and M.K.A.S.; Investigation: S.-R.K.; Resources: S.-R.K.; Data curation: S.A. and M.K.A.S.; Writing–original draft preparation: S.A. and M.K.A.S.; Writing–review and editing: S.A., M.K.A.S., K.J., D.C.H., C.R.P., and S.-R.K.; Visualization: S.A., and K.J.; Supervision: S.-R.K.; Project administration: S.-R.K.; Funding acquisition, S.-R.K. All authors have read and agreed to the published version of the manuscript.

Funding: This project was funded by the NIH-NIBIB R00EB017729-04 (S.-R.K.), R21EB030370-01 (S.-R.K.) and support of Penn State Cancer Institute (S.-R.K.).

Institutional Review Board Statement: The human volunteer studies were conducted by following the internal imaging protocol of CYBERDYNE, INC (Rotterdam, The Netherlands) for healthy-volunteer imaging experiments, adhering to the guidelines of the Declaration of Helsinki.

Informed Consent Statement: Informed consent was obtained from all the subjects involved in this study.

Data Availability Statement: Data available on request from the authors.

Acknowledgments: We gratefully acknowledge the support from Gary Meyers and Eugene Gerber, for their help with the 3-D printing work and the machining of experimental setup parts.

Conflicts of Interest: M.K.A.S. is employed by CYBERDYNE, INC. The author(s) declare no potential conflicts of interest with respect to the research, authorship and/or publication of this article.

Abbreviations

PA	Photoacoustic
US	Ultrasound
PAI	Photoacoustic imaging
LED	Light-emitting diode
SNR	Signal to noise ratio
PRF	Pulse repetition frequency
OPO	Optical parametric oscillator
FWHM	Full width half maximum
MIP	Maximum intensity projection

References

1. Norrving, B.; Kissela, B. The global burden of stroke and need for a continuum of care. *Neurology* **2013**, *80*, S5–S12. [CrossRef] [PubMed]
2. American Diabetes Association. Peripheral arterial disease in people with diabetes. *Diabetes Care* **2003**, *26*, 3333–3341. [CrossRef] [PubMed]
3. Stapleton, P.A.; James, M.E.; Goodwill, A.G.; Frisbee, J.C. Obesity and vascular dysfunction. *Pathophysiology* **2008**, *15*, 79–89. [CrossRef] [PubMed]
4. Zhang, Y.; Xia, H.; Wang, Y.; Chen, L.; Li, S.; Hussein, I.A.; Wu, Y.; Shang, Y.; Yao, S.; Du, R. The rate of missed diagnosis of lower-limb DVT by ultrasound amounts to 50% or so in patients without symptoms of DVT: A meta-analysis. *Medicine* **2019**, *98*, e17103. [CrossRef] [PubMed]
5. Pollak, A.W.; Norton, P.T.; Kramer, C.M. Multimodality imaging of lower extremity peripheral arterial disease: Current role and future directions. *Circ. Cardiovasc. Imaging* **2012**, *5*, 797–807. [CrossRef]
6. Ena, J.; Argente, C.R.; González-Sánchez, V.; Algado, N.; Verdú, G.; Lozano, T. Use of pocket pulse oximeters for detecting peripheral arterial disease in patients with diabetes mellitus. *J. Diabetes Mellit.* **2013**, *3*, 79–85. [CrossRef]
7. Ntziachristos, V.; Razansky, D. Molecular Imaging by Means of Multispectral Optoacoustic Tomography (MSOT). *Chem. Rev.* **2010**, *110*, 2783–2794. [CrossRef]
8. Bungart, B.L.; Lan, L.; Wang, P.; Li, R.; Koch, M.O.; Cheng, L.; Masterson, T.A.; Dundar, M.; Cheng, J.X. Photoacoustic tomography of intact human prostates and vascular texture analysis identify prostate cancer biopsy targets. *Photoacoustics* **2018**, *11*, 46–55. [CrossRef]
9. Hoelen, C.G.A.; De Mul, F.F.M.; Pongers, R.; Dekker, A. Three-dimensional photoacoustic imaging of blood vessels in tissue. *Opt. Lett.* **1998**, *23*, 648–650. [CrossRef]
10. Pan, D.; Pramanik, M.; Senpan, A.; Allen, J.S.; Zhang, H.; Wickline, S.A.; Wang, L.V.; Lanza, G.M. Molecular photoacoustic imaging of angiogenesis with integrin-targeted gold nanobeacons. *FASEB J.* **2010**, *25*, 875–882. [CrossRef]
11. Laufer, J.; Elwell, C.; Delpy, D.; Beard, P. In vitro measurements of absolute blood oxygen saturation using pulsed near-infrared photoacoustic spectroscopy: Accuracy and resolution. *Phys. Med. Biol.* **2005**, *50*, 4409. [CrossRef] [PubMed]
12. Lee, C.; Jeon, M.; Jeon, M.Y.; Kim, J.; Kim, C. In vitro photoacoustic measurement of hemoglobin oxygen saturation using a single pulsed broadband supercontinuum laser source. *Appl. Opt.* **2014**, *53*, 3884–3889. [CrossRef] [PubMed]
13. Lan, B.; Liu, W.; Wang, Y.-C.; Shi, J.; Li, Y.; Xu, S.; Sheng, H.; Zhou, Q.; Zou, J.; Hoffmann, U.; et al. High-speed widefield photoacoustic microscopy of small-animal hemodynamics. *Biomed. Opt. Express* **2018**, *9*, 4689–4701. [CrossRef] [PubMed]
14. Mallidi, S.; Luke, G.P.; Emelianov, S.Y. Photoacoustic imaging in cancer detection, diagnosis, and treatment guidance. *Trends Biotechnol.* **2011**, *29*, 213–221. [CrossRef] [PubMed]
15. Lv, J.; Li, S.; Zhang, J.; Duan, F.; Wu, Z.; Chen, R.; Chen, M.; Huang, S.; Ma, H.; Nie, L. In vivo photoacoustic im-aging dynamically monitors the structural and functional changes of ischemic stroke at a very early stage. *Theranostics* **2020**, *10*, 816. [CrossRef]
16. Kratkiewicz, K.; Manwar, R.; Rajabi-Estarabadi, A.; Fakhoury, J.; Meiliute, J.; Daveluy, S.; Mehregan, D.; Avanaki, K. Photoacoustic/Ultrasound/Optical Coherence Tomography Evaluation of Melanoma Lesion and Healthy Skin in a Swine Model. *Sensors* **2019**, *19*, 2815. [CrossRef]
17. Guggenheim, J.A.; Allen, T.J.; Plumb, A.; Zhang, E.; Rodriguez-Justo, M.; Punwani, S.; Beard, P.C. Photoacoustic imaging of human lymph nodes with endogenous lipid and hemoglobin contrast. *J. Biomed. Opt.* **2015**, *20*, 050504. [CrossRef]
18. Kole, A.; Cao, Y.; Hui, J.; Bolad, I.A.; Alloosh, M.; Cheng, J.-X.; Sturek, M. Comparative Quantification of Arterial Lipid by Intravascular Photoacoustic-Ultrasound Imaging and Near-Infrared Spectroscopy-Intravascular Ultrasound. *J. Cardiovasc. Transl. Res.* **2018**, *12*, 211–220. [CrossRef]
19. Xu, Z.; Zhu, Q.; Wang, L.V. In vivo photoacoustic tomography of mouse cerebral edema induced by cold injury. *J. Biomed. Opt.* **2011**, *16*, 066020. [CrossRef]
20. Yao, D.K.; Maslov, K.; Shung, K.K.; Zhou, Q.; Wang, L.V. In vivo label-free photoacoustic microscopy of cell nuclei by excitation of DNA and RNA. *Opt. Lett.* **2010**, *35*, 4139–4141. [CrossRef]
21. Weber, J.; Beard, P.C.; Bohndiek, S.E. Contrast agents for molecular photoacoustic imaging. *Nat. Methods* **2016**, *13*, 639–650. [CrossRef] [PubMed]
22. Luke, G.P.; Yeager, D.; Emelianov, S. Biomedical Applications of Photoacoustic Imaging with Exogenous Contrast Agents. *Ann. Biomed. Eng.* **2012**, *40*, 422–437. [CrossRef] [PubMed]
23. Dragulescu-Andrasi, A.; Kothapalli, S.R.; Tikhomirov, G.A.; Rao, J.; Gambhir, S.S. Activatable oligomerizable imaging agents for photoacoustic imaging of furin-like activity in living subjects. *J. Am. Chem. Soc.* **2013**, *135*, 11015–11022. [CrossRef] [PubMed]
24. Cheng, K.; Kothapalli, S.-R.; Liu, H.; Koh, A.L.; Jokerst, J.V.; Jiang, H.; Yang, M.; Li, J.; Levi, J.; Wu, J.C.; et al. Construction and Validation of Nano Gold Tripods for Molecular Imaging of Living Subjects. *J. Am. Chem. Soc.* **2014**, *136*, 3560–3571. [CrossRef] [PubMed]
25. Zemp, R.J.; Li, L.; Wang, L.V. Photoacoustic Imaging of Gene Expression in Small Animals in Vivo. In *Photoacoustic Imaging and Spectroscopy*; CRC Press: Boca Raton, FL, USA, 2017; pp. 363–370.
26. Ahn, S.; Kang, J.; Kim, P.; Lee, G.; Jeong, E.; Jung, W.; Park, M.; Song, T.-K. Smartphone-based portable ultrasound imaging system: Prototype implementation and evaluation. In Proceedings of the 2015 IEEE International Ultrasonics Symposium (IUS), Taipei, Taiwan, 21–24 October 2015.

27. Jawahar, D.; Rachamalla, H.R.; Rafalowski, A.; Ilkhani, R.; Bharathan, T.; Anandarao, N.; Jawahar, D. Pulse oximetry in the evaluation of peripheral vascular disease. *Angiology* **1997**, *48*, 721–724. [CrossRef] [PubMed]
28. Regensburger, A.P.; Fonteyne, L.M.; Jüngert, J.; Wagner, A.L.; Gerhalter, T.; Nagel, A.M.; Heiss, R.; Flenkenthaler, F.; Qurashi, M.; Neurath, M.F.; et al. Detection of collagens by multispectral optoacoustic tomography as an imaging biomarker for Duchenne muscular dystrophy. *Nat. Med.* **2019**, *25*, 1905–1915. [CrossRef]
29. Kothapalli, S.-R.; Sonn, G.A.; Choe, J.W.; Nikoozadeh, A.; Bhuyan, A.; Park, K.K.; Cristman, P.; Fan, R.; Moini, A.; Lee, B.C.; et al. Simultaneous transrectal ultrasound and photoacoustic human prostate imaging. *Sci. Transl. Med.* **2019**, *11*, eaav2169. [CrossRef]
30. Dutta, R.; Mandal, S.; Lin, H.-C.A.; Raz, T.; Kind, A.; Schnieke, A.; Razansky, D. Brilliant cresyl blue enhanced optoacoustic imaging enables non-destructive imaging of mammalian ovarian follicles for artificial reproduction. *J. R. Soc. Interface* **2020**, *17*, 20200776. [CrossRef]
31. Oraevsky, A.A.; Clingman, B.; Zalev, J.; Stavros, A.T.; Yang, W.T.; Parikh, J.R. Clinical optoacoustic imaging combined with ultrasound for coregistered functional and anatomical mapping of breast tumors. *Photoacoustics* **2018**, *12*, 30–45. [CrossRef]
32. Allen, T.J.; Beard, P.C. Pulsed near-infrared laser diode excitation system for biomedical photoacoustic imaging. *Opt. Lett.* **2006**, *31*, 3462–3464. [CrossRef]
33. Yao, Q.; Ding, Y.; Liu, G.; Zeng, L. Low-cost photoacoustic imaging systems based on laser diode and light-emitting diode excitation. *J. Innov. Opt. Health Sci.* **2017**, *10*, 1730003. [CrossRef]
34. Kolkman, R.G.M.; Steenbergen, W.; Van Leeuwen, T.G. In vivo photoacoustic imaging of blood vessels with a pulsed laser diode. *Lasers Med Sci.* **2006**, *21*, 134–139. [CrossRef] [PubMed]
35. Upputuri, P.K.; Pramanik, M. Fast photoacoustic imaging systems using pulsed laser diodes: A review. *Biomed. Eng. Lett.* **2018**, *8*, 167–181. [CrossRef] [PubMed]
36. Dangi, A.; Agrawal, S.; Kothapalli, S.R. Lithium niobate-based transparent ultrasound transducers for photoacoustic imaging. *Opt. Lett.* **2019**, *44*, 5326–5329. [CrossRef]
37. Dangi, A.; Cheng, C.Y.; Agrawal, S.; Tiwari, S.; Datta, G.R.; Benoit, R.R.; Pratap, R.; Trolier-McKinstry, S.; Kothapalli, S.-R. A Photoacoustic Imaging Device Using Piezoelectric Micromachined Ultrasound Transducers (PMUTs). *IEEE Trans. Ultrason. Ferroelectr. Freq. Control* **2020**, *67*, 801–809. [CrossRef]
38. Chen, M.-C.; Perez, A.P.; Kothapalli, S.-R.; Cathelin, P.; Cathelin, A.; Gambhir, S.S.; Murmann, B. A Pixel Pitch-Matched Ultrasound Receiver for 3-D Photoacoustic Imaging With Integrated Delta-Sigma Beamformer in 28-nm UTBB FD-SOI. *IEEE J. Solid-State Circuits* **2017**, *52*, 2843–2856. [CrossRef]
39. Kothapalli, S.R.; Ma, T.J.; Vaithilingam, S.; Oralkan, Ö.; Khuri-Yakub, B.T.; Gambhir, S.S. Deep tissue photoacous-tic imaging using a miniaturized 2-D capacitive micromachined ultrasonic transducer array. *IEEE Trans. Biomed. Eng.* **2012**, *59*, 1199–1204. [CrossRef]
40. Liu, S.; Tang, K.; Feng, X.; Jin, H.; Gao, F.; Zheng, Y. Toward Wearable Healthcare: A Miniaturized 3D Imager With Coherent Frequency-Domain Photoacoustics. *IEEE Trans. Biomed. Circuits Syst.* **2019**, *13*, 1417–1424. [CrossRef]
41. Dangi, A.; Agrawal, S.; Datta, G.R.; Srinivasan, V.; Kothapalli, S.R. Towards a Low-Cost and Portable Photoacoustic Microscope for Point-of-Care and Wearable Applications. *IEEE Sens. J.* **2020**, *20*, 6881–6888. [CrossRef]
42. Hariri, A.; Fatima, A.; Mohammadian, N.; Mahmoodkalayeh, S.; Ansari, M.A.; Bely, N.; Avanaki, M.R. Development of low-cost photoacoustic imaging systems using very low-energy pulsed laser diodes. *J. Biomed. Opt.* **2017**, *22*, 075001. [CrossRef]
43. Zhong, H.; Duan, T.; Lan, H.; Zhou, M.; Gao, F. Review of Low-Cost Photoacoustic Sensing and Imaging Based on Laser Diode and Light-Emitting Diode. *Sensors* **2018**, *18*, 2264. [CrossRef] [PubMed]
44. Allen, T.J.; Beard, P.C. High power visible light emitting diodes as pulsed excitation sources for biomedical photoacoustics. *Biomed. Opt. Express* **2016**, *7*, 1260–1270. [CrossRef] [PubMed]
45. Hansen, R.S. Using high-power light emitting diodes for photoacoustic imaging. In Proceedings of the Medical Imaging 2011: Ultrasonic Imaging, Tomography, and Therapy, Lake Buena Vista (Orlando), FL, USA, 13–14 February 2011; International Society for Optics and Photonics: Bellingham, WA, USA, 2011; Volume 7968, p. 79680A.
46. Dai, X.; Yang, H.; Jiang, H. In vivo photoacoustic imaging of vasculature with a low-cost miniature light emitting diode excitation. *Opt. Lett.* **2017**, *42*, 1456–1459. [CrossRef] [PubMed]
47. Zhu, Y.; Feng, T.; Cheng, Q.; Wang, X.; Du, S.; Sato, N.; Yuan, J.; Kuniyil Ajith Singh, M. Towards Clinical Translation of LED-Based Photoacoustic Imaging: A Review. *Sensors* **2020**, *20*, 2484. [CrossRef] [PubMed]
48. Zhu, Y.; Xu, G.; Yuan, J.; Jo, J.; Gandikota, G.; Demirci, H.; Agano, T.; Sato, N.; Shigeta, Y.; Wang, X. Light emitting diodes based photoacoustic imaging and potential clinical applications. *Sci. Rep.* **2018**, *8*, 9885. [CrossRef] [PubMed]
49. Agrawal, S.; Fadden, C.; Dangi, A.; Yang, X.; AlBahrani, H.; Frings, N.; Zadi, S.H.; Kothapalli, S.-R. Light-Emitting-Diode-Based Multispectral Photoacoustic Computed Tomography System. *Sensors* **2019**, *19*, 4861. [CrossRef] [PubMed]
50. Joseph, F.K.; Boink, Y.E.; Dantuma, M.; Ajith, S.M.; Manohar, S.; Steenbergen, W. Tomographic imaging with an ultrasound and LED-based photoacoustic system. *Biomed. Opt. Express* **2020**, *11*, 2152–2165. [CrossRef]
51. Zhu, Y.; Lu, X.; Dong, X.; Yuan, J.; Fabiilli, M.L.; Wang, X. LED-based photoacoustic imaging for monitoring angiogenesis in fibrin scaffolds. *Tissue Eng. Part C: Methods* **2019**, *25*, 523–531. [CrossRef]
52. Xavierselvan, M.; Singh, M.K.A.; Mallidi, S. In Vivo Tumor Vascular Imaging with Light Emitting Diode-Based Photoacoustic Imaging System. *Sensors* **2020**, *20*, 4503. [CrossRef]

53. Jo, J.; Xu, G.; Zhu, Y.; Burton, M.; Sarazin, J.; Schiopu, E.; Gandikota, G.; Wang, X. Detecting joint inflammation by an LED-based photoacoustic imaging system: A feasibility study. *J. Biomed. Opt.* **2018**, *23*, 110501–110504. [CrossRef]
54. Maneas, E.; Aughwane, R.; Huynh, N.; Xia, W.; Ansari, R.; Kuniyil Ajith Singh, M.; Hutchinson, J.C.; Sebire, N.J.; Arthurs, O.J.; Deprest, J.; et al. Photoacoustic imaging of the human placental vasculature. *J. Biophotonics* **2020**, *13*, e201900167. [CrossRef] [PubMed]
55. Xia, W.; Maneas, E.; Huynh, N.T.; Singh, M.K.A.; Brown, N.M.; Ourselin, S.; Gilbert-Kawai, E.; West, S.J.; Desjardins, A.E. Imaging of human peripheral blood vessels during cuff occlusion with a compact LED-based photoacoustic and ultrasound system. In *Photons Plus Ultrasound: Imaging and Sensing*; International Society for Optics and Photonics: Bellingham, WA, USA, 2019; Volume 10878, p. 1087804.
56. Agano, T.; Singh, M.K.A.; Nagaoka, R.; Awazu, K. Effect of light pulse width on frequency characteristics of photoacoustic signal—An experimental study using a pulse-width tunable LED-based photoacoustic imaging system. *Int. J. Eng. Technol.* **2018**, *7*, 4300–4303.
57. Jaeger, M.; Schüpbach, S.; Gertsch, A.; Kitz, M.; Frenz, M. Fourier reconstruction in optoacoustic imaging using truncated regularized inverse k -space interpolation. *Inverse Probl.* **2007**, *23*, S51–S63. [CrossRef]
58. Marquez, G.; Wang, L.V.; Lin, S.-P.; Schwartz, J.A.; Thomsen, S.L. Anisotropy in the absorption and scattering spectra of chicken breast tissue. *Appl. Opt.* **1998**, *37*, 798–804. [CrossRef]
59. American National Standards Institute. *American National Standard for Safe Use of Lasers*; Laser Institute of America: Orlando, FL, USA, 2007.
60. Anas, E.M.A.; Zhang, H.K.; Kang, J.; Boctor, E. Enabling fast and high quality LED photoacoustic imaging: A recurrent neural networks based approach. *Biomed. Opt. Express* **2018**, *9*, 3852–3866. [CrossRef]
61. Farnia, P.; Najafzadeh, E.; Hariri, A.; Lavasani, S.N.; Makkiabadi, B.; Ahmadian, A.; Jokerst, J.V. Dictionary learning technique enhances signal in LED-based photoacoustic imaging. *Biomed. Opt. Express* **2020**, *11*, 2533–2547. [CrossRef]
62. Johnstonbaugh, K.; Agrawal, S.; Durairaj, D.A.; Fadden, C.; Dangi, A.; Karri, S.P.K.; Kothapalli, S.R. A deep learning approach to photoacoustic wavefront localization in deep-tissue medium. *IEEE Trans. Ultrason. Ferroelectr. Freq. Control* **2020**, *67*, 2649–2659. [CrossRef]

sensors

Article

Remote Photoacoustic Sensing Using Single Speckle Analysis by an Ultra-Fast Four Quadrant Photo-Detector

Benjamin Lengenfelder [1,2,*], Martin Hohmann [1,2], Moritz Späth [1,2], Daniel Scherbaum [1], Manuel Weiß [3], Stefan J. Rupitsch [4], Michael Schmidt [1,2], Zeev Zalevsky [2,5] and Florian Klämpfl [1,2]

1 Institute of Photonic Technologies, University Erlangen-Nürnberg (FAU), Konrad-Zuse-Straße 3/5, 91052 Erlangen, Germany; Martin.Hohmann@lpt.uni-erlangen.de (M.H.); Moritz.Spaeth@lpt.uni-erlangen.de (M.S.); danielscherbaum@gmail.com (D.S.); Michael.Schmidt@lpt.uni-erlangen.de (M.S.); Florian.Klaempfl@lpt.uni-erlangen.de (F.K.)
2 Erlangen Graduate School in Advanced Optical Technologies (SAOT), Paul-Gordan-Straße 6, 91052 Erlangen, Germany; zeev.zalevsky@biu.ac.il
3 Sensor Technology, University Erlangen-Nürnberg (FAU), Paul-Gordan-Straße 3/5, 91052 Erlangen, Germany; manuel.weiss@fau.de
4 Laboratory for Electrical Instrumentation and Embedded Systems, Department of Microsystems Engineering—IMTEK, University of Freiburg, Georges-Koehler-Allee 106, 79110 Freiburg, Germany; stefan.rupitsch@fau.de
5 Faculty of Engineering, Bar-Ilan University, Ramat-Gan 52900, Israel
* Correspondence: ben.lengenfelder@fau.de

Citation: Lengenfelder, B.; Hohmann, M.; Späth, M.; Scherbaum, D.; Weiß, M.; Rupitsch, S.J.; Schmidt, M.; Zalevsky, Z.; Klämpfl, F. Remote Photoacoustic Sensing Using Single Speckle Analysis by an Ultra-Fast Four Quadrant Photo-Detector. *Sensors* **2021**, *21*, 2109. https://doi.org/10.3390/s21062109

Academic Editors: Mithun Kuniyil Ajith Singh and Wenfeng Xia

Received: 22 January 2021
Accepted: 11 March 2021
Published: 17 March 2021

Abstract: The need for tissue contact makes photoacoustic imaging not applicable for special medical applications like wound imaging, endoscopy, or laser surgery. An easy, stable, and contact-free sensing technique might thus help to broaden the applications of the medical imaging modality. In this work, it is demonstrated for the first time that remote photoacoustic sensing by speckle analysis can be performed in the MHz sampling range by tracking a single speckle using a four quadrant photo-detector. A single speckle, which is created by self-interference of surface back-reflection, is temporally analyzed using this photo-detector. Phantoms and skin samples are measured in transmission and reflection mode. The potential for miniaturization for endoscopic application is demonstrated by fiber bundle measurements. In addition, sensing parameters are discussed. Photoacoustic sensing in the MHz sampling range by single speckle analysis with the four quadrant detector is successfully demonstrated. Furthermore, the endoscopic applicability is proven, and the sensing parameters are convenient for photoacoustic sensing. It can be concluded that a single speckle contains all the relevant information for remote photoacoustic signal detection. Single speckle sensing is therefore an easy, robust, contact-free photoacoustic detection technique and holds the potential for economical, ultra-fast photoacoustic sensing. The new detection technique might thus help to broaden the field of photoacoustic imaging applications in the future.

Keywords: photoacoustic; remote sensing; endoscopy; speckle

1. Introduction

Photoacoustic imaging (PA) is a new, rising imaging technique since it combines a high penetration depth with a good image contrast [1]. As PA is based on the absorption of a short light pulse and the subsequent acoustic signal generation, the absorption characteristics define the imaging contrast. As the absorption coefficients for different tissue absorbers differ greatly, it is possible to obtain a high image contrast. Hence, PA is especially attractive for displaying blood vessels due to the high hemoglobin absorption compared to other tissue constituents in the visible and near-infrared range [2]. This high contrast is combined with a high penetration depth of several millimeters since the acoustic scattering is up to a factor of 1000 less than the optical scattering [3]. Thus, higher imaging depths than for purely optical methods can be achieved with ease.

Since piezo element transducers, which are contact based, are the state-of-the-art detection technique, PA requires tissue contact. As a consequence, PA may not be suitable for special medical applications, such as wound imaging or laser surgery. Furthermore, transducer miniaturization is challenging, and small elements suffer from low sensitivity, which limits their usage for photoacoustic endoscopy. Therefore, a remote and stable detection system would be beneficial.

Air-coupled transducers could overcome this issue. However, they suffer from poor sensitivity and are still too big for minimally-invasive medical applications [4]. For the non-contact beam deflection technique, the deflection of a probe beam that is positioned above the tissue is monitored by a position sensitive detector [5,6]. A small fraction of the photoacoustic signal is transmitted to the surrounding air and deflects the probe beam. Since the probe beam needs to be scanned above the whole tissue surface, this modality is unsuitable for surgical usage. In addition, the detection bandwidth is limited due to the probe beam size. Interferometric detection methods offer non-contact sensing and a higher detection bandwidth. Here, the surface displacement after photoacoustic signal generation is monitored for the initial pressure reconstruction inside the object [7–10]. Nevertheless, interferometric systems are expensive and require a complicated setup that is noise sensitive [11]. In addition, interferometric setups only monitor the surface displacement correctly, if the surface is properly tilted, which cannot be guaranteed for clinical application. There are also non-interferometric approaches for remote acoustic signal detection. Clark et al. detected surface acoustic waves by tracking the movement of multiple speckles using simple diode detection systems [12,13]. For this approach, however, the speckles are imaged in the near-field, and thus, Fresnel diffraction applies. In this case, the movement of the speckles is dependent on three types of movement, which cannot be separated and occur simultaneously: transverse, axial, and tilt [14]. Due to this disadvantage, the acoustic sensing approach of Clark et al. may not be suitable for photoacoustic imaging. Hajireza et al. sensed the photoacoustic signal directly at its origin by a non-interferometric system [15]. This system monitors the reflection of a probe beam that is sensitive to the elasto-optic index modulation induced by the photoacoustic initial pressure transients. This approach allows non-interferometric, non-contact photoacoustic sensing. However, it provides only penetration depths in the mm-range due to the high optical attenuation for the probe beam. In addition, it is only applicable for photoacoustic microscopy and not suited for photoacoustic tomography since it does not provide temporally resolved acquisition of the acoustic signal and thus requires depth scanning for image acquisition.

Remote speckle analysis is an easy, robust, and non-interferometric vibration sensing technique whose potential for photoacoustic tomography we already demonstrated in previous publications [16–18]. By tracking the movement of multiple speckles, it was possible to remotely reconstruct the photoacoustic signal. However, in these proof-of-concept studies, the acoustic sensing bandwidth and thus resolution were limited to the frame rate of the expensive, high-resolution camera used at approximately 800 kHz.

In the present publication, it is demonstrated for the first time that remote photoacoustic sensing with a sampling rate of 8 MHz can be performed by tracking the movement of a single speckle with only four diodes. The speckle tracking by only four diodes proves the applicability of an economical position sensitive diode for speckle analysis. Polymer phantoms and skin tissue samples are measured in transmission mode and reflection mode. In addition, the capability of easy miniaturization and endoscopic usage of the single speckle analysis is shown by fiber bundle measurements. The new technique is an essential step for the implementation of a remote photoacoustic imaging system using speckle analysis. Therefore, this work might help to broaden the applications of PA in special applications like wound imaging, endoscopy, or guiding laser surgeries.

2. Materials and Methods

2.1. Speckle Sensing by Multiple Speckles' Tracking

The speckle sensing technique is based on the time-resolved detection of the position of a speckle pattern. It is possible to extract the tilt change of a laser illuminated surface by tracking the speckle pattern movement [14]. The left side of Figure 1 shows a tilting object surface, and the right side illustrates the speckle sensing technique.

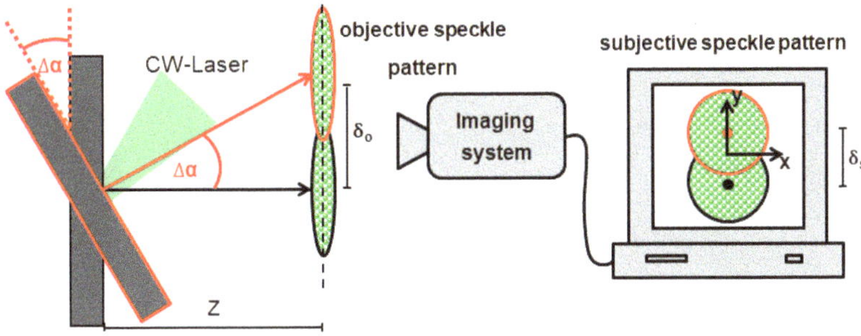

Figure 1. Speckle sensing theory: An object tilt results in a speckle pattern movement, which can be reconstructed with an imaging system. Figure after [19].

If the observation plane distance Z for the generated objective speckle pattern fulfills the far-field approximation ($Z > (D_{ill}^2)/(4\lambda)$, illuminated diameter D_{ill}), the objective speckle pattern movement (δ_o) is only dependent on the surface tilt change $\Delta\alpha$ and Z. By imaging the objective speckle pattern using an imaging system with the magnification M, a subjective speckle pattern is created on the imaging sensor whose lateral movement (δ_s) is also linearly proportional to $\Delta\alpha$ assuming small angle approximation. Equation (1) explains the relation between the speckle movements, M and Z [19].

$$\delta_s = \delta_o \cdot M = tan(\Delta\alpha) \cdot Z \cdot M \tag{1}$$

By acquiring a speckle pattern video sequence with the imaging system, it is thus possible to measure the surface tilt. Correlation analysis of the image frames allows the reconstruction of the speckle pattern movement along the sensor axes x and y, which are representative of the surface tilts along these two axes' directions [19]. In this article, it is shown for the first time that speckle sensing is feasible by tracking the movement of a single speckle in contrast to the described correlation analysis of a multiple speckle image. Therefore, it is proven that a single speckle contains all necessary information for photoacoustic signal detection.

2.2. *Imaging Systems and Setups*

Two imaging systems are established for this work and described here in detail: the free-space setup, which is capable of far distance PA sensing, and the fiber based setup, which is suitable for endoscopy.

2.2.1. Free-Space Single Speckle Sensing

Figure 2 shows the established diode based, free-space imaging system. The imaging system consists of a microscope objective (10×), a beamsplitter and a lens (f = 100 mm) used for image magnification on a camera (DCC1545M, Thorlabs, Newton, NJ, USA), and an avalanche-photodiode sensor (APS; APDcam, Fusion Instruments, Budapest, Hungary). A bandpass filter for the speckle wavelength is placed after the objective in order to block the photoacoustic excitation and room light. The magnification of the system is calculated at 5 with a microscope test target (M = 5). The camera is used as a reference for image calibration and alignment. The calibration is done with a multi-mode fiber (AFS105/125Y, Thorlabs, Newton, NJ, USA) to which a halogen light source (HL-2000, Ocean Optics, Ostfildern, Germany) is coupled. The fiber tip is then imaged on the camera and APS (Figure 2e), and the co-alignment of the camera image and the APS can be verified. The APS consists of a 4 × 8 avalanche-diode array (pixel size: 1.6 mm) and therefore offers 32 pixels.

Of these 32 pixels, only the four central pixels are used for the tracking of a single speckle in order to maximize the acquisition rate.

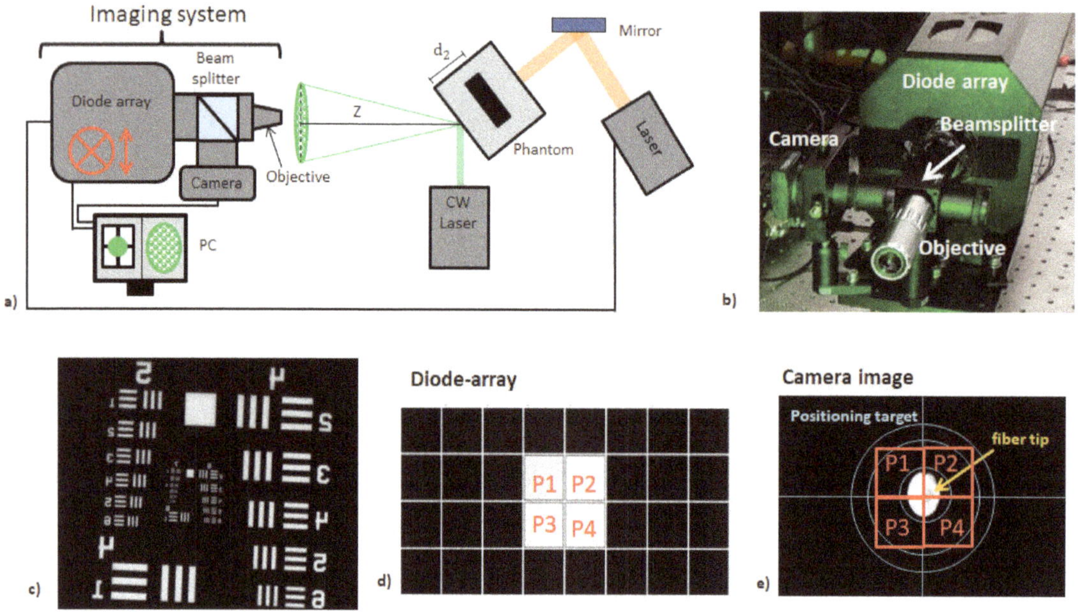

Figure 2. (**a**) Experimental setup (transmission mode) for the remote photoacoustic measurements using the diode array sensing system. The imaging system can be moved mechanically in lateral directions, which is indicated by the red marks. (**b**) Picture of the sensing system. (**c**) The magnification of the sensing system is calculated at 5 using a USAF 1951 Test Target. (**d**,**e**) The camera position of the sensing system is calibrated using a multi-mode fiber. The diode signals and the camera image are shown when an illuminated fiber is placed in their center. Figures after [20].

The speckles are generated by cw illumination (532 nm, 80 mW, diameter: 0.75 mm) of the sample surface and imaged at $Z = 20$ cm. First, a convenient speckle is found by manually moving the imaging system using mechanical stages and visually tracking the camera image. A speckle is considered as convenient for the measurement if it is in the center of the camera and therefore in the center of the diode array (see Figure 3). Furthermore, the speckle size needs to be in the range of the pixel size of the diode array (1.6 mm). This is ensured by comparing the tracked speckle size to the inner circle (diameter 1.4 mm) of the illustrated target in Figure 3, which can be displayed by the camera software.

For the photoacoustic measurements, the samples are excited with a short laser pulse (Q-Smart 450, Quantel laser, Les Ulis, France). The laser parameters are as follows: $\lambda = 1064$ nm, pulse duration 5 ns, beam diameter 7 mm, pulse energy 90 mJ. These parameters result in an exposure of 230 $\frac{\text{mJ}}{\text{cm}^2}$, which is above the maximum permitted exposure (MPE) for single pulse excitation at 1064 nm (100 $\frac{\text{J}}{\text{cm}^2}$). This, however, is desired to achieve a high signal amplitude for the demonstrated proof-of-concept experiments. The laser pulse triggers the acquisition start of the imaging system with a sampling rate of 8 MHz.

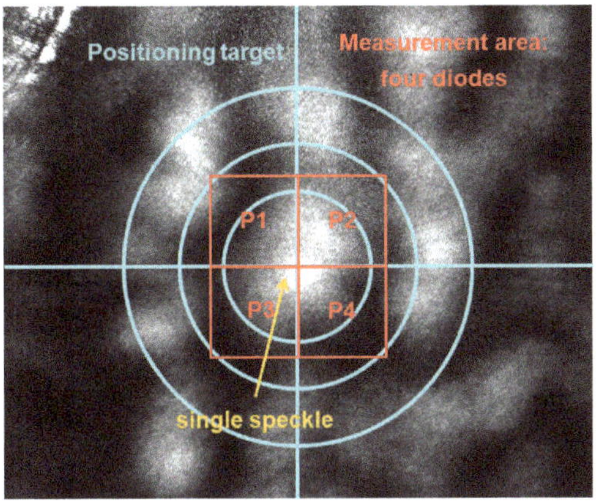

Figure 3. The camera image of an exemplary speckle used for photoacoustic measurement is shown. The speckle is placed in the center of the reference camera (blue target) and thus in the center of four measurements diodes P1/P2/P3/P4 (red) by manually moving the complete imaging system.

In order to prove the safe applicability of the sensing system, ex vivo skin measurements are performed with a total exposure below the MPE. This is achieved by replacing the described cw laser with a temporally pulsed laser diode (IBEAM-SMART-405-S-HP, Toptica Photonics, Gräfelfing, Germany) and by reducing the photoacoustic excitation energy. The laser diode pulse (λ = 405 nm) is temporally triggered by the short laser pulse for photoacoustic excitation and illuminates the tissue only for a duration of 30 µs after photoacoustic excitation at a peak power of 35 mW with an illumination radius of 300 µm at the tissue surface. This temporal speckle illumination together with a reduced excitation energy of 35 mW results in a total exposure that is below the MPE for soft tissue [21].

2.2.2. Fiber-Based Single Speckle Sensing

Figure 4 shows the imaging unit and setup for the fiber based approach and camera images of a USAF 1951 Test Target and a selected speckle.

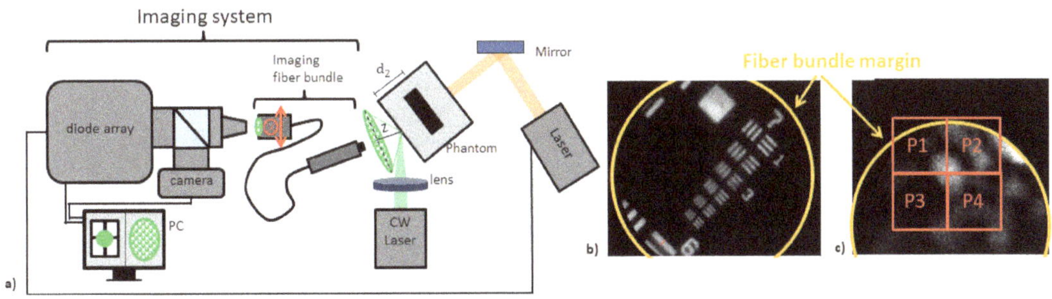

Figure 4. Experimental setup and imaging unit for the remote photoacoustic measurements using the fiber based approach (**a**). The proximal fiber bundle end can be moved in lateral directions, which is indicated by the red marks. The magnification for the camera arm is determined at 50 using a USAF 1951 Test Target (**b**). A convenient speckle that is centralized inside the photodiode measurement area is illustrated (**c**).

In contrast to the free-space approach, an imaging fiber bundle (30,000 fibers, imaging resolution 1 µm, working distance 30 µm, field of view diameter 240 µm) that can be used in endoscopy is used for speckle pattern imaging. At the proximal fiber bundle end, the speckles are imaged by the diode based imaging system mentioned in Section 2.2.1.

In contrast to the free-space system previously described, a 20× magnification objective (WC95248318, Mitutoyo, Japan, 20×) is used. Furthermore, the imaging lens in the diode array arm is changed to f = 200 mm. Together with the magnification of the imaging fiber bundle ($M = 2.5$), these changes result in a magnification of $M = 50$ for the diode array arm and $M = 25$ for the camera arm. These higher optical magnifications allow speckle sensing at a near imaging distance of Z = 2 mm, which is desired for endoscopic usage.

The speckles are generated by focused cw illumination (532 nm, 100 mW, $D_{ill} = 50$ µm) of the sample surface and imaged by the fiber bundle and APS system at $Z = 2$ mm. A convenient speckle is found and automatically centered inside the measurement area of the four photodiodes by analyzing the camera image and moving the proximal fiber bundle end in lateral directions. For the photoacoustic measurements, the phantoms are excited with a short laser pulse (Q-Smart 450, Quantel laser, Les Ulis, France), and the laser parameters are the following: $\lambda = 1064$ nm, pulse duration 5 ns, beam radius 7 mm, pulse energy 110 mJ. These parameters result in an exposure dose of 285 $\frac{\mathrm{mJ}}{\mathrm{cm}^2}$, which is above the MPE for single pulse excitation at 1064 nm (100 $\frac{\mathrm{J}}{\mathrm{cm}^2}$). This, however, is desired to achieve a high signal amplitude for the proof-of-concept experiments. The laser pulse triggers the acquisition start of the APS with a sampling rate of 8 MHz.

2.3. Measurement Modes and Samples

Using the described free-space and fiber based imaging systems, the samples are measured in transmission mode and reflection mode. For transmission mode, the photoacoustic excitation and speckle sensing take place at opposite sample sides, whereas they are on the same side for reflection mode.

The phantoms used in this work are made of the soft polymer PVCP (polyvinyl chloride plastisol; Standard Lure flex (medium), Lure Factors, Great Britain) and consist of two parts: absorber and scattering matrix. In order to adjust the optical properties for these parts, additives are used during the plastisol preparation process. A black plastic color changes the absorption coefficient μ_a, and TiO_2-particles (titanium(IV)-oxide, Sigma Aldrich, Taufkirchen, Germany) adjust the reduced scattering coefficient μ_s'. In this work, a color concentration of 7-vol-% and a TiO_2-concentration of 4 $\frac{\mathrm{mg}}{\mathrm{mL(PVCP)}}$ is used for the absorbing and scattering phantom parts, respectively. The optical properties for these concentrations were determined at the excitation wavelength 1064 nm using spectrophotometric measurements and inverse adding doubling. The absorption coefficient for the absorbing phantom part is 106 cm^{-1}, and the reduced scattering coefficient for the scattering part is 21 cm^{-1}. The scattering coefficient for the absorbing part and the absorbing coefficient for the scattering part can be neglected.

For the measurements using the free-space setup, three different PVCP phantoms (PhAT1, PhAT2, PhAT3) with increasing distances between the absorber surface and detection surface (d) are measured in transmission mode. The "Ph" in the phantom name stands for phantom, "A" for the free-space setup, and "T" for transmission mode. Two PVCP phantoms are measured in reflection mode (PhAR4, PhAR5). The "R" in the sample name stands for reflection mode. Two skin tissue samples (skinAT1, skinAT2p) are measured in transmission mode. The "p" in the sample name stands for the pulsed speckle illumination, which ensures a total exposure below the MPE for the experiments with the skin tissue. The skin tissue was obtained from bisected pig heads, which were obtained from the local slaughterhouse (Unifleisch GmbH, Erlangen, Germany). Therefore, the approval of the Ethics Committee was not necessary. The tissue sample was prepared manually using a scalpel. For skinAT1, a PVCP-absorber was placed at the sample bottom by cutting out a hole in the tissue. For skinAT2, a PVCP-absorber was placed between a fat layer, which was obtained from a local supermarket, and a skin layer. In order to ensure good contact between the sample constituents, ultrasound gel was used, and for skinAT2p, a metallic sample holder gently pressed the tissues.

The speed of sound for the prepared skin sample was measured at 1300 $\frac{\mathrm{m}}{\mathrm{s}}$ and for the PVCP phantoms at 1330 $\frac{\mathrm{m}}{\mathrm{s}}$ with an ultrasound thickness measurement device (Mini

Test 430, Elektro Physik, Germany). For each sample measured with the free-space setup, fifteen measurements were analyzed in order to ensure statistical relevance.

For the measurements using the fiber based imaging setup, two PVCP phantoms (PhBT1, PhBT2) were measured in transmission mode. The "B" in the sample name stands for the fiber based setup. The speed of sound for these samples were measured at 1349 $\frac{\mathrm{m}}{\mathrm{s}}$ with an ultrasound thickness measurement device (Mini Test 430, Elektro Physik, Germany). For these two phantoms, ten measurements were analyzed in order to ensure statistical relevance.

The density ρ of all PVCP phantoms was measured by the volume displacement of ethanol at 1200 $\frac{\mathrm{kg}}{\mathrm{m}^2\mathrm{s}}$. The resulting acoustic impedance ($Z_{ac} = \rho c$) of the used phantoms in this work was therefore in the range of $1.60 \times 10^6 \frac{\mathrm{kg}}{\mathrm{m}^2\mathrm{s}}$–$1.62 \times 10^6 \frac{\mathrm{kg}}{\mathrm{m}^2\mathrm{s}}$, which is in good agreement with the values of soft tissue: the impedance of fat tissue is $1.4 \times 10^6 \frac{\mathrm{kg}}{\mathrm{m}^2\mathrm{s}}$ and for muscle $1.62 \times 10^6 \frac{\mathrm{kg}}{\mathrm{m}^2\mathrm{s}}$ [22].

Figure 5 sketches the phantom position in regard to the excitation and illumination laser for the two measurement modes and shows the corresponding detection distances d of all samples. The absorber thickness is 3 mm for all samples, and d is varied by adjusting the thickness for the scattering sample part. For all samples, the mean detection times by automated single speckle analysis and their standard deviations are compared to the theoretical detection time.

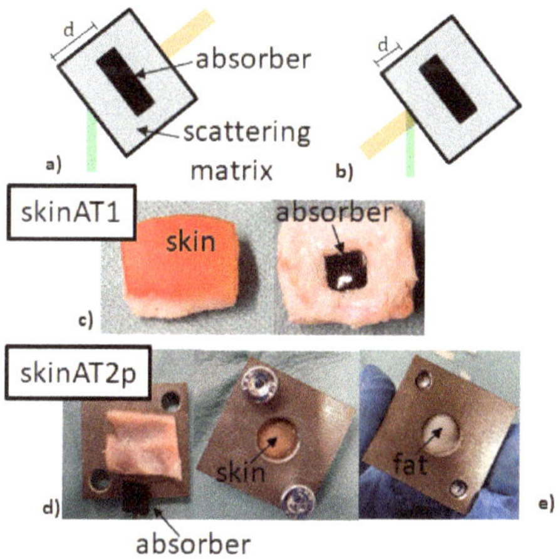

Sample	d [mm]	t_{theo} [µs]	Mode
PhAT1	4	3.01	T
PhAT2	5	3.76	T
PhAT3	6	4.51	T
PhAR4	3	2.26	R
PhAR5	4	3.01	R
skinAT1	10	7.69	T
skinAT2p	4	3.04	T
PhBT1	7.5	5.56	T
PhBT2	5	3.71	T

e)

Figure 5. The transmission setup (T) is sketched, and the phantom parts are marked (**a**). The reflection setup (R) is illustrated (**b**). Pictures of the two skin tissue samples are shown: skinAT1 with an absorber at the bottom (**c**) and skinAT2p with an absorber between skin and fat (**d**). The corresponding detection distances, theoretical detection times, and measurement modes are summarized in the table (**e**).

2.4. Data Analysis for Single Speckle Analysis

Figure 6 shows a speckle, which is initially in the center of the four measurement diodes P1, P2, P3, and P4 and moves to a different position due to the PA signal.

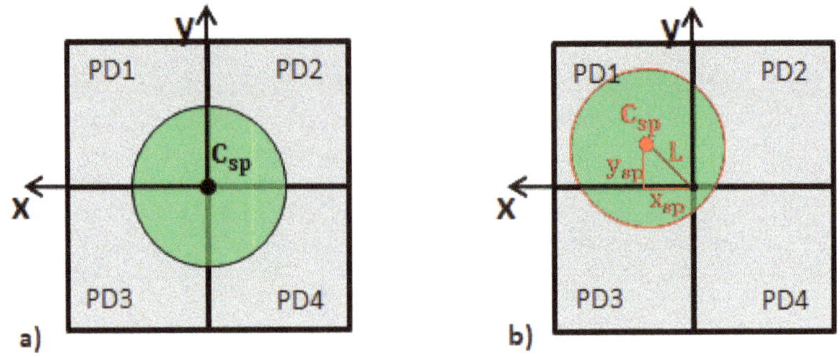

Figure 6. A perfectly round speckle, which is initially in the center of the measurement diodes is shown (**a**). The diode signals (P1, P2, P3, P4) can be used to compute its center of gravity $C_{sp}(x_{sp}, y_{sp})$ and its total vector length L in order to detect speckle movements (**b**).

By using the temporal signals of the four diodes, it is possible to compute the temporal center of gravity of the speckle C_{sp} (x_{sp}, y_{sp}) and its total vector length L by using Equations (2)–(4), similar to a four quadrant position sensitive diode [23]. Since the speckle moves if the surface tilts, this center of gravity is related to the photoacoustic signal, and it is possible to reconstruct the absorber depth d.

$$x_{sp} = \frac{(P1 + P3) - (P2 + P4)}{(P1 + P2 + P3 + P4)} \tag{2}$$

$$y_{sp} = \frac{(P1 + P2) - (P3 + P4)}{(P1 + P2 + P3 + P4)} \tag{3}$$

$$L = \sqrt{x_{sp}^2 + y_{sp}^2} \tag{4}$$

In order to prove the usability of the sensing system for remote photoacoustic detection, the detection times of x_{sp}, y_{sp}, and L are shown for the free-space approach. For L, the mean detection time t_{mean} and its standard deviation σ are computed for all samples and verified to t_{theo} for the acoustic signal, which is calculated using d and c.

2.5. Sensing Parameter Evaluation for Single Speckle Analysis

2.5.1. Sensitivity

The sensitivity in terms of minimal detectable tilt $S_{d,\alpha}$ is limited by the noise floor σ_{nf} of the PA measurements. The noise level σ_{nf} is computed by taking the standard deviation of a PA measurement data set before excitation. For this standard deviation computation, one-hundred fifty data points before PA excitation are used. By determining the noise floor σ_{nf} and taking into account the relevant imaging parameters (Z, pixel size d_{px}, magnification M, illumination diameter D_{ill}), $S_{d,\alpha}$ can be determined. The minimal detectable speckle shift $\delta_{s,min}$ is equal to the multiplication of d_{px} and σ_{nf}. By considering Equation (1), it is then possible to compute the minimal detectable tilt $S_{d,\alpha}$: $\tan(S_{d,\alpha}) = \frac{\delta_{s,min}}{ZM}$. Under the assumption that the investigated speckle pattern or single speckle consists of reflections from the complete illuminated surface area with the diameter D_{ill}, the minimal detectable axial surface deformation is estimated by $S_{d,nm} = \tan(S_{d,\alpha}) \cdot D_{ill}$. The minimal detectable pressure S_d can be determined according to Equation (5) [24].

$$S_d = \pi Z_{ac} S_{d,nm} f \tag{5}$$

The parameter f defines the frequency of the acoustic wave that should be detected, and Z_{ac} is the acoustic impedance. For the established system here, the maximum detectable frequency is half the frame rate. For the established free-space and fiber based sensing system, the described sensitivities are computed and compared to the literature and to contact ultrasound transducers.

2.5.2. Sensing Range

The sensing range, i.e., the tilt interval that is covered by the remote speckle analysis, is dependent on the sensitivity, speckle size, and sensor size. The sensitivity defines the minimal detectable tilt as treated in the previous section. The maximal detectable tilt α_{max} is defined by the sensor and speckle size. The sensing range for the single speckle analysis is discussed.

2.5.3. Linearity

Simulations were carried out in order to evaluate the linearity and robustness against neighboring speckles, meaning speckles that are not situated inside the original image, but appear in the shifted speckle image. Four speckle patterns with a centralized speckle were analyzed. Figure 7 shows these speckle patterns. The images are moved in the horizontal direction with a shift amplitude of -0.26 to 0.26 ($x_{sp,real}$) of the diode pixel size with a shift resolution of 0.025. For each shifted image, the horizontal center of gravity x_{sp} for the fixed sensing diodes region, which is indicated in Figure 7, is computed according to Equation (2). The values of x_{sp} are computed for all shifts and speckle images, compared to $x_{sp,real}$ and compared to linear behavior.

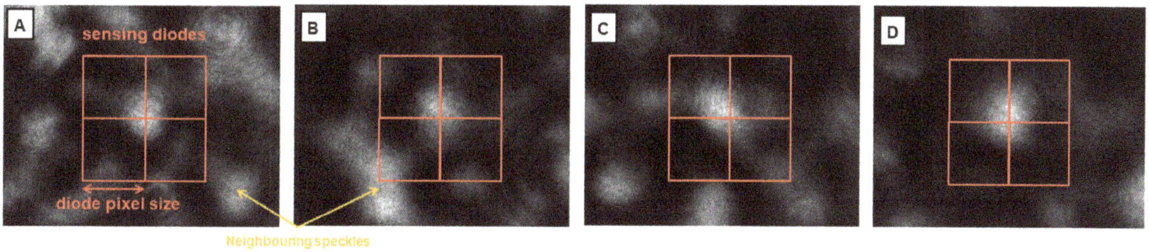

Figure 7. Speckle images (**A–D**) that are evaluated for their single speckle sensing capability. The diode sensing regions are indicated by the red rectangles.

3. Results and Discussion

3.1. Photoacoustic Measurements

Figure 8 shows measurement results in transmission mode of the three phantoms PhAT1, PhAT2, and PhAT3. They illustrate the speckle vector length L, which represents the temporal vibration profile of the surface under investigation. The detection time of the first peak in these profiles is marked with a black circle, because this time point corresponds to the photoacoustic signal. The surface expansion after the photoacoustic excitation results in a surface tilt change and thus in a speckle movement, which can be seen in L. For the phantoms PhAT1-PhAT3, the acquisition times increase as expected with increasing acoustic travel distance d. Considering t_{sp}, the acquisition time increases as follows: 3.13 µs, 3.63 µs, and 4.38 µs. By using the speed of sound (1330 $\frac{\mathrm{m}}{\mathrm{s}}$), the following acoustic travel distances are calculated: 4.163 mm, 4.828 mm, and 5.825 mm. Thus, for each phantom, the acquisition time of the photoacoustic signal by speckle analysis corresponds to the geometrical phantom dimensions (d: 4 mm, 5 mm, 6 mm), taking into account the phantom production uncertainty.

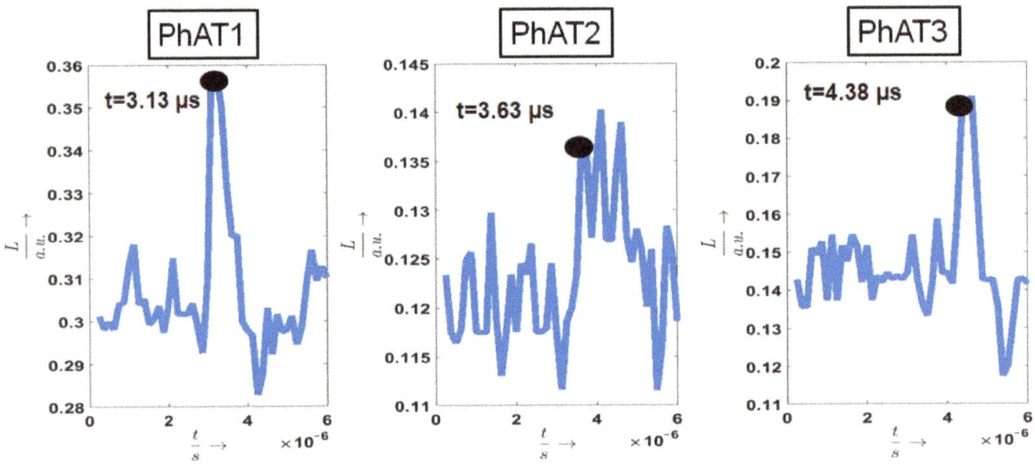

Figure 8. The speckle vector length t_{sp} is illustrated for measurements of phantoms PhAT1–PhAT3. For the three samples, the detection times of the initial generated photoacoustic signal are noted, and the corresponding signal peaks are marked.

Table 1 summarizes t_{mean} for all photoacoustic acquisition times of all samples considering L, their σ, and the theoretical acoustic transit time t_{theo}. For illustration purpose, these results are also plotted in Figure 9.

Table 1. The mean value t_{mean} for all photoacoustic acquisition times of all samples considering, its standard deviation σ, and the theoretical acoustic transit time t_{theo} are listed.

Sample	t_{mean} (μs)	σ (μs)	t_{theo} (μs)
PhAT1	3.19	0.06	3.01
PhAT2	3.71	0.09	3.76
PhAT3	4.48	0.17	4.51
PhAR4	2.15	0.16	2.26
PhAR5	2.99	0.06	3.01
skinAT1	7.72	0.22	7.69
skinAT2p	2.93	0.25	3.04
PhBT1	5.91	0.057	5.56
PhBT2	3.53	0.093	3.71

It is clearly visible that the speckle analysis mean detection times increase with bigger phantom dimensions and that they match the acoustic transit times. The small differences between the mean and theoretical detection time can be explained with inaccuracies for the phantom manufacturing and measuring process. The low standard deviations prove the repeatability and the fact that single speckle sensing allows precise photoacoustic sensing compared to the previous high-speed camera experiments. The standard deviation is in the range of approximately 0.1 μs, which results in a precision of 0.13 mm considering the speed of sound.

Based on the repeatability and successful verification with the theoretical transit time t_{theo} of the transmission mode and reflection mode measurements, it can be concluded that free-space and fiber based single speckle sensing is a reliable technique for the photoacoustic detection on phantoms and on skin tissue samples. Furthermore, the less expensive low-resolution diode sensor, in contrast to the previously used high-speed camera, reaches a high sampling rate of 8 MHz, which allows precise photoacoustic sensing.

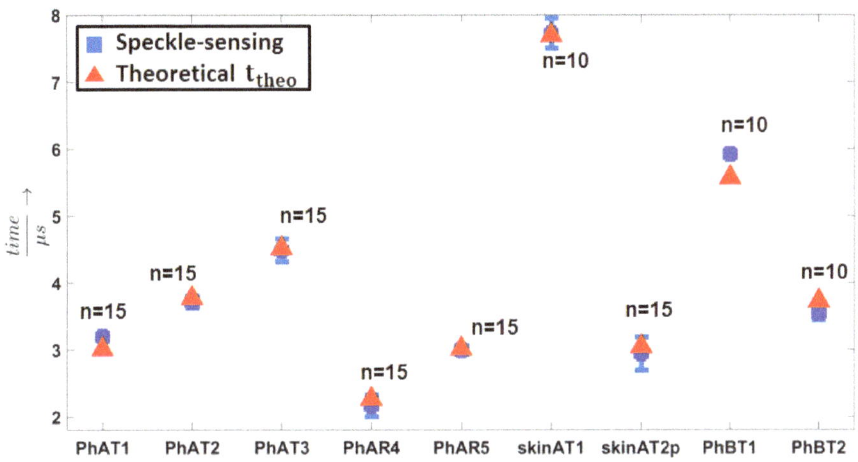

Figure 9. Mean detection times (t_{mean}) and their standard deviation σ for the photoacoustic measurements. The sample names can be explained as follows: Ph stands for a phantom, A for the free-space setup, B for the fiber based setup, T for transmission mode, R for reflection mode, and p for pulsed speckle illumination below the MPE. The theoretical transit time of the photoacoustic signal is used for verification.

3.2. Sensing Parameters

3.2.1. Sensitivity

Table 2 summarizes the relevant parameters for the determination of $S_{d,nm}$ and S_d for the established sensing systems. For the purpose of comparison, f is selected at 1 MHz and 4 MHz, which represents the maximal detectable acoustic frequency for the established sensing system in this work.

Table 2. Relevant parameters for the determination of $S_{d,nm}$ and S_d for the established sensing systems.

	Free-Space		Fiber-Guided
	PVCP	skin	PVCP
σ_{nf}	0.006	0.007	0.0057
d_{px} in μm	1600	1600	1600
M	5	5	50
Z in cm	20	20	0.2
$S_{d,\alpha}$ in $10^{-5\circ}$	55.0	64.2	522
D_{ill} in mm	0.75	0.75	0.05
$S_{d,nm}$ in nm	7.2	8.4	4.56
Z_{ac} in $10^6\,\frac{\mathrm{kg}}{\mathrm{m^2 s}}$	1.4	1.99	1.4
$S_{d,1\,\mathrm{MHz}}$ in kPa	31.67	52.52	20.06
$S_{d,4\,\mathrm{MHz}}$ in kPa	127	210	80

With the systems developed in this work, axial deformations of approximately 5 nm can be detected, which results in a pressure sensitivity of approximately 20 kPa for a 1 MHz acoustic wave. Horstmann et al. reached the following sensitivity parameters with a full-field speckle interferometry approach: $S_{d,\text{nm}} = 1$ nm and $S_{d,1\text{MHz}} = 1.5$ kPa with a sensing bandwidth of 80 MHz [10]. These values, however, were achieved for measurements on silicone, which has a lower impedance ($0.94 \times 10^6 \frac{\text{kg}}{\text{m}^2\text{s}}$) than PVCP, which results in low S_d. Piezoelectric contact transducers that are especially designed and optimized for broadband PA detection achieve high sensitivities, which are dependent on the size and detection bandwidth. For a detection of acoustic frequencies in the range of 10 MHz to 50 MHz with an element size of 30 mm^2, S_d can be estimated to lie between 1.5 Pa and 3.5 Pa [25]. Arrays have a smaller active area per detector element and thus lower sensitivity. An optimized ultrasonic line array can have a sensitivity of 110 Pa for a single element [26]. These sensitivities are better than the sensitivity for the established sensing system in this work. However, as explained in the Introduction, the interferometric setup is more complicated than the speckle analysis applied in the present investigation, and in comparison to the transducer, the single speckle analysis is contact-free. In addition, the speckle analysis sensitivity might even be improved by the usage of smaller photodiodes, tighter focusing of the cw illumination, or new data analysis techniques.

3.2.2. Sensing Range

The single speckle analysis tracks the center of gravity of a single speckle that is positioned in the center of the four measurement diodes (Equations (2) and (3)). In general, the maximal detectable center of gravity coordinates ($x_{sp,max}$, $y_{sp,max}$) are defined by the single speckle diameter l_s and the diode size (d_{px}) by Equation (6). For a larger speckle shift in regards to the sensor center, the single speckle would not be on the sensor completely, which leads to measurement errors. The maximum allowable speckle diameter is defined by twice the diode size. When assuming a speckle diameter of the pixel size, the maximum detectable center of gravity coordinates ($x_{sp,max}$, $y_{sp,max}$) are thus half the diode size.

$$x/y_{sp,max} = (1 - \frac{l_s}{2d_{px}})d_{px} \tag{6}$$

Table 3 gives an overview of the maximal detectable tilts in the horizontal and vertical direction ($\alpha_{max,x}$, $\alpha_{max,y}$). For the single speckle analysis, $x/y_{sp,max}$ can be determined according to Equation (6) with $l_s = 0.6d_{px}$.

Table 3. Overview of the sensor size and surface tilt sensing ranges for the single speckle sensing approach.

	Single Speckle Analysis
d_{px} in µm	1600
$x/y_{sp,max}$ in µm	1120
$\alpha_{max,x/y}$ in °	0.0642

The detectable tilt interval can be defined at [$55 \times 10^{-5°}$; 0.0642°] for the diode system. These intervals can be converted according to the computations from Table 2 to pressure interval [31.67 kPa; 3690 kPa] for a PVCP surface and a 1 MHz acoustic wave. These values result in a dynamic range factor for the pressure detection of approximately 116, which is convenient for PA sensing.

3.2.3. Linearity

Figure 10 illustrates the computed value for x_{sp} for all shifts and speckle images and compares the results to linear behavior.

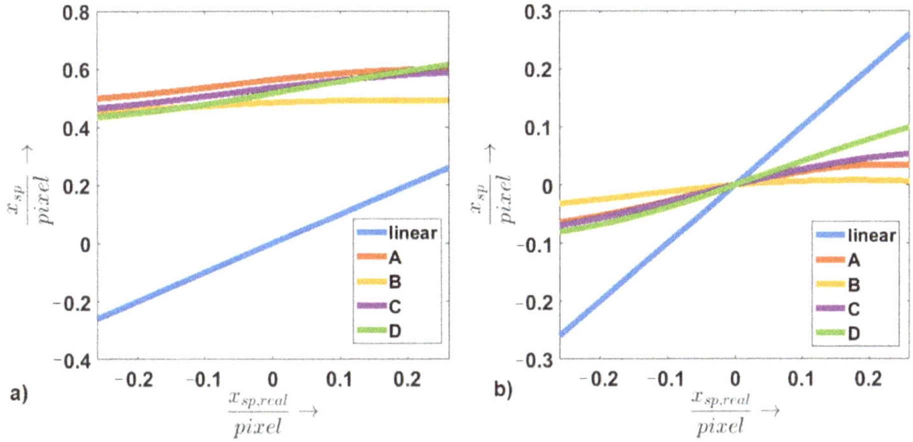

Figure 10. The computed values for x_{sp} are shown over the $x_{sp,real}$ without (**a**) and with zero offset correction (**b**).

Though we tried to centralize the speckle inside the camera image, x_{sp} is not zero for a zero shift (see Figure 10a). This can be explained by the speckle surrounding signal, which also falls into the diode sensing regions. Due to this effect, there is an offset for the computed x_{sp} compared to linear behavior. This offset can be corrected, although there is still a clear difference between the computed x_{sp} and linear behavior (see Figure 10b), which depends on the central speckle size and intensity distribution. Furthermore, the connection between $x_{sp,real}$ and x_{sp} might be surjective. This means that there can be multiple values of x_{sp} for one $x_{sp,real}$. This effect occurs for Speckle Image B, as neighboring speckles that are shifted into the diode sensing region lower the signal strongly for higher shift magnitudes. The non-linearity problem can be corrected since each extracted shift can be mapped to a real shift when assuming a non-surjective behavior for the connection between x_{sp} and $x_{sp,real}$. This could be assured by an automated speckle finding software that analyzes the camera image and potential values for $x_{sp,real}$ and identifies a suitable speckle automatically.

4. Conclusions and Outlook

In previous studies, we already demonstrated the feasibility of remote photoacoustic sensing by multiple speckles analysis [16–18,20]. However, in these previous works, expensive detector systems were used and multiple speckles were analyzed, which limits the achievable sensing rate and thus also the resolution. Furthermore, the detection systems used were not fiber based and thus not suitable for endoscopic applications.

This study reports on a new, purely optical, non-interferometric modality for PA signal acquisition in the MHz range by analyzing a single speckle with four diodes and demonstrates its suitability for endoscopy. Based on the repeatability and successful verification of the transmission mode and reflection mode measurements, it can be concluded that a single speckle provides the information required for reliable PA detection on phantoms that mimic the optical and mechanical properties of tissue and skin samples. The successful fiber based measurements demonstrate the usability of the approach for endoscopic applications. These results are essential steps toward the future application of the technique in a potential imaging device or as a smart feedback system for laser procedures.

However, several challenge arise for the implementation of a future imaging system. Higher sensing rates need to be achieved in order to provide better sensing resolutions. For this work, the frame rate was limited to 8 MHz by the APS, which could be replaced by an economical four quadrant position sensitive diode [27]. With appropriate hardware, this PSD allows higher acquisition rates and thus better sensing resolutions. An array of four quadrant diodes could be developed, and a measurement would be started if speckles were in a convenient position. Therefore, this sensor array would make a positioning unit obsolete and lower the costs of the sensing system significantly since a standard camera could be used for speckle position and shape tracking. Furthermore, the sensing sensitivity needs to be improved, which could be achieved by new hardware and a more precise speckle shift extraction algorithm [23]. This improved sensing sensitivity would reduce the required PA excitation exposure, which was above the MPE for soft tissue in most of the experiments for this investigation. A scan pattern of 10×10, which is considered to be sufficient for imaging, and a single photoacoustic measurement time of 10 µs result in a total measurement time of 1 ms per image, thus allowing imaging rates of 1 kHz. These mentioned steps together with appropriate reconstruction algorithms will allow PA imaging in the future [28].

Author Contributions: Conceptualization, B.L. and F.K.; formal analysis, B.L.; investigation, B.L. and D.S.; methodology, B.L., M.H., and Z.Z.; supervision, M.S. (Michael Schmidt), Z.Z., and F.K.; visualization, B.L.; writing—original draft, B.L.; writing—review and editing, M.H., M.S. (Moritz Späth), D.S., M.W., S.J.R., M.S. (Michael Schmidt), Z.Z., and F.K. All authors read and agreed to the published version of the manuscript.

Funding: This work was funded by the Deutsche Forschungsgemeinschaft (DFG, German Research Foundation)—Project Number 397972545. In addition, funding of the Erlangen Graduate School in Advanced Optical Technologies (SAOT) by the DFG in the framework of the German excellence initiative is gratefully acknowledged.

Institutional Review Board Statement: Not applicable.

Informed Consent Statement: Not applicable.

Data Availability Statement: The data that supports the findings of the study are provided within the article.

Conflicts of Interest: The authors declare no conflict of interest.

Abbreviations

The following abbreviations are used in this manuscript:

APS	avalanche-photodiode sensor
MPE	maximum permitted exposure
PA	Photoacoustic
PhAT	phantom Setup A transmission mode
PhAR	phantom Setup A reflection mode
PhBT	phantom Setup B transmission mode
PVCP	polyvinyl chloride plastisol
skinAT1	skin sample Setup A transmission mode
skinAT2p	skin sample Setup A transmission mode pulsed illumination

References

1. Yao, J.; Xia, J.; Wang, L.V. Multiscale functional and molecular photoacoustic tomography. *Ultrason. Imaging* **2016**, *38*, 44–62. [CrossRef]
2. Jacques, S.L. Optical properties of biological tissues: A review. *Phys. Med. Biol.* **2013**, *58*, R37. [CrossRef]
3. Yao, J.; Wang, L.V. Photoacoustic microscopy. *Laser Photonics Rev.* **2013**, *7*, 758–778. [CrossRef]
4. Kolkman, R.G.; Blomme, E.; Cool, T.; Bilcke, M.; van Leeuwen, T.G.; Steenbergen, W.; Grimbergen, K.A.; den Heeten, G.J. Feasibility of noncontact piezoelectric detection of photoacoustic signals in tissue-mimicking phantoms. *J. Biomed. Opt.* **2010**, *15*, 055011. [CrossRef] [PubMed]
5. Barnes, R.A.; Maswadi, S.; Glickman, R.; Shadaram, M. Probe beam deflection technique as acoustic emission directionality sensor with photoacoustic emission source. *Appl. Opt.* **2014**, *53*, 511–519. [CrossRef] [PubMed]
6. Johnson, J.L.; van Wijk, K.; Caron, J.N.; Timmerman, M. Gas-coupled laser acoustic detection as a non-contact line detector for photoacoustic and ultrasound imaging. *J. Opt.* **2016**, *18*, 024005. [CrossRef]
7. Carp, S.A.; Guerra, A., III; Duque, S.Q., Jr.; Venugopalan, V. Optoacoustic imaging using interferometric measurement of surface displacement. *Appl. Phys. Lett.* **2004**, *85*, 5772–5774. [CrossRef]

8. Hochreiner, A.; Bauer-Marschallinger, J.; Burgholzer, P.; Jakoby, B.; Berer, T. Non-contact photoacoustic imaging using a fiber based interferometer with optical amplification. *Biomed. Opt. Express* **2013**, *4*, 2322–2331. [CrossRef]
9. Rousseau, G.; Gauthier, B.; Blouin, A.; Monchalin, J.P. Non-contact biomedical photoacoustic and ultrasound imaging. *J. Biomed. Opt.* **2012**, *17*, 061217. [CrossRef] [PubMed]
10. Horstmann, J.; Spahr, H.; Buj, C.; Münter, M.; Brinkmann, R. Full-field speckle interferometry for non-contact photoacoustic tomography. *Phys. Med. Biol.* **2015**, *60*, 4045. [CrossRef]
11. Wissmeyer, G.; Pleitez, M.A.; Rosenthal, A.; Ntziachristos, V. Looking at sound: Optoacoustics with all-optical ultrasound detection. *Light. Sci. Appl.* **2018**, *7*, 1–16. [CrossRef] [PubMed]
12. Clark, M. Optical detection of ultrasound on rough surfaces using speckle correlated spatial filtering. *J. Phys. Conf. Ser. IOP Publ.* **2011**, *278*, 012025. [CrossRef]
13. Sharpies, S.; Light, R.; Achamfuo-Yeboah, S.; Clark, M.; Somekh, M.G. The SKED: Speckle knife edge detector. *J. Phys. Conf. Ser. IOP Publ.* **2014**, *520*, 012004. [CrossRef]
14. Zalevsky, Z.; Beiderman, Y.; Margalit, I.; Gingold, S.; Teicher, M.; Mico, V.; Garcia, J. Simultaneous remote extraction of multiple speech sources and heart beats from secondary speckles pattern. *Opt. Express* **2009**, *17*, 21566–21580. [CrossRef]
15. Hajireza, P.; Shi, W.; Bell, K.; Paproski, R.J.; Zemp, R.J. Non-interferometric photoacoustic remote sensing microscopy. *Light. Sci. Appl.* **2017**, *6*, e16278. [CrossRef]
16. Lengenfelder, B.; Mehari, F.; Hohmann, M.; Heinlein, M.; Chelales, E.; Waldner, M.J.; Klämpfl, F.; Zalevsky, Z.; Schmidt, M. Remote photoacoustic sensing using speckle analysis. *Sci. Rep.* **2019**, *9*, 1–11. [CrossRef] [PubMed]
17. Lengenfelder, B.; Mehari, F.; Hohmann, M.; Löhr, C.; Waldner, M.J.; Schmidt, M.; Zalevsky, Z.; Klämpfl, F. Contact-free endoscopic photoacoustic sensing using speckle analysis. *J. Biophotonics* **2019**, *12*, e201900130. [CrossRef]
18. Lengenfelder, B.; Mehari, F.; Hoppe, L.; Klämpfl, F.; Tenner, F.; Zalevsky, Z.; Schmidt, M. Remote photoacoustic tomography using speckle sensing with a high-speed camera. In *Optics and the Brain*; Optical Society of America: Washington, DC, USA, 2016; p. JM3A–16.
19. Lengenfelder, B.; Hohmann, M.; Röhm, M.; Schmidt, M.; Zam, A.; Zalevsky, Z.; Klämpfl, F. Image reconstruction for remote photoacoustic tomography using speckle analysis. *Tissue Opt. Photonics* **2020**, *11363*, 113631F.
20. Lengenfelder, B.; Jarkas, H.; Shabairou, N.; Hohmann, M.; Schmidt, M.; Zalevsky, Z.; Klämpfl, F. Remote photoacoustic tomography using diode-array and speckle analysis. *Tissue Opt. Photonics* **2020**, *11363*, 1136308.
21. Horstmann, J. *Kontaktlose Photoakustische Tomographie: Realisierung und Evaluation einer Optisch-Holographischen Detektionsmethode*; Infinite Science Publishing: Lübeck, Germany, 2016.
22. Culjat, M.O.; Goldenberg, D.; Tewari, P.; Singh, R.S. A review of tissue substitutes for ultrasound imaging. *Ultrasound Med. Biol.* **2010**, *36*, 861–873. [CrossRef]
23. Wu, J.; Chen, Y.; Gao, S.; Li, Y.; Wu, Z. Improved measurement accuracy of spot position on an InGaAs quadrant detector. *Appl. Opt.* **2015**, *54*, 8049–8054. [CrossRef]
24. Shabairou, N.; Lengenfelder, B.; Hohmann, M.; Klämpfl, F.; Schmidt, M.; Zalevsky, Z. All-optical, an ultra-thin endoscopic photoacoustic sensor using multi mode fiber. *Sci. Rep.* **2020**, *10*, 1–8. [CrossRef] [PubMed]
25. Winkler, A.M.; Maslov, K.I.; Wang, L.V. Noise-equivalent sensitivity of photoacoustics. *J. Biomed. Opt.* **2013**, *18*, 097003. [CrossRef] [PubMed]
26. Wang, X.; Fowlkes, J.B.; Cannata, J.M.; Hu, C.; Carson, P.L. Photoacoustic imaging with a commercial ultrasound system and a custom probe. *Ultrasound Med. Biol.* **2011**, *37*, 484–492. [CrossRef] [PubMed]
27. Position Sensitive Diode, First Sensor. Available online: https://www.first-sensor.com/en/products/optical-sensors/detectors/quadrant-apds-qa/ (accessed on 31 August 2020).
28. Huang, C.; Wang, K.; Nie, L.; Wang, L.V.; Anastasio, M.A. Full-wave iterative image reconstruction in photoacoustic tomography with acoustically inhomogeneous media. *IEEE Trans. Med. Imaging* **2013**, *32*, 1097–1110. [CrossRef] [PubMed]

sensors

MDPI

Letter

An Automatic Unmixing Approach to Detect Tissue Chromophores from Multispectral Photoacoustic Imaging

Valeria Grasso [1,2], Joost Holthof [1] and Jithin Jose [1,*]

1 FUJIFILM VisualSonics, 1114 AB Amsterdam, The Netherlands; valeria.grasso@fujifilm.com (V.G.); joost.holthof@fujifilm.com (J.H.)

2 Institute for Animal Science, Hannover Medical School, 30625 Hannover, Germany

* Correspondence: jithin.jose@fujifilm.com

Received: 29 April 2020; Accepted: 4 June 2020; Published: 6 June 2020

Abstract: Multispectral photoacoustic imaging has been widely explored as an emerging tool to visualize and quantify tissue chromophores noninvasively. This modality can capture the spectral absorption signature of prominent tissue chromophores, such as oxygenated, deoxygenated hemoglobin, and other biomarkers in the tissue by using spectral unmixing methods. Currently, most of the reported image processing algorithms use standard unmixing procedures, which include user interaction in the form of providing the expected spectral signatures. For translational research with patients, these types of supervised spectral unmixing can be challenging, as the spectral signature of the tissues can differ with respect to the disease condition. Imaging exogenous contrast agents and accessing their biodistribution can also be problematic, as some of the contrast agents are susceptible to change in spectral properties after the tissue interaction. In this work, we investigated the feasibility of an unsupervised spectral unmixing algorithm to detect and extract the tissue chromophores without any a-priori knowledge and user interaction. The algorithm has been optimized for multispectral photoacoustic imaging in the spectral range of 680–900 nm. The performance of the algorithm has been tested on simulated data, tissue-mimicking phantom, and also on the detection of exogenous contrast agents after the intravenous injection in mice. Our finding shows that the proposed automatic, unsupervised spectral unmixing method has great potential to extract and quantify the tissue chromophores, and this can be used in any wavelength range of the multispectral photoacoustic images.

Keywords: photoacoustic; optoacoustic; spectral imaging; blind source separation; unsupervised unmixing

1. Introduction

The accurate detection and quantification of tissue chromophores is vital in molecular imaging, as it can facilitate the early detection, prediction, and monitor the disease conditions. In recent years, multispectral photoacoustic imaging has emerged as a noninvasive tool to visualize the tissue chromophores [1–4]. The underlying principle of photoacoustic (PA) imaging is based on the conversion of absorbed nanosecond laser pulses into acoustic waves that can be detected just as conventional ultrasound [5,6]. Based on this approach, PA images combine the peculiar optical absorption contrast of the tissue chromophores and the spatial resolution of ultrasound imaging (US). Being a hybrid imaging modality of ultrasound and optical, this multimodal imaging technology can provide anatomical, functional, and molecular information several centimeters deep in the tissues with a resolution up to tens of micrometers. The potential of PA imaging has been demonstrated in various preclinical applications, such as tumor progression and the prediction of tumor recurrence, therapy monitoring, imaging of vasculature, and the biodistribution of the contrast agents [7–13].

Apart from the preclinical applications, PA imaging is also used in clinical research. In addition to breast cancer monitoring [14,15] and sentinel lymph node imaging [16,17], the PA approach is also used to examine inflammatory bowel disease (IBD) [18] and the temporal arteries in patients with suspected giant cell arteritis (GCA) [19]. To expedite the clinical applications of PA imaging, recently, there has been a lot of focus on developing affordable light sources and the use of this technology in low-resource settings. Xia et al. demonstrated the feasibility of a Light-Emitting Diode (LED) based PA imaging system for the visualization of superficial vasculatures and needle guidance for minimally invasive procedures [20]. Zhu et al. used the LED-based approach to explore more clinical applications, such as diagnosing inflammatory arthritis and assessing peripheral microvascular function in patients [21–23].

Although there has been a lot of emphasis on PA hardware development, in the field of affordable settings, data analysis and reconstruction algorithms also play a crucial role in increasing the utility of the technology. Multiwavelength acquisition and spectral image processing is one of the commonly used techniques in PA. Since the optical absorption coefficient of the tissue chromophores varies over the spectrum, multispectral image processing approach can be applied to distinguish and characterize the molecules present in the tissues [24].

In general, the pixel intensity of the multispectral photoacoustic image is proportional to the absorption value of the respective tissue at a specific wavelength. In reality, due to the finite dimension of the pixel (partial volume effect) and the presence of instrumental noise, each spectrum can be a combination of different tissue chromophores. Therefore, it is a challenging task to unmix these signals spectrally and estimate their concentrations. The most common solution to detect the tissue chromophores from multispectral PA imaging is the supervised unmixing [25]. Although this technique yields acceptable results, it requires user interaction to provide the expected source spectral curves as an input to unmix the signals. For translational research with patients, these types of supervised spectral unmixing can be challenging, as the spectral signature of the tissues differs with respect to the disease condition. Imaging exogenous contrast agents and accessing their biodistribution can also be problematic, as some of the contrast agents are susceptible to change in spectral properties after the tissue interaction; thus, the algorithm can forfeit the sensitivity and specificity of imaging.

Hence, an unsupervised unmixing algorithm that can automatically detect the tissue chromophores, without any a-priori knowledge and user interaction, will be optimal, as this can facilitate and improve sensitivity and specificity. Generally, this class of algorithms is referred to as blind source separation (BSS) algorithms, as no a-priori information is required. The study reported by Glatz et al. [26] demonstrated the potential of these approaches to "blindly" extract the oxygenated and deoxygenated hemoglobin absorption spectra from the multispectral photoacoustic images. In the study, they evaluated different unsupervised algorithms, such as multivariate curve resolution analysis (MCR), principal component analysis (PCA), and independent component analysis (ICA) [27–31]. PCA yields an orthogonal transformation that decorrelates the variables. This approach relies upon the hypothesis that the source components are uncorrelated. On the other hand, ICA is based on a different assumption that the source components are maximally independent and non-Gaussian. Recently, Arabul et al. [32] used a similar approach to explore human carotid plaques, in which the ICA blind unmixing approach was constrained non-negatively. Although these approaches demonstrated the potential to detect the tissue chromophores, they suffer from limitations related to the interpretability of the mixed-sign values of their outcomes. Indeed, these aim to fit the training data well but often do not generalize the real and positive data sets.

In this paper, we investigated the possibility of using another blind source separation approach, which is based on non-negative matrix factorization (NNMF) [33]. The concept of NNMF has been widely used in a variety of applications, such as image recognition [34], text classification [35,36], and recommender systems [37]. This approach uses only the non-negative matrices to estimate the prominent components and their spatial distribution, from a linear mixture model. Montcuquet et al. [38] used this approach for in-vivo fluorescent imaging and demonstrated that the positivity condition enhances the convergence and, thus, improves the sensitivity of spectral unmixing [39].

Here, we examine the performance of non-negative matrix factorization (NNMF) to unmix tissue chromophores from multispectral PA images. The algorithm has been optimized to extract the tissue chromophores in the wavelength range of 680–900 nm. We tested the NNMF on synthetic data and on experimental data that mimic the blood vessels. Further, we validated the potential of the approach on an in-vivo study to detect and quantify the endogenous absorbers and exogenous contrast agent accumulation. To our knowledge, this is the first time the NNMF algorithm has been used for PA imaging.

2. Non-Negative Matrix Factorization (NNMF)

NNMF is a data decomposition approach, and it is based on the linear mixing model [40]. In this algorithm, the acquired mixed pixel spectra are differentiated into a collection of constituent spectra (called endmembers) and a set of fractional abundance maps. The endmembers represent the pure molecule absorption spectra present in the imaged sample, and these are extracted from the mixed pixel spectra. The maps of abundance at each pixel represent the percentage of each endmember present in that pixel.

Since the acquired spectral images are known, and the rest has to be estimated, the mixed data (multispectral PA images) can be arranged as a matrix $\boldsymbol{X} \in \mathbb{R}^{n \times m}$, where $\boldsymbol{n}$ represents the number of observations (pixels), and $\boldsymbol{m}$ corresponds to the number of variables per object (different wavelengths). In particular, the unmixing problem can be formulated as a matrix factorization:

$$X \approx WS \tag{1}$$

where X represents the mixed multispectral PA images, W the abundance maps, and S the source spectra. The dimensions of the matrices W and S are $\boldsymbol{n} \times \boldsymbol{k}$ and $\boldsymbol{k} \times \boldsymbol{m}$, respectively, where $\boldsymbol{k}$ is the hyperparameter which represents the number of prominent components.

The NNMF constrained cost function of the optimization problem can be formulated as follows:

$$[W, S] = \min_{W,S} \frac{1}{2} \parallel X - WS \parallel_F^2 \tag{2}$$

$$W_{ij} \geq 0, \;\; S_{ij} \geq 0 \tag{3}$$

where, in Equation (2), the defined cost function considers a Frobenius distance between the acquisition X and the model WS [33]. W and S are iteratively obtained until both matrices satisfy Equation (1), where the distance defined in the cost function (2) is constrained non-negatively (3). The NNMF learns a parts-based representation of the data, and the whole image is formed as a combination of additive components. The non-negativity constraint is computationally expensive to implement but it can lead to more interpretable data.

To solve the iterative optimization, the multiplicative update rules [41] can be used, and the steps can be defined as:

$$S\,(p+1) = S(p) \otimes \frac{\left(W^T \cdot X\right)}{\left(W^T W \cdot S\right)} \tag{4}$$

$$W(p+1) = W(p) \otimes \frac{\left(X \cdot S^T\right)}{\left(W \cdot S \cdot S^T\right)} \tag{5}$$

where $\boldsymbol{p}$ is the iteration step, and the operations of $\otimes$ and division in (4) and (5) are considered element by element. Each component is estimated only up to a multiplying scale factor. Hence, the factorization problem does not have a unique solution, where WS is a lower-rank approximation of X.

3. Experimental Methods

3.1. Simulated Multispectral PA Images

To evaluate the performance of the NNMF algorithm, simulation studies were performed by using a synthetic data set. The synthetic data set contains implemented photoacoustic spectral images within the wavelength range of 680–900 nm, with a step size of 5 nm. Each image at the respective wavelength contains 400 × 600 pixels. From the photoacoustic signal generation, it is evident that the PA signal is not only proportional to the absorption coefficient but also depends on the local fluence. The light fluence generally decreases with depth, and thus degrades the image uniformity, causing spatial fluence variations within the tissue. Consequently, the fluence compensation is significant for quantitative spectral imaging. Since the main focus of the simulation was to test the unmixing algorithm and its accuracy to detect the spectral signature of the prominent components, in this study, the local fluence was assumed to be constant.

Figure 1a depicts the schematic of the 2-dimensional (2-D) data set with six homogeneous inclusions. The inclusions mimic the cross section of blood vessels with different concentrations of oxyhemoglobin and deoxyhemoglobin. The inclusions 1, 2, and 3 contain oxyhemoglobin at 100%, 70%, and 30% of the intensity, respectively. Conversely, the inclusions 4, 5, and 6 include deoxyhemoglobin at 30%, 70%, and 100% of the intensity. Figure 1b shows the theoretical absorption spectra [42] expected from these inclusions, in addition to the background tissue absorption. To mimic the experimental conditions, a positive Gaussian distribution of noise ($mean = 0.04$; $std = 0.1\%$; $SNR = 30$ dB) was also added to the respective data set.

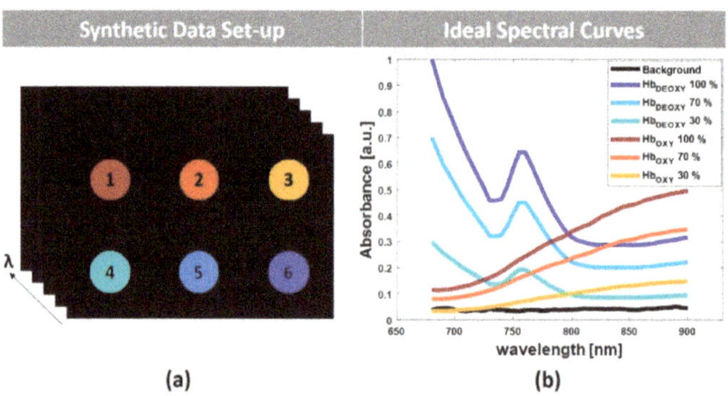

Figure 1. (**a**) Schematic representation of the simulated multispectral photoacoustic (PA) images with six inclusions, (**b**) ideal spectral curves of the inclusions.

3.2. Experimental Set-Up and Tissue-Mimicking Vessel Phantom

In addition to the simulation studies, experiments were performed by using Vevo LAZR-X photoacoustic image technology (FUJIFILM VisualSonics, Inc., Toronto, ON, Canada), as described elsewhere [43]. Vevo Phantom (FUJIFILM VisualSonics, Inc., Toronto, ON, Canada) containing two capillary tubes filled with Indocyanine Green (ICG, PULSION Verwaltungs, GmbH) and Methylene Blue (MB, Sigma-Aldrich), was used to mimic the blood vessels in the tissue. Transparent polyethylene (PE) tubes (SAI Infusion Technologies, Lake Villa, IL, USA), with an inner diameter of 15 µm and an outer diameter of 33 µm, were used. The tubes were positioned at a reciprocal distance of 6 mm and fixed at the same depth of 14 mm from the surface of the transducer. Demineralized water was used as a coupling medium, and the multispectral PA images were obtained in the wavelength range of 680–900 nm, with a step size of 5 nm. Figure 2a shows the schematic of the phantom and the experimental set-up. A 256-element linear array transducer with a central frequency of 21 MHz (MX250), including the integrated light delivery fibers from the sides of the transducer, was used to acquire the PA images. The transducer was aligned perpendicular to the capillary tubes, and the cross-sectional image of the tubes was acquired throughout the wavelength range. 3-Dimensional (3-D) data sets were also collected by linearly translating the transducer with a stepper motor over the capillary tubes while capturing cross-sectional 2-D slices. Figure 2b shows the photoacoustic spectra measured from the capillary tubes by using the Spectro-Mode in the VevoLab software (FUJIFILM VisualSonics, Inc., Toronto, ON, Canada). In the measurement tool, the system allows the user to select the region of interest (ROI) and calculate the average intensity of the photoacoustic signal at different wavelengths. Although this is not quantitative and the values are in arbitrary units, it can provide the spectral absorption trend of the agents in the respective wavelength range.

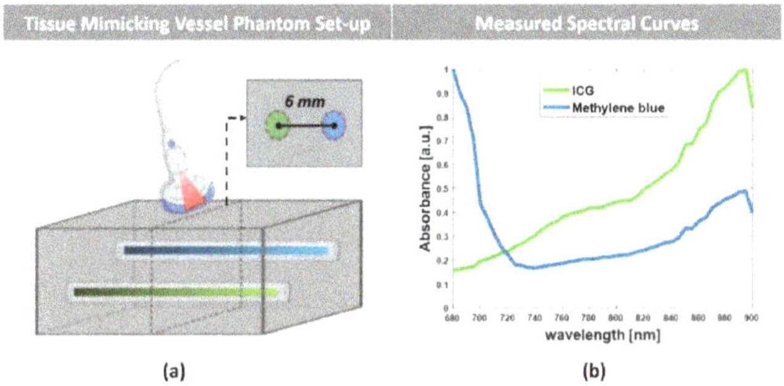

Figure 2. (**a**) Schematic of the tissue-mimicking vessel phantom. (**b**) The PA absorbance spectral graph of the agents measured by using VevoLab.

3.3. In-Vivo Study

Further, in-vivo animal experiments were performed to evaluate the feasibility of the NNMF data analysis on multispectral PA Imaging. The animal experiments were performed at the FUJIFILM Sonosite/VisualSonics facility in Amsterdam. The animal protocols used in this work were evaluated and approved by the Animal Use and Ethics Committee (CEUA) of The Netherlands (Protocol AVD2450020173644). They are in accordance with FELASA guidelines and the National Law for Laboratory Animal Experimentation (Law No. 18.611). The experiments were performed by using the same apparatus (Vevo LAZR-X) used for the phantom studies. A CD-1 female mouse model (Envigo, Horst, the Netherlands) was used for the experiments. The animal was anesthetized with isoflurane and placed on the animal imaging platform of the Vevo LAZR-X system, where temperature, heart rate, and respiration rate were monitored in real time. During the experiments, anesthesia was maintained using a vaporized isoflurane (1 L/min of oxygen and 0.75% isoflurane) gas system. The animal was positioned in right lateral recumbency, and the transducer was aligned perpendicularly. The kidney–spleen area of the animal was imaged before and after the intravenous injection of ICG. To obtain a concentration of 800 µM, 25-mg vial of ICG (PULSION Verwaltungs, GmbH) was resuspended in sterile water. With the help of an infusion pump (flowrate of 15 µL/sec), 80 µL of ICG was injected into the tail vein, and multispectral PA images were acquired in the wavelength range of 680–900 nm.

4. Results and Discussion

4.1. Simulated Multispectral PA Images

Figure 3 shows the main component spectra (a) and the respective abundance maps (b, c, and d) extracted from the synthetic PA data set by using the NNMF algorithm. The obtained spectral graphs (Figure 3a) show that the prominent absorbers present within the inclusions consist of two different endmembers: oxyhemoglobin and deoxyhemoglobin. As expected from the synthetic data, the oxyhemoglobin is mainly distributed on the first row of inclusions, with decreasing intensity from left to right (Figure 3c). The deoxyhemoglobin is present in the second row of inclusions, with increasing intensity from left to right (Figure 3b). Figure 3d displays the spatial distribution of the detected third component (named as background in Figure 3a), which is principally present in the region around the inclusions.

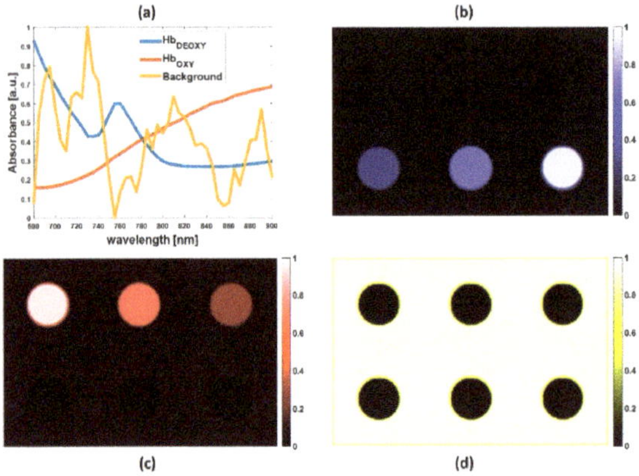

Figure 3. (**a**) Source spectra of oxyhemoglobin, deoxyhemoglobin, and background extracted by the non-negative matrix factorization (NNMF) algorithm. Abundance maps of (**b**) deoxyhemoglobin, (**c**) oxyhemoglobin, and (**d**) background.

NNMF appears to be an accurate method, yielding an explicit unmixing of specific tissue biomarkers, such as oxyhemoglobin and deoxyhemoglobin, and extracting the respective spectral signatures. The calculated Pearson correlation coefficient between the ideal spectra of the oxy and deoxyhemoglobin, and the extracted spectra was equal to 1 (Table 1). This confirms that NNMF can provide encouraging results on extracting tissue chromophores from multispectral PA images.

Table 1. Correlation values between the extracted source components by using the NNMF unmixing approach, and the respective absorption spectral curves used as a reference.

Data Set-Up	Chromophores	Correlation Value
Synthetic data	Oxyhemoglobin	1
	Deoxyhemoglobin	1
Vessel mimicking phantom	ICG	0.9943
	Methylene Blue	0.8344

Further, we investigated the performance of the NNMF approach to quantify the extracted prominent components. Figure 4a shows a graph with the quantitation of the source components (oxyhemoglobin, deoxyhemoglobin) per each circular inclusion of the synthetic data. Considering the synthetic data: inclusions 1, 2, and 3 present approximatively a decreasing content of oxyhemoglobin, and the regions 4, 5, and 6 present an increasing content of deoxyhemoglobin. In the graph, the variations are in the range of zero to one, and it is evident that these normalized intensities are matching the expected proportion of the components. Figure 4b shows the overlapped abundance maps of oxyhemoglobin and deoxyhemoglobin, and it also confirms the expected distribution of the prominent components.

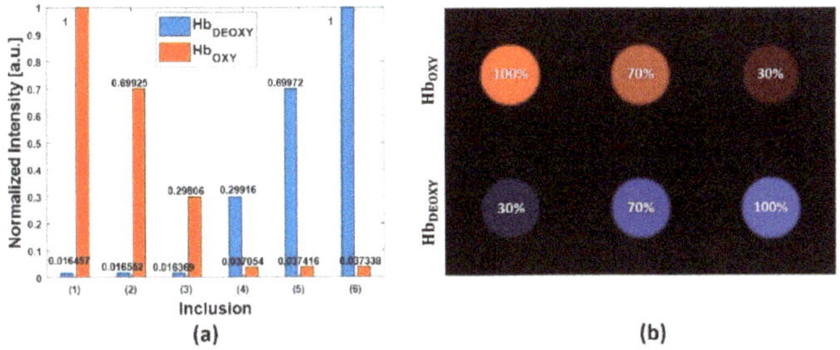

Figure 4. (**a**) Quantitative evaluation of the prominent source components (oxyhemoglobin and deoxyhemoglobin), per each circular region of the synthetic phantom. (**b**) Overlapped abundance maps of oxyhemoglobin and deoxyhemoglobin.

4.2. Tissue-Mimicking Vessel Phantom

Figure 5 shows the prominent spectral curves (a) and the abundance maps of the dyes in the capillary tubes, obtained by using the NNMF algorithm. The abundance maps overlapped in Figure 5b correspond to the 2-D spatial distribution of the contrast agents (ICG in green and MB in blue). Figure 5c shows the unmixing of the capillary tubes in 3-D, where the total imaged range was 6.5 mm with a step size of 200 µm. The Pearson correlation coefficient was evaluated between the extracted spectral curves, by using the NNMF, and the spectra measured by using the Spectro-Mode in the VevoLab.

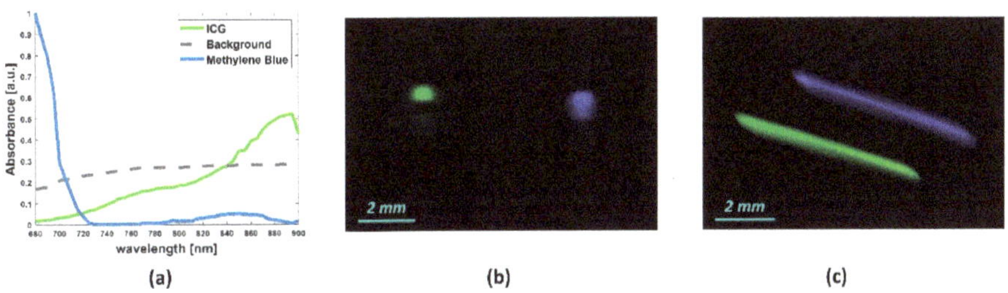

Figure 5. (**a**) Spectral absorption curves of the detected source components by NNMF, from 2-D spectral PA images of the tissue-mimicking vessel phantom. The overlapped abundance 2-D maps (**b**) and 3-D maps (**c**) of the detected source components: ICG and MB.

Table 1 reports the correlation coefficients measured for the ICG and methylene blue. The correlation value obtained for the methylene blue was 0.8344, and it was comparatively lower to the value obtained for the ICG, which was 0.9943. This could be due to the experimental conditions, as the tubes are located adjacently, into the phantom chamber. Hence, due to the short distance, the ICG that has an absorption peak at around 880 nm could influence the absorption spectrum of the MB. The graph reported in Figure 2b supports this assumption, as the MB spectrum shows an additional peak at around 880 nm. This may entail that measured spectral curves in Figure 2b are not the pure agent spectra. On the other hand, the spectra extracted by using NNMF are in accordance with the expected spectral signatures, and it shows promising unmixing performance.

4.3. In-Vivo Study

Figure 6a shows a high-resolution ultrasound (US) image of the kidney–spleen region and (b) is the PA image obtained at 880 nm. Figure 6c shows the oxygen saturation (SO_2) map, obtained before the contrast agent injection. To obtain the oxygen saturation map, we have followed the algorithm reported by Needles et al. [5], where the pixel values of the image are in the range of 0% (lower oxygenation) to 100% (higher oxygenation).

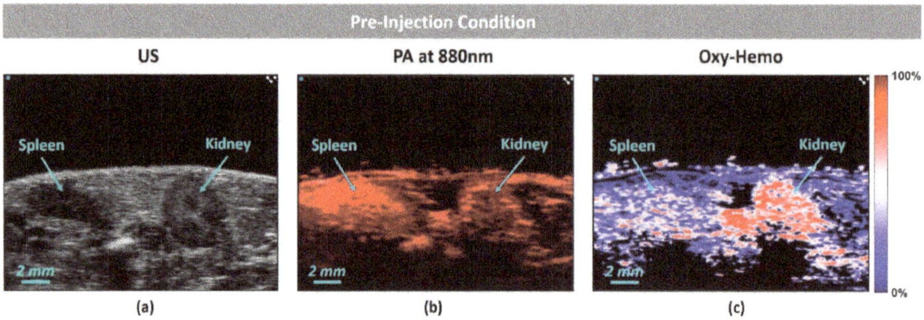

Figure 6. Pre-injection conditions: (**a**) ultrasound (US) image of the kidney–spleen view; respectively (**b**) photoacoustic (PA) image obtained at 880nm, and (**c**) SO_2 map.

Figure 7 shows the post-ICG injection condition. Figure 7a displays the source absorption spectra extracted by the NNMF approach. As expected, in the post-injection condition, the NNMF extracted the typical absorption spectral curve of the ICG, in addition to the endogenous chromophores such as oxy and deoxyhemoglobin spectra. The extracted spectrum of the ICG appears slightly different than the vessel mimicking phantom. Although the peak absorption was at 880 nm, the ICG spectrum at the lower wavelengths was altered. This could be due to the ICG interaction with other chromophores within the tissues. Figure 7 also shows the abundance maps of deoxyhemoglobin (b), oxyhemoglobin (c), and ICG (d). In the ICG map, it is evident that the dye is mostly accumulated in the spleen region. This is in accordance with the biodistribution of the ICG, as the kinetic of the kidney is much faster than the spleen [44].

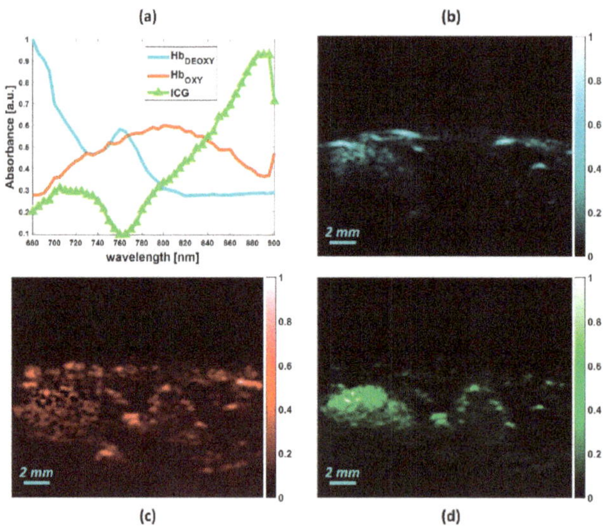

Figure 7. Post-injection conditions: (**a**) spectral signature of the endmembers obtained by using NNMF and abundance distribution maps of (**b**) deoxyhemoglobin, (**c**) oxyhemoglobin, and (**d**) ICG.

5. Conclusions

In summary, we investigated the spectral decomposition of various tissue biomarkers from multispectral PA images. In particular, we have explored the performance of an unsupervised spectral unmixing algorithm, NNMF, in the wavelength range of 680–900 nm. Considering the evidence obtained from the initial results, the NNMF can extract both endogenous and exogenous agents from the multispectral PA data. The unmixing results obtained from the simulation studies performed on synthetic data revealed a high correlation with the expected spectra, and also yielded the quantification of the chromophores. The experiment performed on the tissue-mimicking phantom also supported the results obtained on the synthetic data. Indeed, the NNMF showed promising unmixing performance, allowing the accurate detection of ICG and MB, by eliminating the spectral influence of the other dyes. The in-vivo experiments in the animal model showed the detection of a contrast agent signature, accounting for the spectral variations that may ensue due to tissue–dye interactions. The NNMF can also provide the maps of abundance distribution of contrast agents in different anatomical targets, facilitating in-vivo biodistribution and kinetics studies. Since the algorithm facilitates the automatic unmixing of the tissue chromophores, without any a-priori knowledge about the source components and user interactions, it is easy to adapt, and promising for data-driven studies in multispectral photoacoustic imaging.

In the current algorithm, we have used the time gain compensation (TGC) approach to overcome the PA signal attenuation through depth. Although the approach is not quantitative, it gives improved results. Recent studies demonstrate that taking into account the fluence variance [45,46] can overcome the quantification limits of photoacoustic imaging. Therefore, in future studies, we will investigate the corruption effects of the fluence variation, the finite size, and band-limited frequency response of the detectors, to consider the respective changes on the absorption spectra. Besides, we will expand the current wavelength range of 680–900 nm to the far infrared (FIR), as this may entail the detection of less prominent tissue chromophores, such as melanin, lipids, and collagens.

In conclusion, to the best of our knowledge, this would be the first time that NNMF was used for unmixing multispectral PA imaging. The obtained results confirmed that the NNMF algorithm automatically and accurately detects the component spectra. This proves that the imposed positivity constraints, to the source spectra and abundance distribution maps, are appropriate requirements to unmix tissue chromophores from multispectral PA images.

Author Contributions: Conceptualization, V.G. and J.J.; Methodology, V.G., J.H. and J.J.; Software, V.G.; Validation, V.G. and J.J.; Investigation, J.J.; Resources, J.J.; Data curation, V.G., J.H. and J.J.; Writing—original draft preparation, V.G.; Writing—review and editing, V.G. and J.J.; Visualization, V.G. and J.J.; Supervision, J.J.; Project administration, J.J.; Funding acquisition, J.J. All authors have read and agreed to the published version of the manuscript.

Acknowledgments: This publication is part of a project that has received funding from the European Union's Horizon 2020 research and innovation program under the Marie Skłodowska-Curie grant, agreement No 811226.

Conflicts of Interest: The authors declare no conflict of interest.

References

1. Cox, B.; Laufer, J.G.; Arridge, S.R.; Beard, P.C. Quantitative spectroscopic photoacoustic imaging: A review. *J. Biomed. Opt.* **2012**, *17*, 061202. [CrossRef] [PubMed]
2. Razansky, D.; Vinegoni, C.; Ntziachristos, V. Multispectral photoacoustic imaging of fluorochromes in small animals. *Opt. Lett.* **2007**, *32*, 2891. [CrossRef] [PubMed]
3. Buchmann, J.; Kaplan, B.; Powell, S.; Prohaska, S.; Laufer, J. Quantitative PA tomography of high resolution 3-D images: Experimental validation in a tissue phantom. *Photoacoustics* **2020**, *17*, 100157. [CrossRef] [PubMed]
4. Jnawali, K.; Chinni, B.; Dogra, V.; Rao, N. Automatic cancer tissue detection using multispectral photoacoustic imaging. *Int. J. Comput. Assist. Radiol. Surg.* **2020**, *15*, 309–320. [CrossRef]
5. Needles, A.; Heinmiller, A.; Sun, J.; Theodoropoulos, C.; Bates, D.; Hirson, D.; Yin, M.; Foster, F. Development and initial application of a fully integrated photoacoustic micro-ultrasound system. *IEEE Trans. Ultrason. Ferroelectr. Freq. Control* **2013**, *60*, 888–897. [CrossRef]
6. Xu, M.; Wang, L.V. Photoacoustic imaging in biomedicine. *Rev. Sci. Instrum.* **2006**, *77*, 041101. [CrossRef]
7. Diot, G.; Metz, S.; Noske, A.; Liapis, E.; Schroeder, B.; Ovsepian, S.V.; Meier, R.; Rummeny, E.; Ntziachristos, V. Multispectral Optoacoustic Tomography (MSOT) of human breast cancer. *Clin. Cancer Res.* **2017**, *23*, 6912–6922. [CrossRef]

8. Wilson, K.E.; Wang, T.Y.; Willmann, J.K. Acoustic and photoacoustic molecular imaging of cancer. *J. Nucl. Med.* **2013**, *54*, 1851–1854. [CrossRef]
9. Heijblom, M.; Piras, D.; van den Engh, F.M.; van der Schaaf, M.; Klaase, J.M.; Steenbergen, W.; Manohar, S. The state of the art in breast imaging using the Twente Photoacoustic Mammoscope: Results from 31 measurements on malignancies. *Eur. Radiol.* **2016**, *26*, 3874–3887. [CrossRef]
10. Bauer, D.R.; Olafsson, R.; Montilla, L.G.; Witte, R.S. 3-D photoacoustic and pulse echo imaging of prostate tumor progression in the mouse window chamber. *J. Biomed. Opt.* **2011**, *16*, 026012. [CrossRef]
11. Margolis, R.; Wessner, C.; Stanczak, M.; Liu, J.B.; Li, J.; Nam, K.; Forsberg, F.; Eisenbrey, J.R. Monitoring Progression of Ductal Carcinoma In Situ Using Photoacoustics and Contrast-Enhanced Ultrasound. *Transl. Oncol.* **2019**, *12*, 973–980. [CrossRef] [PubMed]
12. Gargiulo, S.; Albanese, S.; Mancini, M. State-of-the-Art Preclinical Photoacoustic Imaging in Oncology: Recent Advances in Cancer Theranostics. *Contrast Media Mol. Imaging* **2019**, *2019*, 5080267. [CrossRef] [PubMed]
13. Jose, J.; Manohar, S.; Kolkman, R.G.M.; Steenbergen, W.; Leeuwen, T.G. Van Imaging of tumor vasculature using Twente photoacoustic systems. *J. Biophotonics* **2009**, *717*, 701–717. [CrossRef] [PubMed]
14. Manohar, S.; Dantuma, M. Current and future trends in photoacoustic breast imaging. *Photoacoustics* **2019**, *16*, 100134. [CrossRef] [PubMed]
15. Rao, A.P.; Bokde, N.; Sinha, S. Photoacoustic imaging for management of breast cancer: A literature review and future perspectives. *Appl. Sci.* **2020**, *10*, 767. [CrossRef]
16. Liu, S.; Wang, H.; Zhang, C.; Dong, J.; Liu, S.; Xu, R.; Tian, C. In Vivo Photoacoustic Sentinel Lymph Node Imaging Using Clinically-Approved Carbon Nanoparticles. *IEEE Trans. Biomed. Eng.* **2019**. [CrossRef]
17. Jose, J.; Grootendorst, D.J.; Vijn, T.W.; van Leeuwen, T.G.; Steenbergen, W.; Manohar, S.; Wouters, M.; Ruers, T.J.; van Boven, H. Initial results of imaging melanoma metastasis in resected human lymph nodes using photoacoustic computed tomography. *J. Biomed. Opt.* **2011**, *16*, 096021. [CrossRef]
18. Alaeian, M.; Orlande, H.R.B.; Machado, J.C. Temperature estimation of inflamed bowel by the photoacoustic inverse approach. *Int. J. Numer. Methods Biomed. Eng.* **2020**, *36*, e3300. [CrossRef]
19. Sheikh, R.; Cinthio, M.; Dahlstrand, U.; Erlov, T.; Naumovska, M.; Hammar, B.; Zackrisson, S.; Jansson, T.; Reistad, N.; Malmsjo, M. Clinical Translation of a Novel Photoacoustic Imaging System for Examining the Temporal Artery. *IEEE Trans. Ultrason. Ferroelectr. Freq. Control* **2019**, *66*, 472–480. [CrossRef]
20. Xia, W.; Singh, M.K.A.; Maneas, E.; Sato, N.; Shigeta, Y.; Agano, T.; Ourselin, S.; West, S.J.; Desjardins, A.E. Handheld real-time LED-based photoacoustic and ultrasound imaging system for accurate visualization of clinical metal needles and superficial vasculature to guide minimally invasive procedures. *Sensors* **2018**, *18*, 1394. [CrossRef]
21. Zhu, Y.; Lu, X.; Dong, X.; Yuan, J.; Fabiilli, M.L.; Wang, X. LEd-based photoacoustic imaging for monitoring angiogenesis in fibrin scaffolds. *Tissue Eng. Part C Methods* **2019**, *25*, 523–531. [CrossRef] [PubMed]
22. Jo, J.; Xu, G.; Zhu, Y.; Burton, M.; Sarazin, J.; Schiopu, E.; Gandikota, G.; Wang, X. Detecting joint inflammation by an LED-based photoacoustic imaging system: A feasibility study. *J. Biomed. Opt.* **2018**, *23*, 110501. [CrossRef] [PubMed]
23. Zhu, Y.; Xu, G.; Yuan, J.; Jo, J.; Gandikota, G.; Demirci, H.; Agano, T.; Sato, N.; Shigeta, Y.; Wang, X. Light emitting diodes based photoacoustic imaging and potential clinical applications. *Sci. Rep.* **2018**, *8*, 1–12. [CrossRef] [PubMed]
24. Razansky, D.; Distel, M.; Vinegoni, C.; Ma, R.; Perrimon, N.; Köster, R.W.; Ntziachristos, V. Multispectral opto-acoustic tomography of deep-seated fluorescent proteins in vivo. *Nat. Photonics* **2009**, *3*, 412–417. [CrossRef]
25. Cox, B.T.; Arridge, S.R.; Beard, P.C. Estimating chromophore distributions from multiwavelength photoacoustic images. *JOSA A* **2009**, *26*, 443–455. [CrossRef]
26. Glatz, J.; Deliolanis, N.C.; Buehler, A.; Razansky, D.; Ntziachristos, V. Blind source unmixing in multi-spectral optoacoustic tomography. *Opt. Express* **2011**, *19*, 3175. [CrossRef]
27. Bartholomew, D.J. Principal components analysis. *Int. Encycl. Educ.* **2010**, 374–377.
28. Hyvärinen, A. Independent component analysis: Recent advances. *Philos. Trans. R. Soc. A Math. Phys. Eng. Sci.* **2013**, *371*, 20110534. [CrossRef]

29. Le, Q.V.; Karpenko, A.; Ngiam, J.; Ng, A.Y. ICA with reconstruction cost for efficient overcomplete feature learning. In *Advances in Neural Information Processing Systems*; Neural Information Processing Systems Foundation, Inc.: San Diego, CA, USA, 2011; pp. 1017–1025.
30. Ngiam, J.; Koh, P.W.; Chen, Z.; Bhaskar, S.; Ng, A.Y. Sparse filtering. In *Advances in Neural Information Processing Systems*; Neural Information Processing Systems Foundation, Inc.: San Diego, CA, USA, 2011; pp. 1125–1133.
31. Wang, P.; Wang, P.; Wang, H.-W.; Cheng, J.-X. Mapping lipid and collagen by multispectral photoacoustic imaging of chemical bond vibration. *J. Biomed. Opt.* **2012**, *17*, 0960101. [CrossRef]
32. Arabul, M.U.; Rutten, M.C.M.; Bruneval, P.; van Sambeek, M.R.H.M.; van de Vosse, F.N.; Lopata, R.G.P. Unmixing multi-spectral photoacoustic sources in human carotid plaques using non-negative independent component analysis. *Photoacoustics* **2019**, *15*, 100140. [CrossRef]
33. Lee, D.D.; Seung, H.S. Learning the parts of objects by non-negative matrix factorization. *Nature* **1999**, *401*, 788–791. [CrossRef] [PubMed]
34. Lil, S.; Hou, X.; Zhangl, H.; Cheng, Q. Learning Spatially Localized, Parts-B ased Representation. In Proceedings of the 2001 IEEE Computer Society Conference on Computer Vision and Pattern Recognition. CVPR 2001, Kauai, HI, USA, 8–14 December 2001.
35. Buciu, I. Non-negative matrix factorization, a new tool for feature extraction: Theory and applications. *Int. J. Comput. Commun. Control* **2008**, *3*, 67–74.
36. Li, L.; Yang, J.; Xu, Y.; Qin, Z.; Zhang, H. Documents clustering based on max-correntropy nonnegative matrix factorization. *Proc. Int. Conf. Mach. Learn. Cybern.* **2014**, *2*, 850–855.
37. Luo, X.; Zhou, M.; Xia, Y.; Zhu, Q. An efficient non-negative matrix-factorization-based approach to collaborative filtering for recommender systems. *IEEE Trans. Ind. Inform.* **2014**, *10*, 1273–1284.
38. Montcuquet, A.-S.; Herve, L.; Garcia, F.P.N.Y.; Dinten, J.-M.; Mars, J.I. Nonnegative matrix factorization: A blind spectra separation method for in vivo fluorescent optical imaging. *J. Biomed. Opt.* **2010**, *15*, 56009.
39. Lipovetsky, S. PCA and SVD with nonnegative loadings. *Pattern Recognit.* **2009**, *42*, 68–76. [CrossRef]
40. Keshava, N.; Mustard, J.F. Spectral unmixing. *IEEE Signal Process. Mag.* **2002**, *19*, 44–57. [CrossRef]
41. Lee, D.D.; Seung, H.S. Algorithms for non-negative matrix factorization. In *Advances in Neural Information Processing Systems*; The MIT Press: Cambridge, MA, USA, 2001; pp. 556–562.
42. Prahl, S. Tabulated Molar Extinction Coefficient for Hemoglobin in Water. Available online: https://omlc.org/index.html (accessed on 5 June 2020).
43. Toumia, Y.; Cerroni, B.; Trochet, P.; Lacerenza, S.; Oddo, L.; Domenici, F.; Paradossi, G. Performances of a Pristine Graphene—Microbubble Hybrid Construct as Dual Imaging Contrast Agent and Assessment of Its Biodistribution by Photoacoustic Imaging. *Part. Part. Syst. Charact.* **2018**, *35*, 1800066. [CrossRef]
44. Song, W.; Tang, Z.; Zhang, D.; Burton, N.; Driessen, W.; Chen, X. RSC Advances Comprehensive studies of pharmacokinetics and biodistribution of indocyanine green and liposomal indocyanine green by multispectral optoacoustic. *RSC Adv.* **2014**, *5*, 3807–3813. [CrossRef]
45. Gröhl, J.; Kirchner, T.; Adler, T.; Maier-Hein, L. Estimation of blood oxygenation with learned spectral decoloring for quantitative photoacoustic imaging (LSD-qPAI). *arXiv* **2019**, arXiv:1902.05839.
46. Brochu, F.M.; Joseph, J.; Tomaszewski, M.; Bohndiek, S.E. Light fluence correction for quantitative determination of tissue absorption coefficient using multi-spectral optoacoustic tomography. In Proceedings of the European Conference on Biomedical Optics 2015, Munich, Germany, 21–25 June 2015; p. 95390Z.

sensors

MDPI

Article

Multiangle Long-Axis Lateral Illumination Photoacoustic Imaging Using Linear Array Transducer

João H. Uliana [1], Diego R. T. Sampaio [1], Guilherme S. P. Fernandes [1], María S. Brassesco [2], Marcello H. Nogueira-Barbosa [3], Antonio A. O. Carneiro [1] and Theo Z. Pavan [1,*]

[1] Department of Physics, FFCLRP, University of São Paulo, Ribeirão Preto 14040-901, SP, Brazil; joaouliana@usp.br (J.H.U.); diegothomaz@gmail.com (D.R.T.S.); guilherme.santos.fernandes@usp.br (G.S.P.F.); adilton@usp.br (A.A.O.C.)

[2] Department of Biology, FFCLRP, University of São Paulo, Ribeirão Preto 14040-901, SP, Brazil; solbrassesco@usp.br

[3] Department of Medical Images, Hematology and Clinical Oncology, Ribeirão Preto Medical School, University of São Paulo, Ribeirão Preto 14040-901, SP, Brazil; marcello@fmrp.usp.br

* Correspondence: theozp@usp.br

Received: 19 June 2020; Accepted: 19 July 2020; Published: 21 July 2020

Abstract: Photoacoustic imaging (PAI) combines optical contrast with ultrasound spatial resolution and can be obtained up to a depth of a few centimeters. Hand-held PAI systems using linear array usually operate in reflection mode using a dark-field illumination scheme, where the optical fiber output is attached to both sides of the elevation plane (short-axis) of the transducer. More recently, bright-field strategies where the optical illumination is coaxial with acoustic detection have been proposed to overcome some limitations of the standard dark-field approach. In this paper, a novel multiangle long-axis lateral illumination is proposed. Monte Carlo simulations were conducted to evaluate light delivery for three different illumination schemes: bright-field, standard dark-field, and long-axis lateral illumination. Long-axis lateral illumination showed remarkable improvement in light delivery for targets with a width smaller than the transducer lateral dimension. A prototype was developed to experimentally demonstrate the feasibility of the proposed approach. In this device, the fiber bundle terminal ends are attached to both sides of the transducer's long-axis and the illumination angle of each fiber bundle can be independently controlled. The final PA image is obtained by the coherent sum of subframes acquired using different angles. The prototype was experimentally evaluated by taking images from a phantom, a mouse abdomen, forearm, and index finger of a volunteer. The system provided light delivery enhancement taking advantage of the geometry of the target, achieving sufficient signal-to-noise ratio at clinically relevant depths.

Keywords: photoacoustic imaging; illumination scheme; in vivo; mouse; Monte Carlo; linear array

1. Introduction

Photoacoustic imaging (PAI) is a technique based on the photoacoustic (PA) effect, which consists of pressure waves generation due to the absorption of light [1–5]. Currently, laser-based PAI systems use short-duration laser pulses (i.e., ~10^{-9} s) ensuring thermal and stress confinement. As pulsed-light propagates within the target material, its absorption increases the local temperature, causing a thermal-elastic expansion [6] and generating a pressure wave. Thus, PAI encodes the optical absorption information into pressure waves, therefore combining optical contrast with ultrasound spatial resolution [4].

PAI can provide physiological and anatomical information of tissues by accessing their optical, thermal, and mechanical proprieties [5,7]. Since PA signal magnitude is temperature-dependent,

PAI has been used, for example, to map temperature variation within tissues during hyperthermia procedures [8–11]. Moreover, PA magnitude is proportional to the optical absorption of a chromophore; therefore, multi-wavelength PAI is capable of identifying structures with different optical absorption profiles [12,13]. In this context, a typical application of multispectral PAI is to estimate blood oxygen saturation (sO_2) from the relative concentrations of oxyhemoglobin (HbO_2) and deoxyhemoglobin (Hb) [14–19]. In addition, exogenous contrast agents, for example, nanoparticles and organic dyes, can be accessed to obtain molecular PAI and for drug delivery studies [20]. PAI is frequently combined with clinical ultrasound arrays. This approach allows simultaneously displaying conventional ultrasound images of different modalities (e.g., B-mode and Doppler) with PAI. Different preclinical and clinical, e.g., breast cancer [21] and joint arthritis [22,23], applications of PAI integrated with ultrasound scanners are under investigation.

Hand-held PAI systems usually operate in reflection mode (also known as epi-mode), where illumination and PA wave detection are arranged on the same side [24–28]. For linear array transducers, it is common to illuminate the tissue using a rectangular optical fiber output, which is attached to both sides of the elevation plane (short-axis) of the transducer, see Figure 1a (here this strategy will be referred to as standard dark-field illumination, following the terminology used in [29–32]). For this dark-field illumination scheme, when the transducer face is in contact with the skin, light is delivered obliquely and relies on light scattering within the target to illuminate the whole field-of-view (FOV) of the transducer. In addition, light absorption outside the imaging plane generates PA waves that can reach the ultrasound transducer. These signals are a source of clutter, which is also an important limiting factor to obtain PA images at deeper regions [33–35]. Different studies have investigated, for this standard dark-field illumination scheme, light delivery optimization to enhance PA image contrast and signal-to-noise ratio (SNR) by varying the distance between the fiber output and the transducer and incidence angle between the light beam and the imaging plane [30,33,36–38]. In [39], the authors verified, through Monte Carlo simulations, that the optimal illumination configuration depends on the optical properties of the tissues under investigation. They observed that thickness and optical scattering of skin play a major role for this optimization. Another possible strategy to optimize light delivery is to accommodate an optically transparent spacer between the transducer and target's surface to deliver light directly to the tissue underneath the transducer [25,27,28,32]. Improvements in light delivery could also be achieved by using a concave-shaped light catcher that redirects the light reflected by the skin surface back to the tissue, improving the PA signal magnitude at higher depths [40,41]. The aforementioned studies evaluated laser-based PAI systems. More recently, pulsed light-emitting diodes (LED)-based PAI technique has been proposed as an interesting and cost-efficient option [42,43]. LED-based PAI with linear array usually operates using a similar setup as shown in Figure 1a [42,44]. The study [44] suggested that the high divergence of LED illumination decreases the source direction dependency on PA signal compared to laser.

Since optimal light delivery to the tissue is essential to increase image depth and SNR, custom transducers, new materials, and new strategies have been developed to explore different illumination geometries to improve the quality of the PA image. For example, an ultrasonic transducer fabricated on a glass substrate has an improved transparency allowing the laser beam to propagate through the transducer's material with low absorption, resulting in overlapped optical excitation and acoustic detection [45]. An ultrasound transducer with a hollow central bore [46] or an optically transparent acoustic transducer [47] could be other options to provide reflection mode illumination. However, these approaches require an extensive redesign of the ultrasound probe and cannot be easily integrated into standard clinical scanners. In epi-mode PAI using standard linear array transducers, optical and acoustic fields can be coaxially arranged (see Figure 1b) by redirecting the laser beam using an optical/acoustic coupler [48] or by using a single or double acoustic reflector to redirect the acoustic waves [29,31,49]. Another strategy for coaxial illumination consists of a custom linear array transducer where the optical fiber outputs and piezoelectric elements are linearly and alternately arranged [50]. These studies [29,31,48–50] showed that this illumination strategy improved light delivery when compared with the standard dark-field approach. In the present paper, this strategy will be referred to as bright-field illumination, following the terminology used in [30–32,49].

An alternative illumination approach, not yet investigated in the literature, would consist of attaching the fiber bundle terminal ends to both sides of the transducer's long-axis (from now on we will refer to this technique as long-axis lateral illumination), see Figure 1c. In the present paper, we propose a long-axis lateral illumination scheme as a new epi-mode PAI strategy, where the light is delivered within the imaging plane similarly to the coaxial arrangement. In the first part of the paper, Monte Carlo simulations of photon propagation were used to compare light delivery for different illumination strategies; i.e., standard dark-field, long-axis lateral, and bright-field illumination. A transparent spacer positioned between the transducer and the tissue surface was considered for all cases. Tissues with three different geometries were simulated; i.e., targets larger and smaller than the lateral dimension of the imaging plane simulating the human forearm and index finger, respectively and an intermediate situation simulating the cross section of a mouse torso where the abdomen was smaller than the image width and the lower limbs fitted the transducer FOV. The simulations demonstrate that the lateral illumination strategy can provide remarkably improved fluence distribution for targets smaller than the imaging plane.

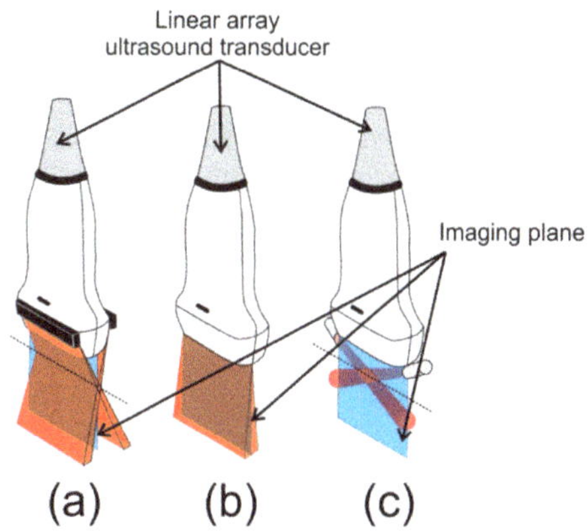

Figure 1. (**a**) Standard dark-field illumination scheme to acquire photoacoustic (PA) image in reflection mode; a rectangular optical fiber terminal illuminates the surface of the target. (**b**) Bright-field illumination where the laser beam and the acoustic field are coaxially aligned. (**c**) Proposed long-axis lateral illumination architecture; the variation of light incidence angle provides wide illumination to the surface of the target.

In the second part of this paper, the development of a simple and easy way to construct a device for long-axis lateral illumination PAI is described. This device employed a nonexpensive commercially available bifurcated optical fiber bundle for light delivery, where no other optical components were required. Since the setup, as shown in Figure 1c, would irradiate only a limited area within FOV, the optical fiber bundle outputs were mounted on movable sockets arranged parallel to the imaging plane to provide multiangle long-axis lateral illumination. The final PA image, covering the full scan area, is then obtained by combining the PA sub-images acquired at different angles. This is a similar strategy as described in [51], where a narrow laser beam scanning approach was proposed for a combined real-time PA-ultrasound imaging system. Then the final PA image was the summation of the sub-images obtained at each laser beam scanning position.

Therefore, this paper presents a novel PAI light delivery where light and sound are coaxially illuminated. Different from other approaches with similar capability [29,31,48,49], the proposed technique does not require an acoustic/optical coupling device. This is an advantage because these coupling modules usually induce important phase distortion to the PA wavefront which can reduce image quality [32]. The light delivery device was prototyped to provide freedom to independently choose the illumination angle, at each side of the transducer, allowing multiangle illumination planning. To show the feasibility of the multiangle long-axis lateral illumination PAI device, images taken from phantom, mouse abdomen, forearm, and index finger of a volunteer were analyzed.

2. Materials and Methods

2.1. Monte Carlo Simulation of Illumination Schemes

The spatial energy deposition may vary depending on the illumination scheme and target shape. To evaluate the performance of the illumination schemes depicted in Figure 1, light transport was simulated using the MCXLAB Matlab toolbox, which is a 3D voxel-based Monte Carlo model [52], for three different target geometries: (i) a cylindrical target shape simulating a situation similar to what was observed for the human index finger (lateral dimension smaller than transducer's width); (ii) geometry similar to the human forearm (lateral dimension larger than transducer's width); (iii) mouse torso as an intermediate case, i.e., part of the target was smaller (mouse abdomen), while the lower limbs of the animal was larger than the transducer's width.

The standard dark-field illumination scheme shown in Figure 1a, based on the setup described in [53], was composed of two optical fiber terminals (38 mm × 1.25 mm) with the same width as the ultrasound linear array used in the experiments of the present paper. Each terminal was positioned so that the light beam incident angle was 20° and the light beams overlapped at the upper surface of the target. For all three illumination schemes, an optically transparent spacer of 19.5 mm (i.e., water) was positioned between the transducer and the target. For the bright-field illumination scheme shown in Figure 1b, the laser beam was coaxial with acoustic detection. In this case, the illumination dimension hitting the target was 38 mm × 5 mm, which is in accordance with [48]. For the long-axis lateral illumination, Figure 1c, the fiber optic bundle terminals were circular in shape with 5 mm diameter. To illuminate the entire transducer FOV, the same multiangle illumination strategy used for the experiments (see next sections for a detailed description) were adopted in the simulations. We verified that at least 5 laser beam incident angles were necessary to ensure a complete illumination. In this case, all simulation parameters were the same as the experimental setup. A total of (5.0×10^6) photons were used to simulate each situation.

The volume dimension for all simulations was 89 mm × 60 mm × 30 mm with a voxel size of 0.25 mm. The volume consisted of two different materials, the background (water) and an inclusion (target) to simulate the tissue. The optical proprieties of the background were: absorption coefficient $\mu_a^{bkg} = 3.5640 \times 10^{-5}$ mm^{-1}, scattering coefficient $\mu_s^{bkg} = 1.0$ mm^{-1}, $g_{bkg} = 1$, and $\eta_{bkg} = 1.37$; where g denotes the anisotropic factor and η denotes the refraction index. For the target, the optical scattering and optical absorption coefficients were chosen for a generic tissue, following the equations [54]:

$$\mu'_s = a(\lambda/500nm)^{-b}, \quad (1)$$

$$\mu_a = BO\mu_a^{HbO_2} + B(1-O)\mu_a^{Hb} + W\mu_a^{water} + F_a\mu_a^{lipid} \quad (2)$$

where μ'_s denotes the reduced scattering coefficient, μ_a denotes the absorption coefficient, B denotes the average blood volume fraction, O is the oxygen saturation of blood, W is the water content, and F_a is the fat content. Selecting a generic tissue composed of 15% of blood at 75% of oxygen saturation, 20% of water, and 10% of fat results in $\mu_a^{tissue} = 0.062$ mm^{-1} at 800 nm. The a and b values were chosen as the mean values estimated for soft tissues ($a = 1.89$ mm^{-1}, $b = 1.286$) [54] resulting in $\mu\prime_s^{tissue} = 1.033$ mm^{-1} at 800 nm. The anisotropic factor (g_{tissue}) and the refraction index (η_{tissue}) were 0.95 and 1.37, respectively.

2.2. Device for Multiangle Long-Axis Lateral Illumination

The PA system is composed by an Nd:YAG Laser (Brilliant B, Quantel Laser, Les Ulis, France) and an Optical Parametric Oscillator (MagicPRISM OPO, Opotek, Carlsbad, CA, USA) connected to a trifurcated optical fiber bundle (Oriel Glass Fiber Optic Bundle; numerical aperture 0.56; core diameter: 7.9 mm (common), 5.5 mm (legs); fiber length 36 in; Newport, Irvine, CA, USA). One terminal end of the optical fiber bundle was connected to the sensor of an energy meter (FieldMax II-TOP, Coherent, Santa Clara, CA, USA) providing the measurement of laser fluence in real-time. The other two terminals were used to illuminate the sample. PA and ultrasonic radiofrequency (RF) data were acquired using a commercial ultrasound system (SonixOP, Ultrasonix Medical Corp., Richmond, BC, Canada) connected to a parallel acquisition receiver module (SonixDAQ, Ultrasonix Medical Corp., Richmond, Canada), operating at a sampling frequency of 40 MHz.

The device for multiangle long-axis lateral illumination was developed using a new architecture that differs from most used configurations presented in previous studies [9,23–26,33–38]. The optical fiber bundle terminal ends were attached to movable sockets placed on the lateral sides of the transducer's long-axis (see Figure 1c), so that different focal illumination spots within the transducer FOV could be obtained by varying the incident laser beam angles (see Figures 2 and 3).

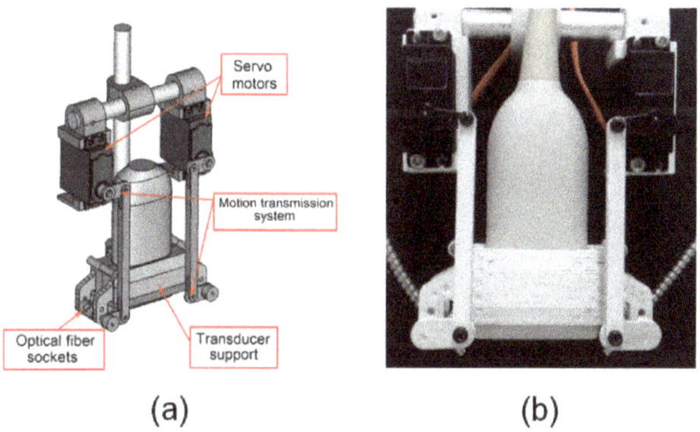

Figure 2. (**a**) Three-dimensional model and (**b**) prototype of the multiangle long-axis lateral illumination device attached to a linear array ultrasound transducer.

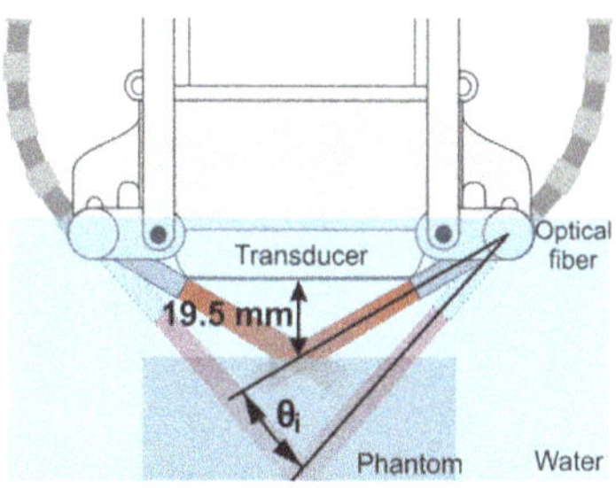

Figure 3. Depiction of the experimental setup used to acquire the multiangle long-axis lateral illumination PA images. The distance of 19.5 mm between the transducer and the phantom surface forces the focal illumination region to be at the phantom surface for $\theta_i = 0°$.

The device consists of three main parts: the ultrasound transducer support, the motion transmission system, and two motors. The support and the motion transmission system were designed using the FreeCAD open-source parametric modeling software, see Figure 2a, and printed with Acrylonitrile Butadiene Styrene (ABS) plastic using a 3D printer (ZMorph 2.0 SX, ZMorph, Wroclaw, Poland), as shown in Figure 2b. The angle of the movable sockets is controlled by the servo motors (MG996R, Tower Pro, Shenzhen, China) connected to the motion transmission systems and controlled by an open-source microcontroller (Arduino UNO, Arduino, Turin, Italy).

A linear L14-5/38 ultrasound transducer (Ultrasonix Medical Corp., Richmond, Canada) with 128 piezoelectric elements and a nominal center frequency of 7.2 MHz, was positioned inside the support, which was then attached to the 3D linear stage (HSC-103, Sigmakoki, Tokyo, Japan). A LabVIEW virtual interface (National Instruments Corp., Austin, TX) was developed to control the position of the device as well as the illumination angles. The timing sequence of the synchronous RF data acquisition and multiangle illumination consists in acquiring a pair of PA and B-mode images for each laser pulse (laser repetition rate is 10 Hz). After K laser pulses ($K{\cdot}100$ ms) the illumination angle is incremented, the process is then repeated for n illumination angles. A 3D volume is obtained by moving the transducer along the elevation axis with a 3-axes translational stage.

2.3. Coherently Summing the PA Subframes

The RF data were acquired intercalating the laser pulse with the pulse-echo transmission for obtaining a prebeamformed PA subframe and then a prebeamformed B-mode frame. The PA sub-image and B-mode image were generated with the delay and sum technique. In PA images, the RF signal $s(x_i, t)$ represents the pressure waves generated by the light absorbers and detected by the i-th transducer element. The PA wave time of flight from the absorber position to each element of the array is

$$\delta(x, x_i, y) = \sqrt{y^2 + (x - x_i)^2}/c \tag{3}$$

where x and y are the lateral and axial position of the pressure wave source, respectively, while x_i denotes the lateral distance of the i-th element of the array to the central element. The delay and sum technique for PA subframe reconstruction consists in applying a delay $\delta(x, x_i, y)$ to the RF signal $s(x_i, t)$ detected by the elements of the transducer and adding coherently [51]

$$S(x, y) = \sum_{x-\alpha}^{x+\alpha} s(x_i, \delta(x, x_i, y)) \tag{4}$$

where α is the aperture of receive beamforming, i.e., the number of adjacent elements summed.

The coherent sum of reconstructed PA subframes acquired using each illumination angle, without other processing steps, is equivalent to a reconstructed PA image acquired using a wide illumination due to the linear behavior of the delay and sum operation (Huygens–Fresnel principle) [51]. Thus, PA signals reconstructed using the delay and sum technique can be coherently added to gather the contribution of each illumination angle as:

$$S_C(x, y) = \sum_{\theta_0}^{\theta_n} S_{\theta_i}(x, y), \tag{5}$$

where $S_c(x, y)$ is the reconstructed RF signal coherently summed (PA final image), $S_{\theta_i}(x, y)$ is the reconstructed RF signal acquired at the i-th illumination angle (PA subframe), θ_n is the maximum illumination angle.

2.4. Phantom Experiments: Evaluation of Multiangle Long-Axis Lateral Illumination PAI

A cubic phantom with a homogeneous distribution of light absorbers (magnetic nanoparticles) was used to evaluate the multiangle illumination and the PA images. The phantom dimensions were 8.0 cm × 8.0 cm × 3.5 cm, and it was manufactured using a mixture of gelatin (Bloom 250, Gelita, Eberbach, Germany) and agar powder (RM026; Himedia Laboratories-LLC, Kennett Square, USA), diluted at dry-weight concentrations of 4% and 2% of water mass, respectively. Iron oxide nanoparticles (Fe_3O_4) with dimensions ranging from 20 nm to 30 nm (Nanostructured and Amorphous Materials Inc., Houston, TX, USA) in the concentration of 0.1% of water mass were added to act as light absorbers.

Formaldehyde in a weight concentration of 0.5% of gelatin mass was added to increase stiffness and melting temperature. The phantom was manufactured according to the description in [55,56].

To avoid any coupling issues, the experiments were performed using targets immersed in water to guarantee that the gap between the ultrasound transducer and the target was completely filled by an optically transparent coupling medium. However, we believe it would be possible to acquire images using a matching layer. For example, this layer could be ultrasound imaging gel (see, for example, [57]) or a gel pad (see, for example, [58]). This is a topic of ongoing research and should appear in future publications.

The phantom was immersed in a water tank with its surface 19.5 mm from the transducer face. For this condition, the focal illumination region was at the phantom surface for $\theta_i = 0^\circ$, considered as the smallest possible angle (see Figure 3). Then, for each illumination angle, two PA subframes were acquired and the angle was varied n times in steps of $\Delta\theta$ until the n-th angle was achieved

$$\theta_n = \theta_{min} + n\Delta\theta. \tag{6}$$

The device moved across the elevation axis to obtain a volume (Table 1), resulting in a total of 360 frames. The PA images were acquired at 720 nm with an average fluence of 15 mJ/cm^2 at the phantom surface. This wavelength was selected for the phantom experiment because one of the energy peaks of the laser is observed at 720 nm; in addition, iron oxide nanoparticles present higher optical absorption at lower wavelengths within the near infrared spectrum region [59]. For the in vivo experiments, 800 nm was selected because it is the isosbestic point of blood [60].

Table 1. Multiangle illumination and acquisition parameters.

Parameters	Phantom	Finger and Forearm	Balb/C Mouse
Angle step	2°	4°	4°
Elevation step	2 mm	2 mm	2 mm
Image axial	45 mm	35 mm	35 mm
Image lateral	38 mm	38 mm	38 mm
Number of angle steps (n)	9	4	4
Number of elevation steps	19	9	5
Number of frames per angle (K)	2	2	2
Wavelength	720 nm	800 nm	800 nm

The phantom was assumed to have a homogeneous distribution of light absorbers; the optical attenuation coefficient was calculated measuring the fluence of transmitted light through the layers of the phantom with different thickness. The estimated light attenuation of the phantom was $\mu^{att}_{phantom} = (0.133 \pm 0.011)$ mm^{-1}. Therefore, an analysis of the light delivery was performed by evaluating the PA signal as a function of axial and lateral directions of an averaged PA image taken over the elevation axis. Since the magnetic nanoparticles at low concentration, which is the case of the present experiment, mainly absorb the light energy, optical scattering was considered negligible for this analysis [59].

We defined image depth as the axial distance between the position of a RF signal inside the target and the target surface, therefore not considering the distance between the transducer face and target surface. The average RF signal of the PA subframes at depths and in lateral direction were evaluated using distinct regions of interest (ROI). Based on the number of elements of the transducer (i.e., 128 elements), we defined a central ROI-1 within FOV, which included the five central elements (62–66) and extended from the phantom surface to the maximum depth, with dimensions 1.5 mm × 25.5 mm. Also, peripheral ROIs (ROI-2 and ROI-3) were defined including two sets of five elements positioned at opposite sides of transducer elements: 5–9 (ROI-2) and 119–123 (ROI-3). ROI-2 and ROI-3 had the same dimensions as ROI-1. In addition, the average PA signal magnitude, in the lateral direction from 0 mm to 2 mm of depth, was calculated for all elements (1–128).

Since the phantom had a homogenous distribution of light absorbers, the PA signal amplitude was related to the amount of light delivered. The quantitative analysis of illumination, in the ROI-1 region, as a function of illumination angle was performed using the mean square root of the RF signal amplitude (A_{RMS}) calculated in each PA subframe:

$$A_{RMS}(\theta) = \sqrt{\frac{\sum_{j=x_a}^{x_b} \sum_{i=y_a}^{y_b} \left(S_\theta\left(x_j, y_i\right)\right)^2}{(y_b - y_a)(x_b - x_a)}} \tag{7}$$

where x_a, x_b, y_a, and y_b are the limits of the ROI-1.

The spatial light delivery information, in the central region of the final PA image, was estimated taking the mean axial position of the RF signal ($\overline{y}$) as a function of illumination angle in ROI-1

$$\overline{y}(\theta) = \frac{\sum_{j=x_a}^{x_b} \sum_{i=y_a}^{y_b} \left(\left|S_\theta\left(x_j, y_i\right)\right| y_i\right)}{\sum_{j=x_a}^{x_b} \sum_{i=y_a}^{y_b} \left(\left|S_\theta\left(x_j, y_i\right)\right|\right)} \tag{8}$$

Thus, A_{RMS} provides information about the mean amount of light delivered per illumination angle in the central region of the transducer while $\overline{y}$ provides spatial information about the mean axial position of generated pressure waves.

The final multiangle PA image SNR was calculated taking the envelope-detected image amplitude [61]:

$$\text{SNR} = \sum_i \sum_j \left[S_H\left(x_i, y_j\right) - \overline{S}_{nH}\right] / \sigma_{nH} \tag{9}$$

where S_H is the Hilbert transform modulus of the RF signal ($S_H = |H\{S_C\}|$), $\overline{S}_{nH}$ and σ_{nH} are the average and standard deviation of the background noise in S_H, respectively.

2.5. In Vivo Experiments: Human (Finger and Forearm) and Animal (Balb/C Mouse)

The index finger of the left hand and left anterior forearm of a human volunteer were photoacoustically imaged using multiangle PAI with illumination parameters according to Table 1. Figure 4a shows photographs of the index finger and forearm where the dashed lines indicate the position and orientation of the transducer. The volunteer immersed his hand and forearm in a water tank; the distance between the transducer face and the skin was chosen so that the laser beams were focused at the skin surface when illumination angle was minimum ($\theta_i = 0°$). The experiments were performed using an average fluence of 9.0 mJ/cm^2 at 800 nm, obtaining a total of 90 frames.

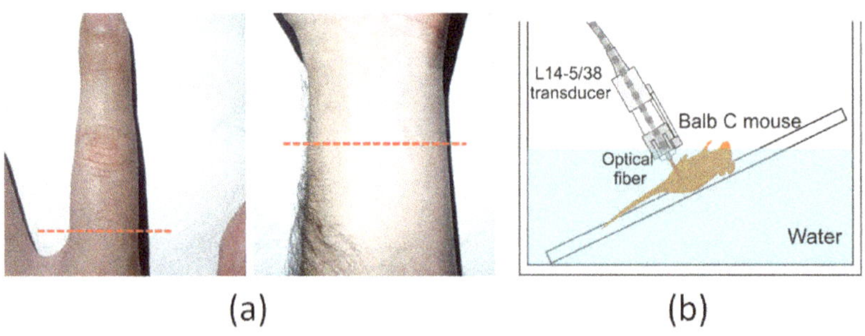

Figure 4. (**a**) Photographs of the index finger and forearm of the volunteer. The dashed lines indicate the position and orientation of the transducer. (**b**) Depiction of the experimental setup used to acquire in vivo PA images of Balb/C mouse.

A male Balb/C mouse, at the age of ten weeks, was anesthetized using vaporized isoflurane (1.0–1.5% isoflurane, Vetflurano, Virbac, São Paulo, Brazil). The animal was positioned in a ramp platform immersed in water with a controlled temperature of 36 °C, as shown in Figure 4b. Using the same wavelength and fluence of the human experiment, PA images of the animal abdomen were acquired.

The experiments involving humans and animals used controlled fluence lower than 20.0 mJ/cm^2, considering the limit for short-pulse lasers at the skin, which is defined by the American National Standards Institute (ANSI). The animal procedures were approved by the Animal Ethical Committee of Ribeirão Preto Medical School, University of São Paulo (process No. 005/2017-1). The experiments with the volunteer were conducted according to the procedure approved by the Research Ethical Committee of Faculty of Philosophy, Science, and Letters of Ribeirão Preto, University of São Paulo (CAAE: 08860819.4.0000.5407).

3. Results and Discussion

3.1. Comparison of Illumination Schemes for Different Target Shapes Using Monte Carlo simulation

Monte Carlo simulations were conducted to analyze the influence of the target shape and illumination scheme on light delivery. Figure 5 shows the normalized fluence maps obtained for all illumination strategies. Figure 5a–c,e–g show the results for the cases where the target is larger and smaller than the image width, respectively. Figure 5i–k show an intermediate situation representing a mouse torso. In this case, the geometry was obtained by segmenting an experimental B-mode image which will be shown in the next sections. Figure 5a,e,i show the results obtained for multiangle long-axis lateral illumination, while in Figure 5b,f,j, the results for the bright-field illumination coaxial with acoustic detection are shown. Finally, Figure 5c,g,k show the results obtained using the standard dark-field illumination scheme. In these images, ROIs were used to compare the light fluence for different spatial locations (white square is ROI-A; black square is ROI-B and magenta square is ROI-C). Figure 5d,h,l show bar graphs comparing the average fluence estimated within ROIs A, B, and C.

In the central region, all illumination schemes presented similar relative fluence at a shallow depth (ROI-A); yet, to some extent it was consistently higher for the bright-field illumination independent of the target shape. For targets with a nonflat surface, the focus region of the two laser beams used in the standard dark-field arrangement can be partially outside of the material and imaging plane as can be seen on the left side of Figure 6. The light delivered outside of the imaging plane contributes less to the PA image generation and can be a clutter source [62]. In this case, both situations, bright-field and long-axis lateral illumination, have the advantage of delivering light inside the imaging plane even when the target's surface is not flat. For this reason, the average fluence within ROIs B and C, located at higher depths, was consistently lower for the standard dark-field illumination scenario. It is important to recall that the light beams used for the standard dark-field were overlapping at the tissue surface. Other studies have shown that light delivery can be increased at higher depths, by using a deeper located illumination focus [30,37,38,53]. However, this strategy can dramatically decrease light delivery at shallow depths. For example, the study [48] showed that by positioning the focus at 13.5 mm depth, the simulated fluence estimated for the standard dark-field optical illumination was considerably lower than what was estimated for the bright-field illumination scheme at depths lower than 10 mm.

For targets smaller than the images' lateral dimension, the long-axis lateral illumination can deliver light to the sides of the target, increasing the penetration of light inside the material, which is depicted at the right side of Figure 6 and can be observed in the Monte Carlo simulation results. Moreover, the long-axis lateral illumination redirects the light to the target while part of the bright-field and standard dark-field illumination schemes do not contribute to PA signal generation. The simulations show that the relative fluence obtained with multiangle long-axis lateral illumination was dramatically improved for the case of a cylindrical geometry with a diameter smaller than the width of the ultrasound probe, which is a similar situation as the human finger as will be described in the next sections. Figure 5h shows that the average fluence for the long-axis lateral illumination scheme, measured within ROI-C, was four times higher than fluence delivered by the bright-field illumination and one order of magnitude higher compared to the dark-field illumination scheme.

For a target that combines parts smaller and parts larger than image width, as the mouse's torso, both long-axis lateral and the bright-field illumination schemes provided a relatively uniform light delivery to the entire target surface. On the other hand, the light delivered by the standard dark-field illumination was considerably higher at the top surface. For this situation, the long-axis lateral illumination provided a little increment of fluence within ROI-C compared to the bright-field illumination scheme.

The next two sections aim to evaluate the feasibility of generating PA images using the long-axis lateral illumination scheme. First, the device and the multiangle imaging strategy are evaluated with a phantom experiment; then the possibility of generating the PA images, in vivo, of targets with similar geometries adopted for the simulations are verified.

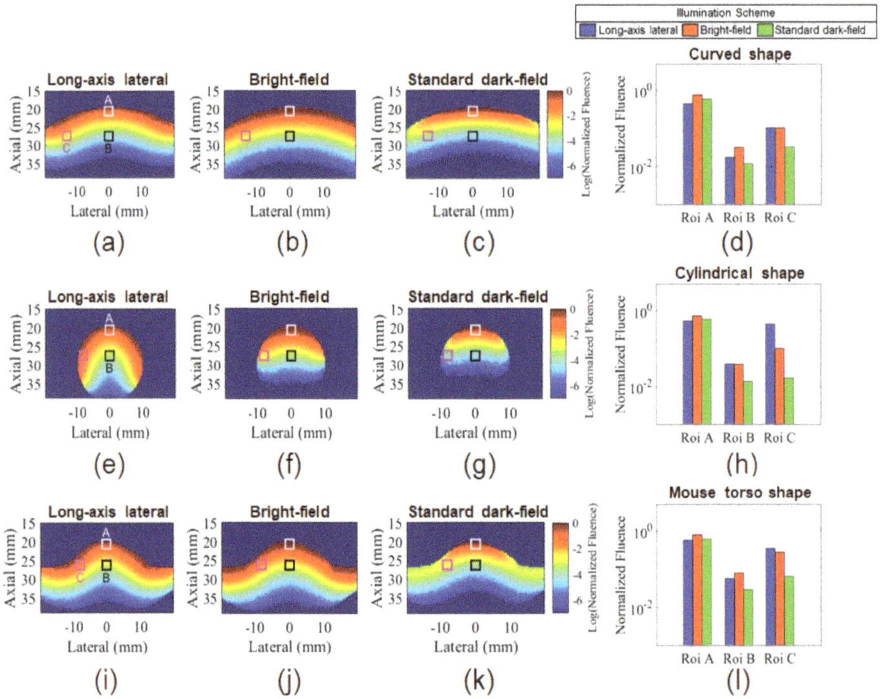

Figure 5. Normalized fluence maps obtained by Monte Carlo simulation for targets larger (**a**–**c**) and smaller (**e**–**g**) than the image width. An intermediate situation representing a mouse torso was also considered (**i**–**k**). All cases were simulated for the bright-field, standard dark-field, and long-axis lateral illumination schemes. Average fluence values were estimated within regions of interest (ROIs) A, B, and C for all cases (**d**,**h**,**l**).

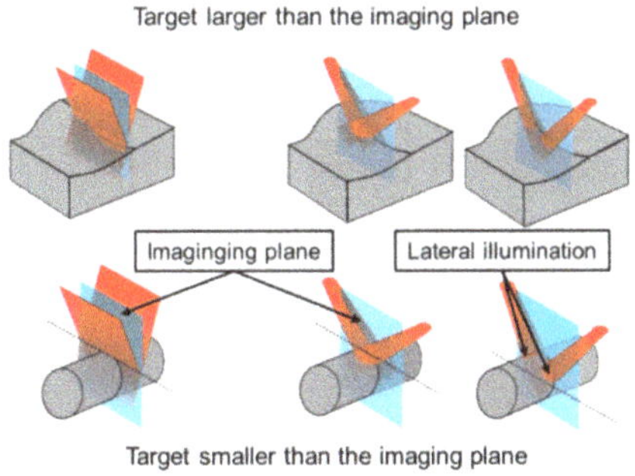

Figure 6. Comparison between the standard dark-field illumination scheme and the multiangle long-axis lateral illumination. For targets larger than the image width with a nonflat surface, the long-axis lateral and the bright-field illumination schemes have the advantage of delivering light within the imaging plane. For targets smaller than the transducer width, the long-axis lateral illumination scheme can deliver light to the sides of the target.

3.2. Analysis of Illumination Angles Contribution to the PA Image of the Phantom

The homogeneous phantom with a flat surface is useful for the characterization of light delivery using different illumination angles. To evaluate the light delivered to the phantom, each PA subframe at θ_i is represented as the average of the 19 PA subframes acquired at different positions of elevation axis (slice) using the same i-th illumination angle. The averaged PA subframes in Figure 7 shows the light propagation along depth. The blue arrows indicate the PA signal generated beyond the laser focal region for the illumination angles 0°, 2°, and 4°. This observation can be understood as an advantage of providing illumination from the laterals of the transducer, therefore generating PA signals within FOV for regions not only at the focus. However, the amplitude of the PA signal is a function of the illumination angle, because the laser focus region moves towards higher depth, while the light path increases, reducing the fluence due to light attenuation.

PA signal magnitude increased for depths greater than 10 mm and illumination angles higher than 8°, as it can be seen in Figure 8a. Although illumination along the peripheral areas of the transducer is mostly achieved by just one of the optical fiber outputs, the incident angle of the laser beam in this region decreased relative to the normal surface, delivering light at higher depths, as shown in Figure 8a,b. Moreover, the average PA signal showed the separation of laser beams along the lateral direction; see Figure 8c.

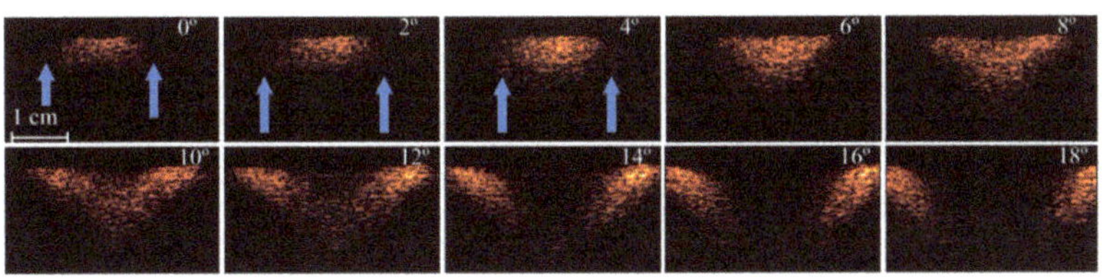

Figure 7. PA subframes of the homogeneous phantom for increasing illumination angles in the range 0°–18°. Each subframe is an average of the phantom's elevational dimension (i.e., 3.8 cm). Blue arrows indicate the generation of PA signals beyond the laser focal region.

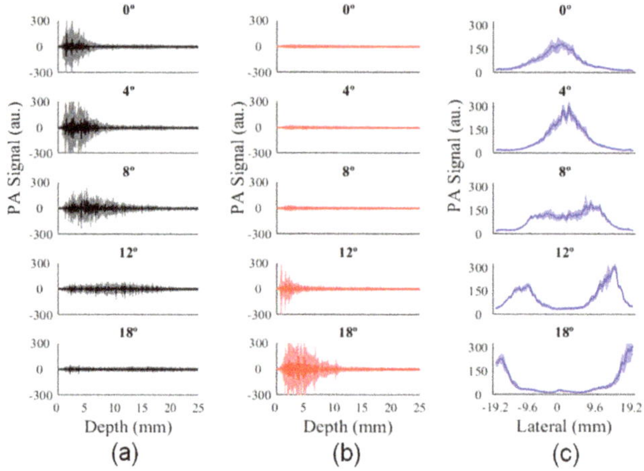

Figure 8. Average PA signal as a function of the illumination angle, along the axial direction at (**a**) central (ROI-1) and (**b**) peripheral ROIs (ROI-2 and ROI-3); (**c**) average PA signal magnitude along lateral direction for depths ranging from 0 mm to 2 mm.

The analysis of the A_{RMS} at the central region revealed a proportional decrease in the amount of light delivered for angles higher than 4°, which is probably related to light attenuation within the phantom (Figure 9a). Besides, a peak of maximum A_{RMS} could be observed for θ_i = 4°, showing that the maximum light delivery to the central area occurred when the laser focus region was completely inside the phantom, where the light was less attenuated (shallow depths). These results show the contribution of illumination using θ_i < 4° was less significant for the image of the phantom. In addition, the mean depth of the PA signal increased as a function of the illumination angle (Figure 9b), which could be qualitatively inferred from the plots in Figure 8a.

The final PA image was obtained from the summation of the PA subframes at different illumination angles. For example, a PA image of a single image slice of the phantom is shown in Figure 10a. The average PA signal calculated at depths ranging from 0 mm to 25.5 mm for all elements of this PA image showed the contribution of all illumination angles; see Figure 10b. Figure 10c shows SNR as a function of depth demonstrating the multiangle long-axis illumination PA imaging could be used to perform studies at relevant imaging depths.

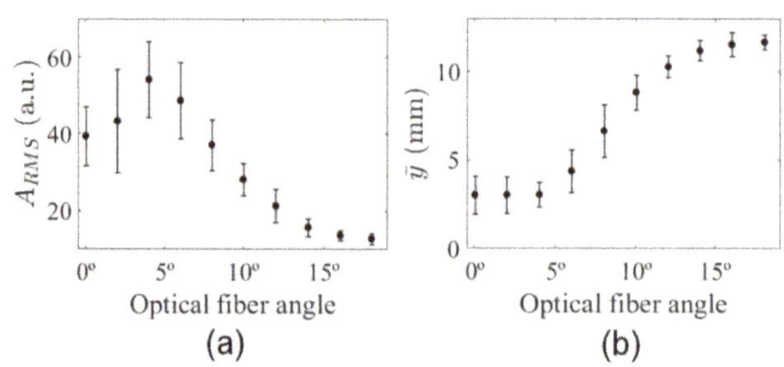

Figure 9. (**a**) Mean square root of PA signal and (**b**) mean depth of PA signal as a function of illumination angle.

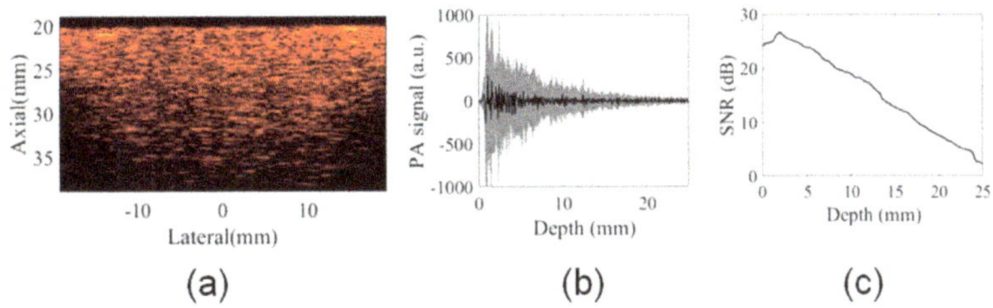

Figure 10. (**a**) Final PA image of a single image slice of the phantom, (**b**) PA signal profile of this PA image shows the contribution of all illumination angles to generate PA signal at depths greater than 10 mm, and (**c**) SNR as a function of image depth.

3.3. In Vivo PA Images

Different in vivo experiments were conducted to evaluate the multiangle long-axis lateral illumination PAI. The first in vivo PA images were acquired from a human forearm. In this case, the shape of the surface was larger than the long-axis dimension of the transducer, providing similarities with the flat surface of the phantom and the simulation study. The anatomical structures of the human forearm such as palmaris longus tendon, subcutaneous blood vessels, and epithelial tissue could be identified in the B-mode image as well as in the PA image, as it can be seen in Figure 11a,b. SNR was calculated within two ROIs of 2.4 mm × 1.6 mm in the axial and lateral dimensions, respectively. For the in vivo evaluation, noise ROIs, with the same dimensions, were positioned within 5 mm distance from the structure under analysis. At a tissue depth of 7.3 mm, ROI overlaid the tendon of flexor digitorum superficialis [63] and provided SNR = 14 dB (green rectangle in Figure 11a), while another ROI overlaid a subcutaneous blood vessel and provided SNR = 25 dB at a 2.5 mm tissue depth (cyan rectangle in Figure 11a). Differences in the SNR values between deep and shallow regions were mostly due to light attenuation; the tendon (collagen) also presents an optical absorption coefficient lower than those for melanin or hemoglobin at 800 nm [64], which reduced its SNR on the PA image.

The second acquisition of PA images was obtained from the human index finger, which has a cylindrical-like shape that differs from the forearm or phantom's surface and has a maximum lateral extension of approximately 20 mm. However, the finger cross-section lateral dimension is smaller than the lateral FOV, allowing illumination from laterals to be more efficient. Consequently, the laser focus becomes deeper while the light path within the tissue is shortened, reducing the light attenuation. SNR was analyzed within ROIs of 2.0 mm × 3.0 mm in the axial and lateral dimensions, respectively. Those ROIs were placed over the location of the dorsal and palmar digital arteries at 3.4 mm (cyan rectangle in Figure 11c) and 10.3 mm tissue depths (green rectangle in Figure 11c), resulting in SNR of 22.5 dB and 22 dB, respectively.

Lastly, a challenging combination of both aforementioned surface shapes was observed when acquiring the PA image of the cross-section of the mouse abdomen. The abdomen had a lateral dimension smaller than the lateral length of the transducer, while the lower limbs fit FOV. In this case, illumination angles provided light delivery to the sides of the mouse abdomen and hit the surface of lower limbs obliquely. Furthermore, the mouse skin has an average thickness of 0.5 mm and optical absorption lower than human skin [65], increasing light penetration. In the B-mode image of the mouse, the bladder, femoral artery, and a branch of the abdominal aorta can be identified [66]. PA signals from the femoral artery were evaluated at 1.3 mm (cyan rectangle in Figure 11e) and the aortic branch at 10.5 mm of depth (green rectangle in Figure 11e) presenting SNR = 21.5 dB for the femoral artery and SNR = 17.5 dB for the aortic branch (ROIs of 2.0 mm × 2.8 mm).

For three among the most studied PA targets in biomedical applications, the multiangle long-axis lateral illumination was able to provide PA images with high SNR at a depth of 10.5 mm for target shapes with lateral extension smaller than the lateral length of transducer (mouse and index finger). In the human forearm, a tendon at depth of 7 mm generated PA signal with sufficient SNR for good visualization of the structure. However, it should be mentioned that the in vivo PA images were acquired using a measured laser fluence of 9 mJ/cm^2, which is much lower than the limit for short-pulse laser at the skin. Increasing the laser fluence to values close to the safety limit can improve SNR of PA images at greater depths.

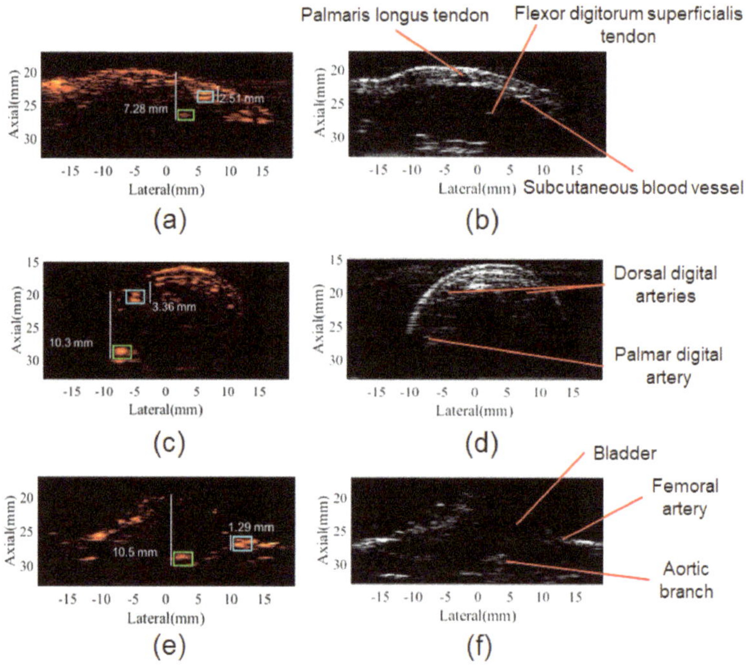

Figure 11. In vivo multiangle long-axis lateral illumination PA and B-mode images. (**a**) PA image and (**b**) B-mode image of the human forearm, anatomical structures such as tendons and subcutaneous blood vessels can be identified; (**c**) PA image and (**d**) B-mode image of the human index finger, the cylindrical shape allows light delivery by the laterals promoting the visualization of the palmar digital artery at depth of 10.3 mm. (**e**) PA image and (**f**) B-mode image of the Balb/Cmouse abdomen, PA signal from an aortic branch at 10.5 mm of depth can be visualized.

Figure S1 shows the subframes acquired at each illumination angle for the in vivo experiments and plots with the corresponding SNR for ROIs positioned at the selected structures (vessels and tendon). The pronounced variation in the SNR values across the subframes acquired at different angles is evident. Clearly, SNR, at each ROI, observed for the final PA image is similar (only slightly higher) to that subframe with the highest SNR where the illumination area comprised the structure of interest. In the case of combining N PA images at the same illumination angle, it is expected that the SNR will be increased by $\sqrt{N}$, which would be higher than SNR obtained with the multiangle approach for a particular ROI. A more concentrated illumination in the proposed approach, when compared to the techniques illustrated in Figure 1a,b, can also help improve SNR at specific locations of the image. This concept has been also explored in other studies [51,53].

Multiangle long-axis lateral illumination could be a useful approach to improve the quality of PA images in preclinical studies with mice, since important anatomic structures are smaller than the FOV. Even for higher frequency linear arrays with a width smaller than the probe used here, murine tumor models still fit in this category [53]. In [53], the authors demonstrated that improving light delivery for this situation can greatly improve PA image quality. Since PA images taken from the human finger joint has shown great potential to evaluate inflammatory arthritis [22,23], it is a possible clinical application where the proposed technique could be beneficial. Future studies will include: (i) evaluating multi-wavelength PA images and fluence correction strategies [57] to monitor blood oxygen saturation and (ii) evaluate the feasibility of using a high divergent source like LED [44] to reduce the number of illumination angles.

Since the pulse repetition frequency of the laser (L_{PRF}) is 10 Hz, the in vivo images of the present study were acquired at a frame rate of 1 frame per second. Each image was composed of two subframes per angle and five different angles were used. The maximum frame rate for this configuration is two frames per second if a single frame was acquired per angle. To acquire the full transducer FOV, a few subframes acquired at different angles are needed; therefore, the frame rate for the proposed technique will be lower than that obtained for the configurations shown in Figure 1a,b. This is a limitation of the proposed approach, especially for the cases where the illumination source operates at low PRF as the laser used in the present study (L_{PRF} = 10 Hz) and monitor fast-changing dynamics is the goal. However, for lasers working at higher repetition rate (L_{PRF} ~ 100 Hz), see, for example, [67]) this limitation can be minimized and the frame rate can be increased. The maximum L_{PRF} supported by the setup depends of the angular velocity of servomotors (ω_s), and the step angle ($\Delta\theta$): $L_{PRF} = \omega_s/\Delta\theta$. The servomotors used in the setup can take 0.2 s to rotate the fibers output from 0° to 60°, resulting in a maximum angular velocity of $\omega_s = 5.2\ rad{\cdot}s^{-1}$ when operating at 5 V. In this case, the setup could acquire PA subframes using = 75 Hz with $\Delta\theta = 4°$ and provide 15 frames per second when five illumination angles are used.

The present paper introduced and evaluated the feasibility of the multiangle long-axis lateral illumination to generate PA images, contributing to the development of new illumination strategies in PAI. An advantage of the setup presented here is that it could be additive to other existing illumination schemes. Light delivery by the laterals of the target could be used together with more conventionally used illumination schemes to improve light delivery to targets with lateral dimension smaller than the transducer's width. The concept of PA images acquired using multiangle illumination can also be applied to setups with similar design as described in [37,38], where the angle of incident light delivered by transducer's short-axis can be controlled.

4. Conclusions

This paper demonstrated the feasibility of using a novel multiangle long-axis lateral illumination PAI. Monte Carlo simulations compared light delivery to tissue for three different illumination schemes: bright-field, standard dark-field, and long-axis lateral illumination. Illumination schemes performance were evaluated for three preclinical and clinically relevant cases for PAI. The shape of the target influenced light delivery for all illumination schemes. Long-axis lateral illumination provided substantial improvement when targets smaller than the lateral width of the transducer were evaluated. The prototype developed to produce multiangle long-axis lateral illumination was evaluated with phantom and in vivo experiments. PA images of good quality were generated from mouse abdomen, forearm, and index finger of a volunteer. Based on the results presented here, a novel PAI system was proposed for preclinical and clinical research. In addition, long-axis lateral illumination could be used together with more conventional illumination schemes to improve light delivery in reflection mode PAI.

Supplementary Materials: The following are available online at http://www.mdpi.com/1424-8220/20/14/4052/s1, Figure S1: Subframes acquired at each illumination angle used to generate the in vivo PA images shown in Figure 11 of the main article for the (a) forearm, (c) index finger, and (e) mouse abdomen. SNR as a function of illumination angle for each subframe of selected structures located at the (b) forearm, (d) index finger, and (f) mouse abdomen.

Author Contributions: Conceptualization and design, J.H.U. and T.Z.P.; Data curation, J.H.U., D.R.T.S. and G.S.P.F.; Analysis and interpretation of data, J.H.U., D.R.T.S., G.S.P.F., M.S.B., M.H.N.-B. and A.A.O.C. and T.Z.P.; Funding acquisition, A.A.O.C. and T.Z.P.; Investigation, J.H.U and T.Z.P.; Methodology, J.H.U., D.R.T.S., G.S.P.F., M.S.B., M.H.N.-B., A.A.O.C. and T.Z.P.; Project administration, T.Z.P.; Resources, M.S.B., A.A.O.C. and T.Z.P.; Software, J.H.U., D.R.T.S. and G.S.P.F.; Supervision, T.Z.P.; Writing—original draft, J.H.U. and T.Z.P.;

Writing—review & editing, J.H.U., D.R.T.S., G.S.P.F., M.S.B., M.H.N.-B. and A.A.O.C. and T.Z.P. All authors critically read and approved the final version of the manuscript.

Funding: This study was supported by Coordenação de Aperfeiçoamento de Pessoal de Nível Superior-Brasil (CAPES)-Finance Code 001, Conselho Nacional de Desenvolvimento Científico e Tecnológico (CNPq) grant 436657/2018-0 and Fundação de Amparo à Pesquisa do Estado de São Paulo (FAPESP) grant 2016/22374-8, 2015/05684-0, 2017/14482-8, and 2013/18854-6.

Conflicts of Interest: The authors declare no conflict of interest.

References

1. Kresge, C.T.; Leonowicz, M.E.; Roth, W.J.; Vartuli, J.C.; Beck, J.S. Generation of ultrasonic waves from a layered photoacoustic source. *Nature* **1992**, *359*, 710–713. [CrossRef]
2. Kruger, R.A. Photoacoustic ultrasound. *Med. Phys.* **1994**, *21*, 127. [CrossRef] [PubMed]
3. Wang, X.; Xu, Y.; Xu, M.; Yokoo, S.; Fry, E.S.; Wang, L.V. Photoacoustic tomography of biological tissues with high cross-section resolution: Reconstruction and experiment. *Med. Phys.* **2002**, *29*, 2799–2805. [CrossRef] [PubMed]
4. Xu, M.; Wang, L.V. Photoacoustic imaging in biomedicine. *Rev. Sci. Instrum.* **2006**, *77*, 041101. [CrossRef]
5. Beard, P. Biomedical photoacoustic imaging. *Interface Focus* **2011**, *1*, 602–631. [CrossRef]
6. Wang, X.; Pang, Y.; Ku, G.; Xie, X.; Stoica, G.; Wang, L.V. Noninvasive laser-induced photoacoustic tomography for structural and functional in vivo imaging of the brain. *Nat. Biotechnol.* **2003**, *21*, 803–806. [CrossRef]
7. Wang, L.V.; Hu, S. Photoacoustic Tomography: In Vivo Imaging from Organelles to Organs. *Science (80-)* **2012**, *335*, 1458–1462. [CrossRef]
8. Larina, I.V.; Larin, K.V.; Esenaliev, R.O. Real-time optoacoustic monitoring of temperature in tissues. *J. Phys. D Appl. Phys.* **2005**, *38*, 2633–2639. [CrossRef]
9. Shah, J.; Park, S.; Aglyamov, S.R.; Larson, T.; Ma, L.; Sokolov, K.V.; Johnston, K.P.; Milner, T.E.; Emelianov, S.Y. Photoacoustic imaging and temperature measurement for photothermal cancer therapy. *J. Biomed. Opt.* **2008**, *13*, 034024. [CrossRef]
10. Yao, J.; Ke, H.; Tai, S.; Zhou, Y.; Wang, L.V. Absolute photoacoustic thermometry in deep tissue. *Opt. Lett.* **2013**, *38*, 5228–5231. [CrossRef]
11. Ke, H.; Tai, S.; Wang, L.V. Photoacoustic thermography of tissue. *J. Biomed. Opt.* **2014**, *19*, 026003. [CrossRef]
12. Razansky, D.; Vinegoni, C.; Ntziachristos, V. Multispectral photoacoustic imaging of fluorochromes in small animals. *Opt. Lett.* **2007**, *32*, 2891. [CrossRef]
13. Laufer, J.; Zhang, E.; Beard, P. Evaluation of absorbing chromophores used in tissue phantoms for quantitative photoacoustic spectroscopy and imaging. *IEEE J. Sel. Top. Quantum Electron.* **2010**, *16*, 600–607. [CrossRef]
14. Laufer, J.G. Pulsed near-infrared photoacoustic spectroscopy of blood. *Proc. SPIE* **2004**, *5320*, 57–68. [CrossRef]
15. Laufer, J.; Elwell, C.; Delpy, D.; Beard, P. In vitro measurements of absolute blood oxygen saturation using pulsed near-infrared photoacoustic spectroscopy: Accuracy and resolution. *Phys. Med. Biol.* **2005**, *50*, 4409–4428. [CrossRef] [PubMed]
16. Zhang, H.F.; Maslov, K.; Stoica, G.; Wang, L.V. Functional photoacoustic microscopy for high-resolution and noninvasive in vivo imaging. *Nat. Biotechnol.* **2006**, *24*, 848–851. [CrossRef] [PubMed]
17. Zhang, H.F.; Maslov, K.; Sivaramakrishnan, M.; Stoica, G.; Wang, L.V. Imaging of hemoglobin oxygen saturation variations in single vessels in vivo using photoacoustic microscopy. *Appl. Phys. Lett.* **2007**, *90*, 5–7. [CrossRef]
18. Chen, Z.; Yang, S.; Xing, D. In vivo detection of hemoglobin oxygen saturation and carboxyhemoglobin saturation with multiwavelength photoacoustic microscopy. *Opt. Lett.* **2012**, *37*, 3414. [CrossRef]
19. Shao, Q.; Ashkenazi, S. Photoacoustic lifetime imaging for direct *in vivo* tissue oxygen monitoring. *J. Biomed. Opt.* **2015**, *20*, 036004. [CrossRef]
20. Fu, Q.; Zhu, R.; Song, J.; Yang, H.; Chen, X. Photoacoustic Imaging: Contrast Agents and Their Biomedical Applications. *Adv. Mater.* **2019**, *31*, 1–31. [CrossRef]
21. Oraevsky, A.A.; Clingman, B.; Zalev, J.; Stavros, A.T.; Yang, W.T.; Parikh, J.R. Clinical optoacoustic imaging combined with ultrasound for coregistered functional and anatomical mapping of breast tumors. *Photoacoustics* **2018**, *12*, 30–45. [CrossRef] [PubMed]

22. Jo, J.; Xu, G.; Cao, M.; Marquardt, A.; Francis, S.; Gandikota, G.; Wang, X. A Functional Study of Human Inflammatory Arthritis Using Photoacoustic Imaging. *Sci. Rep.* **2017**, *7*, 1–9. [CrossRef] [PubMed]
23. Jo, J.; Tian, C.; Xu, G.; Sarazin, J.; Schiopu, E.; Gandikota, G.; Wang, X. Photoacoustic tomography for human musculoskeletal imaging and inflammatory arthritis detection. *Photoacoustics* **2018**, *12*, 82–89. [CrossRef] [PubMed]
24. Niederhauser, J.J.; Jaeger, M.; Lemor, R.; Weber, P.; Frenz, M. Combined ultrasound and optoacoustic system for real-time high-contrast vascular imaging in vivo. *IEEE Trans. Med. Imaging* **2005**, *24*, 436–440. [CrossRef]
25. Daoudi, K.; Van Den Berg, P.J.; Rabot, O.; Kohl, A.; Tisserand, S.; Brands, P.; Steenbergen, W. Handheld probe integrating laser diode and ultrasound transducer array for ultrasound/photoacoustic dual modality imaging. *Opt. Express* **2014**, *22*, 26365. [CrossRef]
26. Hai, P.; Zhou, Y.; Zhang, R.; Ma, J.; Li, Y.; Wang, L.V. Label-free high-throughput detection and quantification of circulating melanoma tumor cell clusters. *J. Biomed. Opt.* **2016**, *22*, 1-041004–1-041005-6. [CrossRef]
27. Zhou, Y.; Liang, J.; Wang, L.V. Cuffing-based photoacoustic flowmetry in humans in the optical diffusive regime. *J. Biophotonics* **2016**, *9*, 208–212. [CrossRef]
28. Zhao, L.; Yang, M.; Jiang, Y.; Li, C. Optical fluence compensation for handheld photoacoustic probe: An in vivo human study case. *J. Innov. Opt. Health Sci.* **2017**, *10*, 1740002-1–1740002-18. [CrossRef]
29. Montilla, L.G.; Olafsson, R.; Bauer, D.R.; Witte, R.S. Real-time photoacoustic and ultrasound imaging: A simple solution for clinical ultrasound systems with linear arrays. *Phys. Med. Biol.* **2013**, *58*. [CrossRef]
30. Wang, G.W.G.; Zhao, H.Z.H.; Ren, Q.R.Q.; Li, C.L.C. Simulation of light delivery for photoacoustic breast imaging using the handheld probe. *Chin. Opt. Lett.* **2014**, *12*, 051703–051706. [CrossRef]
31. Lee, Y.J.; Jeong, E.J.; Song, H.W.; Ahn, C.G.; Noh, H.W.; Sim, J.Y.; Song, D.H.; Jeon, M.Y.; Lee, S.; Kim, H.; et al. Photoacoustic imaging probe for detecting lymph nodes and spreading of cancer at various depths. *J. Biomed. Opt.* **2017**, *22*, 091513. [CrossRef] [PubMed]
32. Bai, Y.; Cong, B.; Gong, X.; Song, L.; Liu, C. Compact and low-cost handheld quasibright-field linear-array probe design in photoacoustic computed tomography. *J. Biomed. Opt.* **2018**, *23*, 1. [CrossRef] [PubMed]
33. Held, G.; Preisser, S.; Akarçay, H.G.; Peeters, S.; Frenz, M.; Jaeger, M. Effect of irradiation distance on image contrast in epi-optoacoustic imaging of human volunteers. *Biomed. Opt. Express* **2014**, *5*, 3765. [CrossRef] [PubMed]
34. Frenz, M.; Jaeger, M. Optimization of tissue irradiation in optoacoustic imaging using a linear transducer: Theory and experiments. In Proceedings of the Photons Plus Ultrasound: Imaging and Sensing 2008: The Ninth Conference on Biomedical Thermoacoustics, Optoacoustics, and Acousto-optics, San Jose, CA, USA, 19–24 January 2008.
35. Preisser, S.; Held, G.; Akarçay, H.G.; Jaeger, M.; Frenz, M. Study of clutter origin in in-vivo epi-optoacoustic imaging of human forearms. *J. Opt.* **2016**, *18*, 094003. [CrossRef]
36. Haisch, C.; Eilert-Zell, K.; Vogel, M.M.; Menzenbach, P.; Niessner, R. Combined optoacoustic/ultrasound system for tomographic absorption measurements: Possibilities and limitations. *Anal. Bioanal. Chem.* **2010**, *397*, 1503–1510. [CrossRef]
37. Sivasubramanian, K.; Periyasamy, V.; Wen, K.K.; Pramanik, M. Optimizing light delivery through fiber bundle in photoacoustic imaging with clinical ultrasound system: Monte Carlo simulation and experimental validation. *J. Biomed. Opt.* **2016**, *22*, 041008. [CrossRef]
38. Sangha, G.S.; Hale, N.J.; Goergen, C.J. Adjustable photoacoustic tomography probe improves light delivery and image quality. *Photoacoustics* **2018**, *12*, 6–13. [CrossRef] [PubMed]
39. Sowers, T.; Yoon, H.; Emelianov, S. Investigation of light delivery geometries for photoacoustic applications using Monte Carlo simulations with multiple wavelengths, tissue types, and species characteristics. *J. Biomed. Opt.* **2020**, *25*, 1. [CrossRef]
40. Wang, Z.; Ha, S.; Kim, K. A new design of light illumination scheme for deep tissue photoacoustic imaging. *Opt. Express* **2012**, *20*, 22649. [CrossRef]
41. Yu, J.; Schuman, J.S.; Lee, J.K.; Lee, S.G.; Chang, J.H.; Kim, K. A Light Illumination Enhancement Device for Photoacoustic Imaging: In Vivo Animal Study. *IEEE Trans. Ultrason. Ferroelectr. Freq. Control* **2017**, *64*, 1205–1211. [CrossRef]

42. Xia, W.; Kuniyil Ajith Singh, M.; Maneas, E.; Sato, N.; Shigeta, Y.; Agano, T.; Ourselin, S.; JWest, S.; EDesjardins, A. Handheld real-time LED-based photoacoustic and ultrasound imaging system for accurate visualization of clinical metal needles and superficial vasculature to guide minimally invasive procedures. *Sensors* **2018**, *18*, 1394. [CrossRef] [PubMed]
43. Zhu, Y.; Feng, T.; Cheng, Q.; Wang, X.; Du, S.; Sato, N. Towards Clinical Translation of LED-Based Photoacoustic Imaging: A Review. *Sensors* **2020**, *20*, 2484. [CrossRef]
44. Kuriakose, M.; Nguyen, C.D.; Singh, M.K.A.; Mallidi, S. Optimizing irradiation geometry in LED-based photoacoustic imaging with 3d printed flexible and modular light delivery system. *Sensors* **2020**, *20*, 3789. [CrossRef] [PubMed]
45. Zhang, X.; Wu, X.; Adelegan, O.J.; Yamaner, F.Y.; Oralkan, O. Backward-Mode Photoacoustic Imaging Using Illumination Through a CMUT with Improved Transparency. *IEEE Trans. Ultrason. Ferroelectr. Freq. Control* **2018**, *65*, 85–94. [CrossRef] [PubMed]
46. Chen, Q.X.; Davies, A.; Dewhurst, R.J.; Payne, P.A. Photo-acoustic probe for intra-arterial imaging and therapy. *Electron. Lett.* **1993**, *29*, 1632–1633. [CrossRef]
47. Laufer, J.; Zhang, E.; Raivich, G.; Beard, P. Three-dimensional noninvasive imaging of the vasculature in the mouse brain using a high resolution photoacoustic scanner. *Appl. Opt.* **2009**, *48*, 299–306. [CrossRef]
48. Li, M.; Liu, C.; Gong, X.; Zheng, R.; Bai, Y.; Xing, M.; Du, X.; Liu, X.; Zeng, J.; Lin, R.; et al. Linear array-based real-time photoacoustic imaging system with a compact coaxial excitation handheld probe for noninvasive sentinel lymph node mapping. *Biomed. Opt. Express* **2018**, *9*, 1408–1422. [CrossRef]
49. Wang, Y.; Lim, R.S.A.; Zhang, H.; Nyayapathi, N.; Oh, K.W.; Xia, J. Optimizing the light delivery of linear-array-based photoacoustic systems by double acoustic reflectors. *Sci. Rep.* **2018**, *8*, 1–7. [CrossRef]
50. Ida, T.; Iwazaki, H.; Kawaguchi, Y.; Kawauchi, S.; Ohkura, T.; Iwaya, K.; Tsuda, H.; Saitoh, D.; Sato, S.; Iwai, T. Burn depth assessments by photoacoustic imaging and laser Doppler imaging. *Wound Repair Regen.* **2016**, *24*, 349–355. [CrossRef] [PubMed]
51. Wei, C.W.; Nguyen, T.M.; Xia, J.; Arnal, B.; Wong, E.Y.; Pelivanov, I.M.; O'Donnell, M. Real-time integrated photoacoustic and ultrasound (PAUS) imaging system to guide interventional procedures: Ex vivo study. *IEEE Trans. Ultrason. Ferroelectr. Freq. Control* **2015**, *62*, 319–328. [CrossRef] [PubMed]
52. Fang, Q.; Boas, D.A. Monte Carlo Simulation of Photon Migration in 3D Turbid Media Accelerated by Graphics Processing Units. *Opt. Express* **2009**, *17*, 20178. [CrossRef] [PubMed]
53. Rich, L.J.; Chamberlain, S.R.; Falcone, D.R.; Bruce, R.; Heinmiller, A.; Xia, J.; Seshadri, M. Performance Characteristics of Photoacoustic Imaging Probes with Varying Frequencies and Light-delivery Schemes. *Ultrason. Imaging* **2019**, *41*, 319–335. [CrossRef] [PubMed]
54. Jacques, S.L. Optical properties of biological tissues: A review. *Phys. Med. Biol.* **2013**, *58*, 5007–5008. [CrossRef]
55. Pavan, T.Z.; Madsen, E.L.; Frank, G.R.; Carneiro, A.A.O.; Hall, T.J. Nonlinear elastic behavior of phantom materials for elastography. *Phys. Med. Biol.* **2010**, *55*, 2679. [CrossRef]
56. Uliana, J.H.; Sampaio, D.R.T.; Carneiro, A.A.O.; Pavan, T.Z. Photoacoustic-based thermal image formation and optimization using an evolutionary genetic algorithm. *Res. Biomed. Eng.* **2018**, *34*, 147–156. [CrossRef]
57. Naser, M.A.; Sampaio, D.R.; Muñoz, N.M.; Wood, C.A.; Mitcham, T.M.; Stefan, W.; Sokolov, K.V.; Pavan, T.Z.; Avritscher, R.; Bouchard, R.R. Improved Photoacoustic-Based Oxygen Saturation Estimation With SNR-Regularized Local Fluence Correction. *IEEE Trans. Med. Imaging* **2019**, *38*, 561–571. [CrossRef] [PubMed]
58. Liu, S.; Feng, X.; Jin, H.; Zhang, R.; Luo, Y.; Zheng, Z.; Gao, F.; Zheng, Y. Handheld Photoacoustic Imager for Theranostics in 3D. *IEEE Trans. Med. Imaging* **2019**, *38*, 2037–2046. [CrossRef]
59. Singh, C.P.; Bindra, K.S.; Bhalerao, G.M.; Oak, S.M. Investigation of optical limiting in iron oxide nanoparticles. *Opt. Express* **2008**, *16*, 8440. [CrossRef]
60. Jacques, S.L.; Prahl, S. Tabulated Molar Extinction Coefficient for Hemoglobin in Water. Available online: https://omlc.org/spectra/hemoglobin/index.html (accessed on 16 July 2020).
61. Bushberg, J.T.; Seibert, J.A.; Leidholdt, E.M.; Boone, J.M. *The Essential Physics of Medical Imaging*, 3rd ed.; Lippincott Williams & Wilkins: Philadelphia, PA, USA, 2011.
62. Alles, E.J.; Jaeger, M.; Bamber, J.C. Photoacoustic clutter reduction using short-lag spatial coherence weighted imaging. *IEEE Int. Ultrason. Symp. IUS* **2014**, 41–44. [CrossRef]
63. Netter, F.H. *Atlas of Human Anatomy*, 6th ed.; Sauders Elsevier: Philadelphia, PA, USA, 2014; Volume 1.

64. Lee, H.D.; Shin, J.G.; Hyun, H.; Yu, B.A.; Eom, T.J. Label-free photoacoustic microscopy for in-vivo tendon imaging using a fiber-based pulse laser. *Sci. Rep.* **2018**, *8*, 1–9. [CrossRef] [PubMed]
65. Sabino, C.P.; Deana, A.M.; Yoshimura, T.M.; da Silva, D.F.; França, C.M.; Hamblin, M.R.; Ribeiro, M.S. The optical properties of mouse skin in the visible and near infrared spectral regions. *J. Photochem. Photobiol. B Biol.* **2016**, *160*, 72–78. [CrossRef] [PubMed]
66. Cook, M.J. *The Anatomy of the Laboratory Mouse*; Academic Press: London, UK; New York, NY, USA, 1965.
67. Choi, W.; Park, E.Y.; Jeon, S.; Kim, C. Clinical photoacoustic imaging platforms. *Biomed. Eng. Lett.* **2018**, *8*, 139–155. [CrossRef] [PubMed]

sensors

MDPI

Article

Enhancement of Photoacoustic Signal Strength with Continuous Wave Optical Pre-Illumination: A Non-Invasive Technique

Anjali Thomas [1], Souradip Paul [1], Joy Mitra [2] and Mayanglambam Suheshkumar Singh [1,*]

1 Biomedical Instrumentation and Imaging Laboratory (BIIL), School of Physics (SoP), Indian Institute of Science Education and Research Thiruvananthapuram (IISER-TVM), Thiruvananthapuram 695551, India; anjalithomas16@iisertvm.ac.in (A.T.); souradip.rkm16@iisertvm.ac.in (S.P.)

2 Scanning Probe Microscopy and Plasmonics Lab, School of Physics (SoP), Indian Institute of Science Education and Research Thiruvananthapuram (IISER-TVM), Thiruvananthapuram 695551, India; j.mitra@iisertvm.ac.in

* Correspondence: suhesh.kumar@iisertvm.ac.in

Abstract: Use of portable and affordable pulse light sources (light emitting diodes (LED) and laser diodes) for tissue illumination offers an opportunity to accelerate the clinical translation of photoacoustic imaging (PAI) technology. However, imaging depth in this case is limited because of low output (optical) power of these light sources. In this work, we developed a noninvasive technique for enhancing strength (amplitude) of photoacoustic (PA) signal. This is a photothermal-based technique in which a continuous wave (CW) optical beam, in addition to short-pulse (∼nsec) laser beam, is employed to irradiate and, thus, raise the temperature of sample material selectively over a pre-specified region of interest (we call the process as pre-illumination). The increase in temperature, in turn enhances the PA-signal strength. Experiments were conducted in methylene blue, which is one of the commonly used contrast agents in laboratory research studies, to validate change in temperature and subsequent enhancement of PA-signal strength for the following cases: (1) concentration or optical absorption coefficient of sample, (2) optical power of CW-optical beam, and (3) time duration of pre-illumination. A theoretical hypothesis, being validated by numerical simulation, is presented. To validate the proposed technique for clinical and/or pre-clinical applications (diagnosis and treatments of cancer, pressure ulcers, and minimally invasive procedures including vascular access and fetal surgery), experiments were conducted in tissue-mimicking Agar phantom and ex-vivo animal tissue (chicken breast). Results demonstrate that pre-illumination significantly enhances PA-signal strength (up to ∼70% (methylene blue), ∼48% (Agar phantom), and ∼40% (chicken tissue)). The proposed technique addresses one of the primary challenges in the clinical translation of LED-based PAI systems (more specifically, to obtain a detectable PA-signal from deep-seated tissue targets).

Keywords: photoacoustic imaging; signal enhancement; pre-illumination; photo-thermal effect; heat capacity

Citation: Thomas, A.; Paul, S.; Mitra, J.; Singh, M.S. Enhancement of Photoacoustic Signal Strength with Continuous Wave Optical Pre-Illumination: A Non-Invasive Technique. *Sensors* **2021**, *21*, 1190. https://doi.org/10.3390/s21041190

Academic Editors: Mithun Kuniyil Ajith Singh and Wenfeng Xia

Received: 25 December 2020
Accepted: 23 January 2021
Published: 8 February 2021

1. Introduction

Photoacoustic imaging (PAI) has been proven as a promising technology for nondestructive recovery of vital patho-physiological parameters (functional, structural, hemodynamics, mechanical, and molecular distribution [1–6] of biological tissues with a microscopic resolution at an unprecedented penetration depth (~cm) [7]. On the other hand, LED-based systems hold great potential in clinical translation because of their portability and affordability [8–10]. In LED-based PAI systems, the pulsed laser source is replaced by LEDs and, thus, the optical power of LEDs is insufficient to induce detectable photoacoustic (PA)-signal at higher penetration depths, which is the major drawback of the system [9]. Similar is true for the cases of laser diode based PAI systems [11]. In this context, enhancement of the achievable PA-signal strength is a longstanding problem in low energy light source based PAI systems and temperature dependent PA-signal enhancement is one of the most focusing areas of research studies in the past few years [12–23]. Enhancement in signal strength enables not only to improve obtainable imaging depth but also to increase accuracy in the quantitative measurement of physiological parameters by improving signal-to-noise-ratio (SNR) [12]. In this regard, an injection-based technique—in which exogenous contrast agents or target specific biomarkers are introduced to tissue sample of interest externally through injection—has been proven successful and remains the only standard technique [13,24,25]. Unfortunately, the technique suffers from serious drawbacks: (1) limited studies of patho-physiological activities due to limited availability or selection of target-specific biomarkers [24,25], (2) bio-compatibility, (3) bio-toxicity, and (4) invasive in nature because of the introduction of exogenous dyes—which are foreign substances—to body as well as breakage of the intervening tissues including skin. To our best knowledge, experimental studies on noninvasive techniques for enhancement of the PA-signal strength (or amplitude) are limited to the articles reported in Refs. [12–23]. In all of these reported studies for enhancement of PA-signal strength, imaging sample was immersed inside a heating bath filled with water that controls the thermodynamic equilibrium temperature (with an exception of the study [20] that controls the temperature of the imaging target by connecting an electrically conducting wire to imaging target of interest). Thus, all of the reported (experimental) techniques are not suitable for the end applications being targeted to clinical settings. Refs. [21,26,27] reported studies on theoretical aspects of the generation of PA-waves and its contribution from thermodynamic equilibrium temperature (T) and temperature raise (ΔT) induced by transient optical illumination. In our present article, we report a unique photothermal-based technique for enhancing the achievable PA-signal strength. Experimentally, enhancement is facilitated by pre-illumination of the sample with a continuous wave (CW)-optical beam (in addition to irradiation of the sample with short-pulse laser (with pulse-width ~ nsec)), thereby raising (thermodynamic equilibrium) temperature (T) of the imaging sample selectively over a pre-specified region of interest. Shortly, enhancement of PA-signal strength is achieved by control of thermodynamic property of imaging sample over a particular region of interest (unlike controlling temperature of entire sample or background as it is the case for the above mentioned or reported techniques). In this way, our proposed technique is noninvasive (without introducing contrast agents) and nondestructive (without damaging tissues).

With a close similarity to our proposed technique, in PAI-guided phototherapy or photothermal therapy [28–30], a CW-optical beam is employed to irradiate target-specific contrast dyes (being introduced to tissue targets), and the subsequent temperature raise (ΔT) is imaged by PAI modality. In the study [16], a CW-laser beam was used for inducing interstitial tissue coagulation while heating of the imaging targets was facilitated by hot air blow from a heat gun or conducting heating. Introduction of photo-absorbing contrast agents—like gold nanoparticles, nano-shells, and plasmonic nanoparticles—in the target region facilitates enhancement in contrast and photoacoustic signal [31–33]. This is an invasive procedure. On the other hand, recovery of the temperature distribution ($T(\vec{r})$) of tissue samples with PA-imaging modality has been studied for more than a decade of years or so [12,14–16,23,34]. However, experimental techniques (including CW-irradiation) for enhancement of PA-signal strength from endogenous signal contrasts of (imaging) tissue sample and its clinical applications are not yet reported in literature. In this article, we propose a theoretical (analytical) hypothesis that derives a mathematical relationship of explicit dependence of PA-signal strength on ambient temperature (T) of the given physical system (tissue sample, in our case). This proposed theoretical hypothesis is validated with numerical simulation studies. For the simulation studies, we adopted k-wave (MATLAB) platform [35–37], which is widely employed as a standard numerical simulation tool for photoacoustic imaging. The experimental results, with experiments being performed in tissue-mimicking Agar phantom, as well as chicken breast, sample, demonstrate that the proposed photothermal-based technique can be adapted for clinical applications. In other words, the proposed technique was validated with experiments being conducted in tissue-mimicking phantom (Agar, in our case) as it is done conventionally in (biomedical) research laboratory settings [3–5,12,15,18] while the feasibility of clinical translation of the proposed technique was validated with experiments being conducted in tissues (chicken breast).

2. Materials and Methods

2.1. Theory

In PA-imaging, a beam of short laser pulses (pulse width ~ nsec) is delivered to the sample surface so as to irradiate a particular target of interest. Thermoelastic expansion occurs [38] due to rapid heating and subsequent cooling of the irradiated sample material. This results in the generation of pressure waves in the sample, which are known as initial photoacoustic pressure waves. One may consider the photoacoustic effect as a thermodynamic process [38]. The initial PA-pressure can be expressed as [3,38]:

$$P_0 = \frac{\beta}{\kappa}\frac{1}{\rho c_V}\mu_a\phi = \Gamma\mu_a\phi. \tag{1}$$

where $\Gamma = \frac{\beta}{\kappa}\frac{1}{\rho c_V}$ is a dimensionless physical quantity, commonly referred as Grüeisen parameter, and it is a measure of thermoelastic efficiency of a given material. Here, in the case of PAI, Γ gives the measure of conversion efficiency from pulse optical energy to acoustic energy. We know that optical absorption coefficient (μ_a) is characterized by optical extinction coefficient (ε) that can be expressed as:

$$\mu_a = \varepsilon[C], \tag{2}$$

where $[C]$ is the concentration.

For a physical system or medium including solution of low concentration (methylene blue solution, as it is the case for our present study) and biological tissue, one can deduce Equation (1) as (a detailed derivation is provided in Appendix A):

$$P_0 = \frac{\beta^{(water)}}{\kappa^{(water)}}\frac{1}{\rho^{(water)}c_V^{(water)}}\varepsilon^{(methy)}[C]^{(methy)}\phi \tag{3}$$

Equation (3) implies that photoacoustic wave generation is dictated by the thermodynamic properties of the surrounding (background) medium/fluid (water, in our study). It is similar to the studies [21–23,26,27,39] that demonstrated PA-wave generation is dictated by thermodynamic properties of the fluid in which optical absorbing targets of vanishingly small size (point source or nanoparticles) were immersed. In view of the arguments (discussed in Appendix A), the above Equation (3) holds true for the case of Agar gel phantom where superscripts, (water) and (methylene blue) are replaced by (agar gel) and (ink), respectively. We adopt mathematical representation corresponding to methylene blue solution.

The reported study [40] gives the variation of thermodynamic parameters with (ambient) thermodynamic equilibrium temperature (T) (as it is given in Table 1). From Table 1, one can conclude that, in the temperature range (~20 °C to 40 °C), $\beta^{(water)}$ changes linearly with T, which can be represented by a straight line ($\beta^{(water)}$ = a + bΔT, where 'a' and 'b' are intercept and slope, respectively). This assumption is in agreement with Taylor expansion of β close to equilibrium temperature (T_0) given *by* $\beta(T_0) \rightarrow \beta(T_0+\Delta T) \approx \beta(T_0) + \Delta T\frac{d\beta}{dT} = \beta_{equil} + \Delta T\frac{d\beta}{dT}$, as it is done in Refs. [21,26,27], where ΔT is the differential change in temperature. β_{equil} $(\equiv \beta(T_0))$ is the thermal expansion coefficient under thermodynamic equilibrium temperature (T_0), which is again dependent on T. For water, thermal expansion coefficient (β) vanishes at T ~ 3.98 °C, i.e., $\beta(T)$ = 0 for $T \sim$ 3.98 °C [13,17]). At this zero-crossing temperature, from Equation (3), one can conclude that PA-waves cannot be generated under any circumstances (including increasing optical absorption coefficient (μ_a) and/or optical fluence (ϕ)), which was validated by experimental studies in the past [21,26,41]. In our present study, we conducted experiments with $T \sim$ 20–30 °C, which is far above the zero-crossing temperature (∼4 °C) and the equilibrium term (β_{equil}) cannot be neglected [21]. In this temperature range of interest (∼20–30 °C) for laboratory as well as clinical studies, from Table 1, we estimated 'b' to be ∼9.64 °C and relative change of β is found to be ∼46.63%. However, other thermodynamic parameters give negligible relative changes compared to β. Under these circumstances, we assume that $\kappa^{(water)}$, $\rho^{(water)}$, and $c_V^{(water)}$ are independent of T or constant in comparison to the dependence of $\beta^{(water)}$ on T [13,42]. Specific heat capacity (c^V) is independent of temperature (for soft tissue) over temperature range $<$ 50 °C [16,43]. From Equation (3), we can obtain the explicit expression for P_0 at an arbitrarily chosen temperature (T) in the neighborhood of thermodynamic equilibrium temperature (T_0), i.e., P_0 at $T_0 \to T = T_0 + \Delta T$ can be expressed as:

$$P_0(T_0) \to P_0(T = T_0 + \Delta T) = P_0(T_0)\ \frac{[C]^{(methy)}}{\alpha^{(water)}} \left((\varepsilon^{(methy)} \left(\frac{\partial \beta^{(water)}}{\partial T} \right)_{T_0} \Delta T + \beta^{water} \left(\frac{\partial \varepsilon^{(methy)}}{\partial T} \right)_{T_0} \Delta T \right) \phi \quad (4)$$

where $\alpha^{(water)}$ $(=\kappa^{(water)} \rho^{(water)} c_V^{(water)})$ and $[C]^{(methy)}$ are considered as independent of T. Equation (4) shows that the strength of initial PA-pressure waves (P_0) is characterized by the ambient temperature ($T = T_0 + \Delta T$) of the medium in addition to μ_a and ϕ. In our present study, the differential raise in thermodynamic equilibrium temperature (ΔT) is facilitated by the pre-illumination of the sample with a CW-laser beam.

Table 1. Dependence of thermodynamic parameters on thermodynamic equilibrium temperature (T) for water.

Temperature (°C or K)	$\beta^{(water)}$ **(10^{-6} K^{-1})**	$\kappa^{(water)}$ **(10^{-10} Pa^{-1})**	$\rho^{(water)}$ **(kg m^{-3})**	$C_P^{(water)}$ **(kJ kg K^{-1})**	$C_V^{(water)}$ **(kJ kg K^{-1})**
20 or 293	206.80	4.5891	998.21	4.1818	4.1545
30 or 303	303.23	4.4770	995.65	4.1784	4.1159
40 or 113	385.30	4.4240	992.22	4.1785	4.0826

Equation (4) gives an explicit representation of P_0 in integral form, into two separate thermal terms. The first term attributes to thermodynamic equilibrium temperature (T_0), while the second term corresponds to transient changes in thermal expansion coefficient (β) and optical extinction coefficient (ε) that are resulted due to thermal perturbation in the physical (imaging) system by an external agency and its associated heating [44]. In our present study, we employed a CW-laser beam to raise the thermodynamic equilibrium temperature $T_0 \to T_0 + \Delta T$ and, thus, the second term in the generation of PA-signals (in Equation (4)) while pulse laser beam, which is contributing to the enhancement from the two terms, is kept fixed. Equation (4) shows that enhancement of strength of initial PA-pressure waves, being generated in mechanical medium (by transient optical illumination), can be achieved (experimentally) by control of ambient temperature (a) directly through heating of the imaging sample or by thermal technique and (b) indirectly through photo-illumination of the specimen raising incident optical power or photo-thermal technique. Direct method demands heating of the imaging specimen as a whole and, thus, it is not of much clinical importance (as it is discussed in Section 1). Indirect method facilitates to raise the temperature (T) of deep-seated tissue targets selectively over a pre-specified region of interest and, thus, this technique is of significant clinical impacts. Here, we focus on the indirect technique where we employed the CW-laser beam (for photo-illumination) in addition to the pulsed-laser beam that is adopted for transient illumination and subsequent generation of initial P_0.

2.2. Enhancement of PA-Signal Strength

One can express relative enhancement in PA-signal strength, due to a raise in thermodynamic equilibrium temperature T_0, as:

$$P_0^{(enh)} = \frac{P_0(T = T_0 + \Delta T) - P(T = T_0)}{P_0(T = T_0)} \times 100 \ (in \ \%) \quad (5)$$

2.3. Numerical Simulation

A circular target of radius $\sim$3 mm having contrast in μ_a in comparison to that of the background is situated at the center of the background of area 21.6 $\times$ 21.6 mm^2. This numerical sample represents the deep-seated inhomogeneous target illuminated with CW-laser (to raise the temperature) in the background tissue medium. A circular sensor (diameter $\sim$20 mm) of 145 identical detectors is collecting the PA-waves for reconstruction as it is done in our previous study [45]. Figure 1a presents a representative image of the distribution of initial PA-pressure waves (P_0) at a temperature T_0 (say, room temperature), while Figure 1c presents the same at a temperature $T' = T_0 + \Delta T$ where ΔT is the temperature raise due to pre-illumination. The values of ΔT corresponding to different concentration $[C]$, absorption coefficient (μ_a), and optical power are taken from experiment. Figure 1b,d present the reconstructed images corresponds to Figure 1a,c. Variation of PA-signal along the marked lines, which are indicated in Figure 1b,d, plotted in Figure 1e. We estimate the strength of initial PA-pressure waves, for the reconstructed pressure distribution, as the average value over full width at the 75% of the maximum of the line plots and it is found to be 130 Pa at room temperature (T_0) and 193 Pa at a temperature $T' = T_0 + \Delta T$. (Here, ΔT is the temperature raise corresponding to a concentration of $\sim$7 mM). From these values, PA-signal enhancement due to temperature raise is calculated as 51%. In similar ways, we study variation PA-signal strength with change in temperature due to the changes in the physical parameters of the targets, and the results are plotted in Section 3.

2.4. Experimental Set-Up

For experiments, we employed a home-built acoustic resolution photoacoustic microscopy (AR-PAM) imaging system in transmission configuration. Figure 2a depicts a schematic diagram of the experimental set-up. A train of short duration pulses of optical beam—beam diameter ~2.3 mm, power ~25 mW, pulse width ~6 nsecs, pulse repetition frequency ~100 Hz, and wavelength ~670 nm—from a tunable pulsed OPO laser source (SpitLight OPO Evo 150–532, Innolas Lasers, Krailling, Germany) was coupled by an optical fibre (diameter ~6 mm) so as to irradiate tissue sample of interest. A tightly focusing ultrasound transducer (V375-SU, Olympus, Shinjuku City, Tokyo, Japan)—focal spot size ~154 µm (calculated using, $BD = 1.02 \frac{F*v}{f*D}$ [46], where F is focal length; v is speed of sound; f is operating frequency; D is the diameter of US transducer), operating frequency ~30 MHz, and focal length ~19.10 mm—was employed to pick-up transient light-induced PA-signals (P_0) selectively from its narrow focal zone.

The detected PA-signals were amplified using a pulser-receiver (Part No.: 5073PR-40-P, Olympus, Shinjuku City, Tokyo, Japan) and, then, acquired using a data acquisition card (Part No.: 779745-02, NI PCI-5114, 250 MS/s, National Instruments, Austin, Texas, USA) being attached to a computer system. From time-resolved A-scan, depth-resolved data is obtained (as it was done in previous studies [3,5]). We obtained 2D or 3D data representative of PA-signals by raster scanning. We employed a high precession translation stage (Newmark NSC-G series, Newmark Systems Inc., Rancho Santa Margarita, CA, USA) for raster scanning of the transducer with step-size (~100 µm) while the optical pulse beam and the sample was kept fixed in a position. For a scanning area of ~4.8 × 3 mm^2—that can be achieved by ~48 × 30 raster scanning steps—we can acquire an image at the frame rate ~0.69 Hz (~4 frames per min). In addition to pulse (nsec) laser beam, as it is shown in Figure 2a, the imaging sample is illuminated with a collimated optical beam (diameter ~3 mm) from a CW-laser source (Stradus-642-110, VORTRAN Laser Technology, Roseville,

CA, USA; wavelength, 642 nm) in such a way that the CW-beam and pulse-laser beam intersect each other at a pre-specified region of interest. Optical wavelengths of CW-illumination (for enhancement of PA-signal strength) and PA-excitation (to induce PA-signals) can be selected independently. CW-laser is continuously switched on during the experiments to maintain the temperature. The entire AR-PAM imaging system is controlled by a LabVIEW-based software. Figure 2b gives a typical 1D PA-signal acquired by our AR-PAM imaging system, while Figure 2c depicts a photograph of the sample holder (transparent cuvette). During experiments, the cuvette filled with sample material (say, methylene blue solution) is completely immersed inside an acoustic-coupling medium (water, in our case) for proper coupling of acoustic signals. We employed AMIRA software for the generation of 3D images from a sequence of 2D images, which are, in turn, reconstructed by using MATLAB.

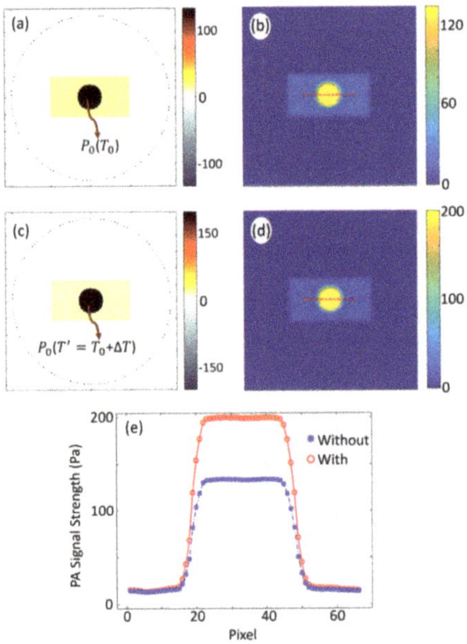

Figure 1. Representative images of initial pressure distribution of the tissue-mimicking numerical sample with a circular target being embedded in the background and sensor array distributed over a circular ring (**a**,**c**). Numerical samples with circular targets at a temperature T_0 (**a**), and temperature $T' = T_0 + \Delta T$ (**c**). The corresponding reconstructed images (**b**) for target at a temperature T_0 and (**d**) for target at temperature T' obtained using k-wave MATLAB toolbox. (**e**) Variation of PA-signal strength along the marked lines shown in (**b**,**d**).

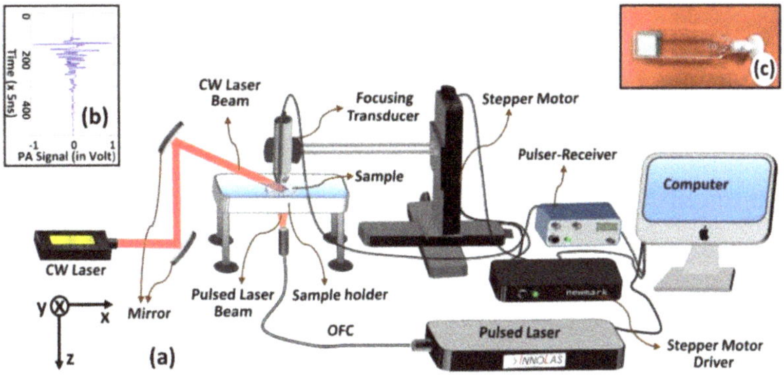

Figure 2. Schematic diagram of the experimental set-up (**a**). A typical 1D PA-signal acquired by our acoustic resolution photoacoustic microscopy (AR-PAM) imaging system (**b**). (**c**) Photograph of transparent cuvette (inner thickness ~1 mm) that is employed as holder for imaging sample (methylene blue solution).

3. Experimental Results and Discussion

3.1. Enhancement of PA-Signal from Methylene Blue

Figure 3 gives the experimental results to compare PA-signal strengths (P_0) with and without pre-illumination of CW-laser beam. Experiments were conducted with methylene blue solution (concentration of $\sim$7 mM using water as a solvent) as an imaging sample. Figure 3a,b present 3D image representatives of P_0 with and without pre-illumination, respectively, while Figure 3c,d give 2D images corresponding to a randomly selected plane (at $z = 7$). For quantitative analysis, the variation of P_0 along the marked lines, which are indicated in Figure 3c,d, are plotted in Figure 3e. In Figure 3c,d, yellow spot appearing in blue background corresponds to P_0 acquired from pulse laser-irradiated spot in sample while the background corresponding to un-illuminated region, which gives no PA-signal. The size and shape of the image, which is characterized by pulse-laser spot, are in agreement with that of cross-section (FWHM) of pulse-laser beam. From Figure 3c,d, it is observed that maximum P_0 achievable with pre-illumination is 0.60 V against 0.40 V (without pre-illumination). Average P_0 over full-width at 75% of maximum is estimated to be 0.50 V (for pre-illumination) against 0.35 V (without pre-illumination) and enhancement ($P_0^{(enh)}$), corresponding to the plane $z = 7$, is found to be 46%. In the similar fashion, $P_0^{(enh)}$ for the planes (corresponding to $z = 6$ and $z = 8$) were obtained as 40% and 38%, respectively. The average value of the measurements from the three consecutive planes (41%) is considered as the $P_0^{(enh)}$ with pre-illumination for a particular power and concentration. The results demonstrate that pre-illumination with CW-optical beam provides a significant enhancement in P_0, which is due to enhancement in temperature change following optical illumination of CW-beam (as it is given by Equation (5)). Similar experiments were conducted for empty cuvette and water (filled in cuvette) being employed as imaging sample. The experimental results, presented in supplementary figures (Figures S1 and S2), demonstrate that cuvette and water give no significant PA-signal for both of the two cases (with and without pre-illumination). These experimental results imply that $P_0^{(enh)}$ is not contributed from sample holder (cuvette) and acoustic-coupling medium (water), or enhancement is contributed only from the sample (methylene blue, in our case).

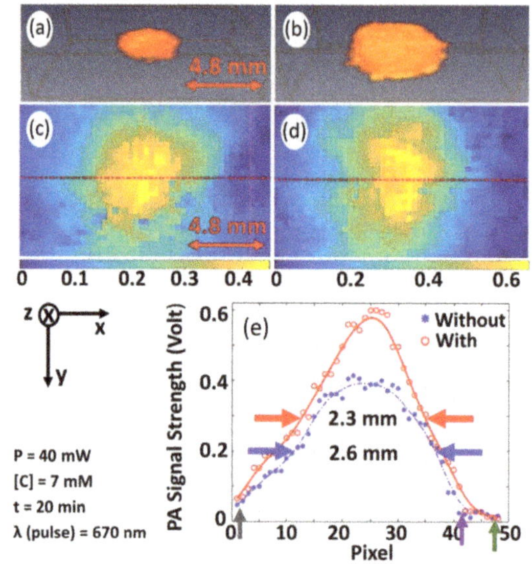

Figure 3. 3D image representatives of P_0 without (**a**) and with (**b**) pre-illumination for methylene blue while 2D image corresponding to a randomly selected plane (at z = 7) ((**c**), without pre-illumination and (**d**), with pre-illumination). (**e**) Variation of P_0 along the marked lines depicted in (**c**,**d**). Arrowheads in both sides of scale bar (being included in (**a**,**c**)) indicate measure of the entire length of the image along x-axis. Arrowheads, marked in (**e**), indicate the points of observations for estimation of size of the illuminated spot.

From Figure 3d, the size of the illuminated spot was estimated—for pre-illumination as well as without pre-illumination—from two different aspects: (a) measuring the full width at half maximum (FWHM) using four cross-sectional profiles (or line plots) (as it is shown in Figure S3). The profile along the x-axis is given in Figure 3e and FWHM corresponding to without (blue color arrow) and with (red color arrow) pre-illumination are estimated to be ~2.3 mm and ~2.6 mm, respectively. The average value of the measured FWHM from all profiles is calculated as ~2.3 mm (with) and ~2.2 mm (without), respectively. This shows that, by our proposed pre-illumination technique, the effective resolution is reduced. (b) Measuring the change in object size across the baseline, i.e., the distance between two points indicated in Figure 3e by arrowheads (grey color and violet color (without pre-illumination); grey color and green color (with pre-illumination)). The width of the spot was found to be ~4.3 mm (without pre-illumination) and ~4.7 mm (with pre-illumination) and the increment in width is obtained as 9.5%. This implies that the obtainable spatial resolution of the imaging system is slightly reduced (~9.5%) by the proposed optical pre-illumination technique. This may be because of thermal excitation and diffusion of heat (resulted from pre-illumination of CW-laser beam) [18,22], thereby enhancing strength of PA-signal from the immediate background of the (pulse-laser) illuminated spot. More elaborately, transient heat generation, which results from transient pulse-laser excitation, is localized [47] (under the physical condition of thermal and stress confinement) while (ambient) temperature raise (and its heat generation) induced by CW-optical beam illumination is diffused (i.e., not localized over the illuminated region perfectly). In this way, the effective size of the illuminated spot becomes wider due to pre-illumination with CW-optical beam.

3.1.1. Variation of PA-Signal Enhancement with Variation in Different Parameters

Figure 4a presents the variation of P_0 with concentration ([C]) of sample for both of the two cases (with pre-illumination and without pre-illumination) where P_0 is measured as an average over full-width at 75% of maximum as it is done in the previous case (shown in Figure 3e). In the experiments, we employed CW-laser with power (P) $\sim$ 60 mW and time duration of pre-illumination (t) ~20 min. Figure 4b depicts the variation of P_0 with respect to μ_a (which is estimated from [C]), where $P = 60$ mW and $t = 20$ min being employed as the experimental parameters. Variation in $P_0^{(enh)}$ with pre-illumination, which is estimated using Equation (5), is also depicted. It is observed that P_0 and $P_0^{(enh)}$ increases linearly with concentration. The observed linearity may be because of an increase in deposition of optical energy as well as mechanical coupling of constituent particles with an increase in $[C]$. Variation in P_0 and $P_0^{(enh)}$ with CW-optical power used for pre-illumination is shown in Figure 4c, (where $[C] = 7$ mM and $t = 20$ min being employed as the experimental parameters), while $P_0^{(enh)}$ with time interval (t) of pre-illumination of CW-optical beam is depicted in Figure 4d (where $[C] = 7$ mM, $P = 60$ mW). Without pre-illumination, the optical power of incident pulse-beam remains the same so that P_0 remains unchanged (which are depicted by blue-color dotted lines in Figure 4c,d). In this case, P_0 and $P_0^{(enh)}$ increases non-linearly, which is followed by saturation at a particular time-point $t \sim 20$ min. This is due to saturation in the deposition of optical energy and, thus, thermo-elastic expansion. Saturation in $P_0^{(enh)}$ suggests that, for practical applications, selection of stability point of signal enhancement is demanded. In our validation study, with experiments being conducted in tissue-mimicking Agar phantom and chicken tissue, pre-illumination of the sample with $t = 20$ min was performed.

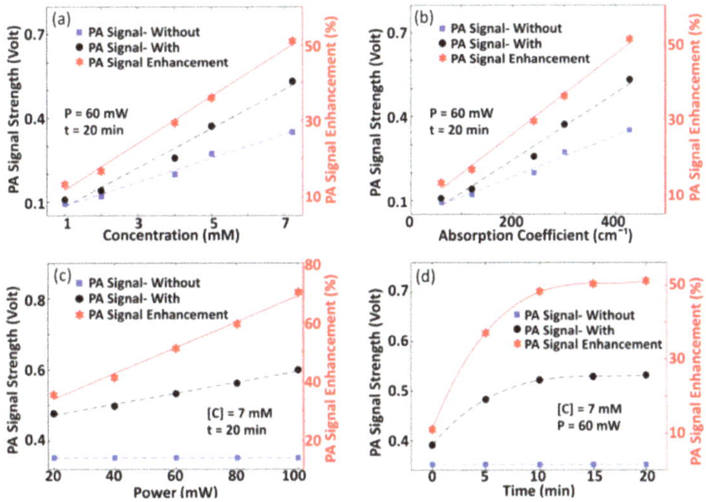

Figure 4. (**a**) Variation of measured P_0 and $P_0^{(enh)}$ with concentration ([C]) of sample solution. In the experiments, we employed a continuous wave (CW)-laser of power (P) $\sim$ 60 mW and time duration of pre-illumination (t) $\sim$ 20 min). (**b**) Variation of measured P_0 and $P_0^{(enh)}$ with optical absorption coefficient (where $P = 60$ mW and $t = 20$ min). (**c**) Variation of measured P_0 and $P_0^{(enh)}$ with optical power of CW-optical beam (where $[C] = 7$ mM and $t = 20$ min), (**d**) Variation of P_0 and $P_0^{(enh)}$ with time-interval of pre-illumination (where $[C] = 7$ mM, $P = 60$ mW). In all of the experiments, we employed pulsed laser beam of wavelength, $\lambda \sim 670$ nm.

3.1.2. Variation of Temperature Raise with Variation in Different Parameters

Figure 5 presents the experimental results for characterization studies of the variation of temperature-rise (ΔT) with various physical parameters of interest. A detailed description of the experimental set-up employed for this study is provided in supplementary (see Figure S4). Figure 5a depicts the variation of ΔT with respect to the time interval of pre-illumination (t) of the CW-optical beam where concentration $([\mathrm{C}])$ of methylene blue is considered as an experimental parameter. Meanwhile, the variation of ΔT with t, where optical power of CW-optical beam is considered as an experimental parameter, is shown in Figure 5b. In these figures, it is observed that the temperature raise increases non-linearly and attains saturation at a certain point of the time interval of pre-illumination ($\sim$20 min). Inset in Figure 5b shows a curve-fit, for 80 mW power of CW-laser, using $\Delta T = \Delta T_{sat}(1 - e^{-\gamma t})$ (where ΔT_{sat} is the maximum temperature raise attained at saturation and γ is a constant that characterizes the growth rate) and it is found that maximum temperature raise is 9.13 °C while growth rate, $\gamma \sim 0.3909$ sec^{-1}. From the figures, it is observed ΔT_{max} and γ are dependent on $[C]$ and optical power of incident CW-beam. Figure 5c shows the variation of temperature raise with $[C]$, which are obtained from Figure 5a with $t = 20$ min. From Figure 5c, we obtained variation of ΔT with μ_a and it is shown in Figure 5d. Similarly, the variation of ΔT with optical power of CW-beam at $t = 20$ min is given in Figure 5e. In the figures (Figure 5a–e), $P_0^{(enh)}$, which is presented in Figure 4, is also presented for comparison of ΔT and $P_0^{(enh)}$. This linearity in ΔT—which is due to increase and subsequent saturation in the absorption of optical energy—is in agreement with that for $P_0^{(enh)}$. From these results (Figures 4 and 5), one may conclude that $P_0^{(enh)}$ is dominantly contributed from ΔT due to CW-optical pre-illumination in support of our theoretical hypothesis given by Equation (4). This is in agreement with the (simulation and experimental) study, reported by Mahmood et al. [20], that demonstrated Γ and, thus, P_0 are dependent on temperature. The time duration for achieving the saturation point of signal enhancement is characteristic of samples, and it can be considered as a preparatory requirement for imaging.

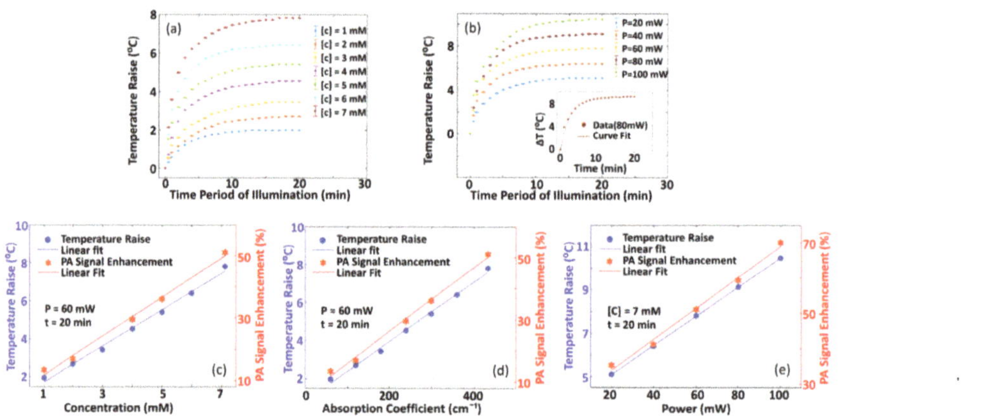

Figure 5. (**a**) Variation of temperature-raise (ΔT) with respect to time-interval (t) of pre-illumination of CW optical beam (with concentration as experimental parameter (for $P = 60$ mW)). (**b**) Variation of ΔT against t with optical power (P) of CW laser beam as an experimental parameter (for $[C] = 7$ mM), (**c**) Variation of ΔT against concentration (where $P = 60$ mW, $t = 20$ min were employed), (**d**) Variation of ΔT against optical absorption coefficient (where $P = 60$ mW and $t = 20$ min were employed), and (**e**) Variation of ΔT against optical power of CW beam for pre-illumination (where $[C] = 60$mW and $t = 20$ min were employed).

3.2. *Variation of Optical Extinction Coefficient of Methylene Blue with Temperature*

Figure 6 gives the experimental results for the characterization study of variation of optical extinction coefficient (ε) with temperature. For this study, we employed a UV-VIS as spectrum analyzer (PerkinElmer, Lambda 750) and ε is estimated from absorbance spectrum using $\varepsilon = \frac{Absorbance}{d\,[C]}$ where d is the (inner) thickness of sample holding cuvette ($\sim$1 cm) where methylene blue solution is filled in. The sample is heated before introduction to the UV-VIS spectrum analyzer. Temperature (T) is measured before and after the experiments, and its mean value is considered as the temperature measurement for the particular spectrum analysis experiment. Figure 6a depicts the variation of ε against wavelength (λ) while Figure 6b gives the variation of ε against T. From the figure, it is observed that ε varies linearly with T where intercept and slope of the linear curve are characteristics of λ. From this experimentally estimated ε, for a given T, we estimated the contribution of optical absorption to the enhancement of photoacoustic signal strength (given in Equations (4) and (5)). However, thermal expansion coefficient at any given temperature T was estimated using a linear fit of β with T given in Table 1. Using these measurements of ε and β (in Equations (4) and (5)), we estimated $P_0^{(enh)}$, which we consider as analytical calculation (in Figure 7), wherein thermal perturbation (ΔT) is adopted from the measurements corresponding to Figure 5.

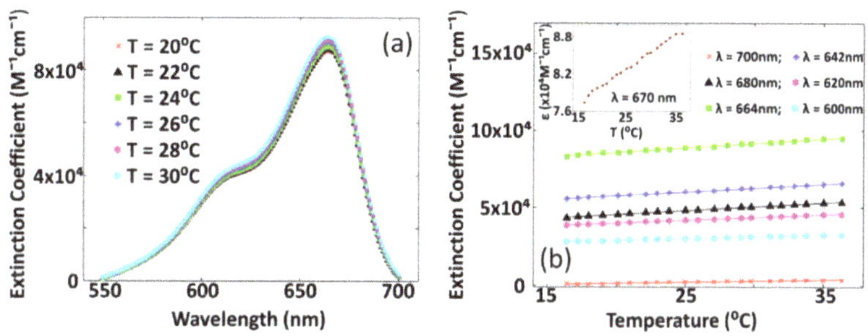

Figure 6. Variation of extinction coefficient (**a**) with respect to optical wavelength (λ) (at various thermodynamic temperature (T)). (**b**) Variation of optical extinction coefficient (ε) against (T) for different λ. Inset in (**b**) presents variation of ε against T for $\lambda = 670$ nm, which is the optical wavelength of pulsed laser beam used for imaging (in our present study). In the experiments, we used methylene blue solution with $[C] \sim 2$ μM.

3.3. Comparison of Numerical Simulation and Experimental Results, and Validation of Theoretical Hypothesis

Figure 7 presents the results of numerical simulation studies—for validation of our proposed theoretical hypothesis (presented in Section 2.1)—in comparison to the experimental results (given in Figure 4). In the meantime, $P_0^{(enh)}$, which is estimated theoretically or analytically using Equations (4) and (5), is also included. Variation of $P_0^{(enh)}$ with respect to $[C]$, μ_a, optical power of CW-beam (used for pre-illumination), and ΔT are shown in Figure 7a–d. We estimate the slopes for variation of $P_0^{(enh)}$ with: (i) concentration as 6.24/mM (from experiments) against 5.52/mM (from analytical calculation) and 5.92/mM (from simulations), (ii) optical absorption coefficient as 0.103 cm (from experiments) against 0.091 cm (from analytical calculation) and 0.098 cm (from simulations), (iii) optical power of CW-beam as 0.439/mW (from experiments) against 0.406/mW (from analytical calculation) and 0.442/mW (from simulations), (iv) temperature raise (due to concentration or optical power) as 5.97 °C^{-1} (from analytical calculation) and 6.37 °C^{-1} (from simulations). In the figures, it is observed that slopes of the variation of $P_0^{(enh)}$ with the physical parameters of interest, obtained from experiments and numerical simulations, are in good agreement. This demonstrates that the proposed hypothesis of the dependence of P_0 and $P_0^{(enh)}$ on T (as it is given in Equations (4) and (5)) is validated. A deviation in absolute measurements of enhancement, as it is observed in Figure 7, may be due to lower in measurement of temperature of the imaging sample at a pre-specified region of interest—over which CW laser beam illuminates—resulting from tips of the (thermocouple) temperature sensor not being situated exactly at the point of illumination while attempting to prevent direct exposure of the tips to the incident light beam. To note, direct exposure of light to thermocouple tips results in the heating of the tips, which gives inaccurate measurements of the temperature of the medium.

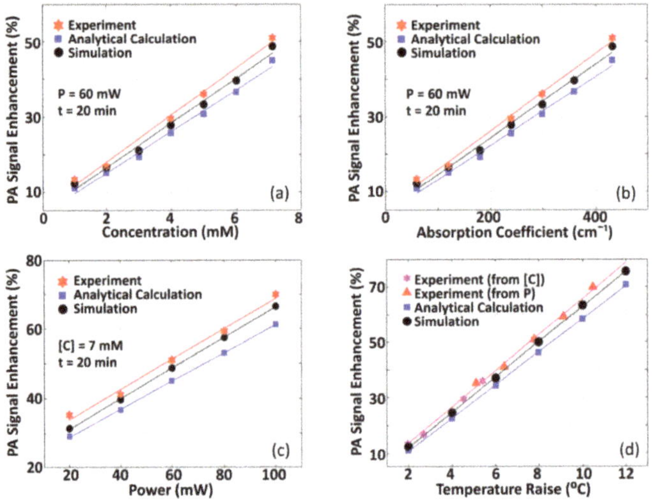

Figure 7. Comparison of the results obtained from experiments, theoretical hypothesis, and numerical simulation. (**a**) Variation of $P_0^{(enh)}$ with concentration (we employed $P = 60$ mW and $t = 20$ min), (**b**) Variation of $P_0^{(enh)}$ with optical absorption coefficient ($P = 60$ mW and $t = 20$ min were employed). (**c**) Variation of $P_0^{(enh)}$ with optical power (*where* $[C] = 7$mM *and* $t = 20$ min). (**d**) Experimental measurements in $P_0^{(enh)}$ with respect to temperature raise (ΔT), with ΔT being obtained from variation of optical power as well as concentration. The wavelength of the pulsed laser used for experiment is 670 nm.

3.4. Validation of the Proposed Technique for Pre-Clinical/Clinical Studies

To validate the proposed technique for enhancement of PA-signal strength, in preclinical and/or clinical studies, experiments were conducted in tissue-mimicking Agar-phantom as well as biological tissue (chicken breast collected from supermarket). Experimental results are depicted in Figure 8 (for Agar phantom) and Figure 9 (for chicken tissue). Figure 8a,b give 3D images representative of P_0 obtained without and with pre-illumination. Figure 8c,d depict 2D images (corresponding to $y = 3$ in 3D images). Line plots, showing the variation of P_0 and $P_0^{(enh)}$ along marked lines indicated in Figure 8b,c, are depicted in Figure 8d. $P_0^{(enh)}$ is found to be 48%, which is in agreement with ~1.5% °C^{-1} increase in the Grüneisen parameter with temperature (for biological tissue) [8,20]. It is observed that the measured P_0 is significantly enhanced with pre-illumination of CW-laser beam. Figure 8f gives a photograph of Agar-sample where a target ($1.2 \times 0.5 \times 2.2$ mm$^3(xyz)$) was embedded in a background phantom ($24 \times 1.5 \times 10$ mm^3). Similar to reported studies [3,5], optical and elastic properties were tailored such that the target has only contrast in μ_a with respect to background (0.02 mm^{-1} (target) against 0.01 mm^{-1} (background)). Figure 9a,b gives 3D images, while Figure 9c,d give 2D images, representative of P_0 obtained without and with pre-illumination. Line plots, showing the variation of P_0 and $P_0^{(enh)}$ along the marked lines indicated in Figure 9b,c, are depicted in Figure 9d. $P_0^{(enh)}$ is estimated and found to be ~40%, which is a significant enhancement with pre-illumination of CW-laser beam. Figure 9f gives a photograph of the chicken breast tissue sample where a transparent (glass) tube of inner diameter (~1 mm) is inserted through the chicken breast sample at depth ~2.5 mm. The tube was filled with methylene blue, whereby mimicking deep-seated blood vessels in (chicken breast) tissue sample. In Figure 8, one observes that PA-representative image of the rectangular-shaped target appears to be circular (against appearing as a line in

Figures 8c and 9c). This is because of the pre-illumination of CW-laser beam, which is circular in cross-section. These results, with experiments being conducted in (chicken) tissue sample, demonstrate the practical applicability of the proposed photo-thermal technique to pre-clinical and/or clinical studies.

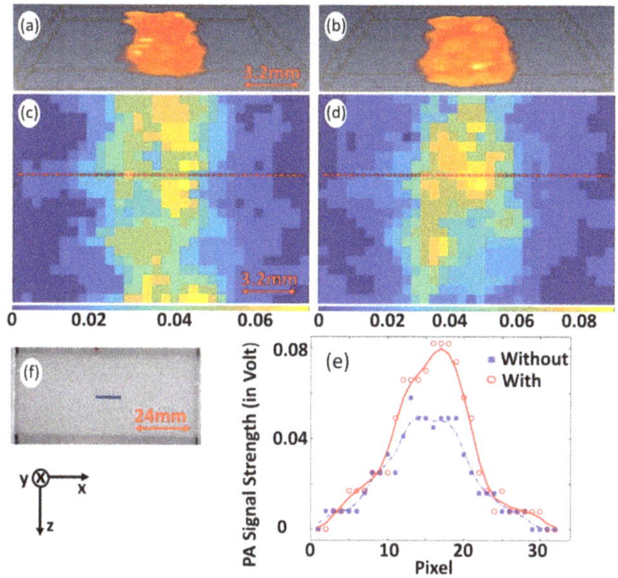

Figure 8. 3D images representative of P_0 obtained without (**a**) and with pre-illumination (**b**), while (**c**,**d**) give corresponding 2D images for plane (at $y = 3$). (**e**) Variation of P_0 and $P_0^{(enh)}$ along marked lines indicated in (**c**,**d**). Photograph of tissue-mimicking Agar-sample (**f**) where marked line indicates region for raster scanning. Arrowheads in both sides of the scale bar (being included in the figure) indicate measure of entire length of the image along x -axis.

3.5. Discussion

The enhancement of PA-signal strength can be contributed from various factors (including change in specific heat capacity and isothermal compressibility), which are neglected in view of the argument described in Section 2.1. In our study, we assume that signal strength enhancement is contributed from a raise in thermodynamic equilibrium temperature (T) and, thus, thermal expansion coefficient (β) and optical extinction coefficient (ε) of the tissue target. Moreover, the CW-laser beam enhances (but does not induce itself) PA-signal strength induced by pulse-laser excitation. Even though the CW-laser illumination increases the optical energy being irradiated to sample, the total optical energy is restricted within this safety limit ($\sim$ 20 mJ/cm^2, FDA, Silver Spring, MD, USA [48]). Experiments were carried out to compare efficiency in the enhancement of PA-signal strength with CW-laser pre-illumination and pulse-laser excitation. Experimental results demonstrate that it is more efficient (46% (for CW-laser pre-illumination) against 98% (for pulsed-laser excitation)) to increase PA-signal strength with an increase of pulsed-laser energy in comparison to a similar amount of increase in CW-laser energy, which is in agreement with the reported study [27]. Even though pulse-laser is more efficient to enhance PA-signal strength, the price of pulse-laser increases drastically as the output pulse energy increase, which then leads to an increase in the overall cost of the PAI system. In this way, the proposed technique paves a cost-effective way to improve PA-signal strengths. Again, from our study, it is observed that a temperature raise due to illumination of CW-optical beam is $\sim$10.5 °C, which is very low in comparison to temperature increase as in the case of photo-therapy (where targeted temperature measures $\sim$40 °C [28,43,49]), which remains as a standard clinical therapy. To the best of our knowledge, this pre-illumination technique may not impose any significant side effects in tissue (in in-vivo study and its applications).

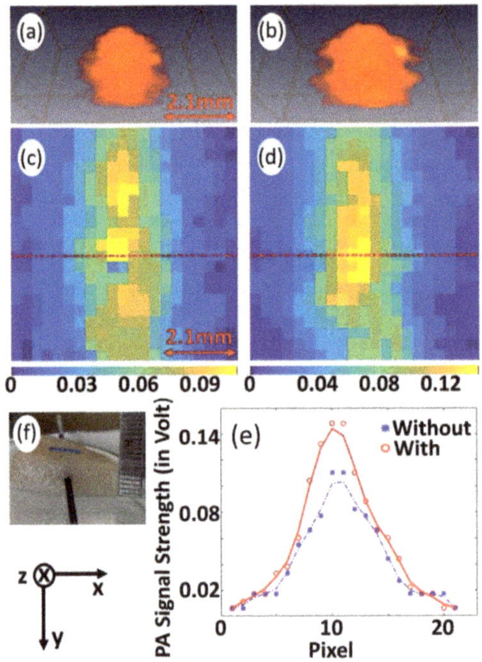

Figure 9. 3D images representative of P_0 obtained without (**a**) and with pre-illumination (**b**), while (**c**,**d**) give corresponding 2D images for plane (at $y = 3$). (**e**) Variation of P_0 and $P_0^{(enh)}$ along marked lines indicated in (**c**,**d**). Photograph of chicken breast tissue (**f**) in which transparent glass of inner diameter (~1 mm) filled with methylene blue, thereby, mimicking deep-seated blood vessels is inserted at depth ~2.5 mm (as it is observed from attached scale). Marked line indicates region for raster scanning. Arrow heads in both sides of scale bar (being included in the figure) indicate measure of entire length of the image along x -axis.

Refs. [14,15] demonstrated that with an increase in temperature ~37 °C (from ~33 °C to ~70 °C), PA- signal strength is enhanced by ~120% (from ~5 mV to ~11 mV) for soft tissue (chicken breast) and ~116% (from ~6 mV to ~13 mV) (for porcine liver) [14,15]. For porcine muscle, enhancement in PA-signal strength is ~42% corresponding to change in temperature ~9 °C [28]. The enhancement of PA-signal strength we achieved in our study is relatively high in comparison to that of Refs. [14–16]. This may be because of different experimental conditions. In our study, a deep-seated target of interest is selectively illuminated by CW-laser beam and, thus, the target is selectively heated while in Ref. [14], the entire sample is immersed inside a heating bath, and the temperature is raised as a whole including background (that also generate PA-waves), i.e., the difference in the enhancement of PA-signal strength may be due to reduce in the obtainable SNR resulted from the increasing background signal [12]. In Ref. [21], an optical fluence ~5 mJ/cm^2 was employed against ~15 mJ/cm^2 in our present study that gives a difference of 6 times in enhancement of signal strength.

One of the drawbacks of our proposed photo-thermal technique is that enhancement of PA-signal strength is at the cost of degradation in the obtainable spatial resolution. However, the degradation in spatial resolution is ~9%, which is much lower than the enhancement in PA-signal strength (~70% (methylene blue), ~48% (Agar phantom)), and ~40% (chicken tissue)). Our study suggests that one can employ a CW-optical beam to perform pre-illumination specifically over a certain point deep inside the tissue without disturbing the intervening tissue medium so as to further enhance the PA-signal strength from a specific deep-seated target reducing the noise from the background or surroundings. LED-based PAI is recently gaining popularity in a wide range of superficial imaging applications, and it holds greater potential in clinical translation because of its portability and affordability. The proposed technique is a promising method to image deep-seated tissues with LED-based systems and accelerates its clinical translation. Strengthening the intensity of boundary measured signals—for improving signal detectability—is a critical factor in imaging (in general) and PA-imaging (in particular). In addition, the obtainable signal contrast between targets (including contrast agents) and the background is more relevant to the improvement of imaging quality. Our proposed photo-thermal technique offers improvement in signal contrast and, thus, the image quality.

4. Conclusions

In conclusion, we demonstrate an optothermal-based technique for enhancement of obtainable PA-signal strength that is, experimentally, facilitated by pre-illumination with CW-optical beam (in addition to pulse-optical beam) in imaging sample. We propose a theoretical hypothesis and it is validated by numerical simulation studies being performed using k-wave toolbox. Experimental studies were conducted in tissue-mimicking Agar phantom and ex-vivo animal tissue (chicken breast) samples. Experimental results demonstrate that a significant enhancement (up to ∼70% (methylene blue), ∼48% (Agar phantom), and ∼40% (chicken tissue)) in the measured PA-signal strength can be achieved by our proposed photo-thermal technique. This unique (non-invasive and non-destructive) technique for enhancement of PA-signal strength will have a significant impact in PA-imaging while improving not only achievable penetration depth but also SNR and, hence, accuracy in quantitative measurement of vital patho-physiological parameters. Particularly, this technique can address the pertaining challenge associated with weak PA-signal from deep-seated tissue regions, which is the major issue associated with the clinical translation of the LED-based PAI systems.

Supplementary Materials: The following are available online at https://www.mdpi.com/1424-8220/21/4/1190/s1, Figure S1: 2D images representative, for empty cuvette as sample specimen for imaging, of PA-signal strength obtained without and with pre-illumination, Figure S2: 2D images representative, for water (filled in cuvette) as sample specimen for imaging, of PA-signal strength obtained without and with pre- illumination, Figure S3: 2D images (corresponding to Figure 3) and line plots along different direction for the calculation of FWHM, Figure S4: A schematic diagram of experimental set-up for the study of variation of temperature rise.

Author Contributions: Conceptualization, A.T., J.M. and M.S.S.; Data curation, A.T. and S.P.; Formal analysis, A.T. and J.M.; Funding acquisition, M.S.S.; Investigation, A.T. and S.P.; Methodology, A.T. and M.S.S.; Project administration, M.S.S.; Resources, S.P. and J.M.; Software, A.T.; Supervision, M.S.S.; Validation, A.T.; Visualization, A.T.; Writing—original draft, M.S.S.; Writing—review & editing, A.T., S.P., J.M. and M.S.S. All authors have read and agreed to the published version of the manuscript.

Funding: This research received no external funding.

Institutional Review Board Statement: Not applicable.

Informed Consent Statement: Not applicable.

Data Availability Statement: The data that supports the findings of the study are provided within the article and the supplementary material.

Acknowledgments: The authors want to acknowledge Rinsa S.R., BII Lab, IISER-TVM, for her technical support and suggestions.

Conflicts of Interest: The authors have no relevant financial interests in this article and no other potential conflicts of interest to disclose.

Appendix A. Dependence of Initial Pressure Waves on Thermodynamic Parameters for Low Concentration Materials

For a physical system or medium (including solution and biological tissue), that is constituted by various individual components, its mass density is given by $\rho \approx \frac{\sum_{i=1}^{N} m_i}{\sum_{i=1}^{N} V_i}$ (where m_i and V_i are mass and volume of ith component, and N is the total number of individual components). Under physical condition of low concentration, as it is the case of our present study (methylene blue solution with concentration of methylene (~mM) being dissolved in water (as a solvent) and tissue-mimicking Agar phantom with concentration of India ink (~3–5 µl) in Agar gel (~100 gms) [3–5]), one can approximate $\sum_{i=1}^{N} m_i \approx m^{(water)}$ and $\sum_{i=1}^{N} V_i \approx V^{(water)}$ (for methylene blue solution), i.e., $\rho \approx \rho^{(water)}$ (for methylene blue solution) and $\rho \approx \rho^{(agar)}$ (for tissue-mimicking Agar phantom added with India ink as an optical absorber). Similarly, specific heat capacity can also be approximated as $c_V = \frac{\sum_{i=1}^{N} Q_i}{\sum_{i=1}^{N} m_i}$ $\approx \frac{Q^{(water)}}{m^{water}\Delta T} = c_V^{(water)}$ (where Q_i is the heat absorbed by i^{th} components of the medium). In the meantime, optical properties of the system is given by collective effects of the individual components as linear combination [50], i.e., $\mu_a = \sum_{i=1}^{N} \mu_a^i = \sum_{i=1}^{N} \varepsilon^i [C^i]$ (where ε^i and $[C^i]$ are extinction (optical absorption) coefficient and concentration of ith component, respectively). For methylene blue solution, we can approximate that $\mu_a = \mu_a^{(water)} + \mu_a^{(methy)}$ $= \mu_a^{(water)} + \varepsilon^{(methy)}[C^{(methy)}] \approx \mu_a^{(methy)}$ (for $\mu_a^{(water)} \ll \mu_a^{(methy)}$ [37]). Similarly for Agar phantom, one can approximate $\mu_a = \mu_a^{(agar)} + \mu_a^{(ink)} = \varepsilon^{(agar)}[C^{(agar)}] + \varepsilon^{(ink)}[C^{(ink)}]$ $\approx \varepsilon^{(ink)}[C^{(ink)}] \approx \mu_a^{(ink)}$ (for $\varepsilon^{(agar)} \ll \varepsilon^{(ink)}$ [51]). Again, from Dalton's law of partial pressures, total pressure of a thermodynamic system is given by an algebraic sum of partial pressures imparted by the individual components, i.e., $P = \sum_{i=1}^{N} P_i$ (where P_i is the partial pressure of ith component) and it can be approximated as $P \approx P^{(water)}$ (for low concentration methylene solution) or $P \approx P^{(agar)}$ (for Agar phantom). Further, under thermodynamic equilibrium condition, there exists a uniform distribution of temperature (T) at any arbitrarily chosen (spatial) point $(\vec{r})$ over the entire space of the thermodynamic system, i.e., T is independent of spatial position $(\vec{r})$ and, thus, measurements of T for all of the individual constituent components are the same. From the thermodynamic relations of κ and β, we can approximate $\kappa = -\frac{1}{V}\left(\frac{\partial V}{\partial P}\right)_T \approx -\frac{1}{V^{(water)}}\left(\frac{\partial V^{(water)}}{\partial P^{(water)}}\right)_T = \kappa^{(water)}$ and $\beta = -\frac{1}{V}\left(\frac{\partial V}{\partial T}\right)_P \approx -\frac{1}{V^{(water)}}\left(\frac{\partial V^{(water)}}{\partial T^{(water)}}\right)_P = \beta^{(water)}$ (for methylene blue solution) while $\kappa = -\frac{1}{V}\left(\frac{\partial V}{\partial P}\right)_T \approx -\frac{1}{V^{(water)}}\left(\frac{\partial V^{(agar)}}{\partial P^{(agar)}}\right)_T = \kappa^{(agar)}$ and $\beta = -\frac{1}{V}\left(\frac{\partial V}{\partial T}\right)_P \approx$ $-\frac{1}{V^{(agar)}}\left(\frac{\partial V^{(agar)}}{\partial T^{(agar)}}\right)_P = \beta^{(agar)}$ (for Agar phantom). Now, for the case of low concentration, we can write Equation (A1) as:

$$P_0 = \frac{\beta^{(water)}}{\kappa^{(water)}} \frac{1}{\rho^{(water)} c_V^{(water)}} \mu_a^{(methy)} \phi \tag{A1}$$

References

1. Strohm, E.M.; Moore, M.J.; Kolios, M.C. Single cell photoacoustic microscopy: A review. *IEEE J. Sel. Top. Quantum Electron.* **2016**, *22*, 6801215. [CrossRef]
2. Wang, L.V.; Hu, S. Photoacoustic tomography: In vivo imaging from organelles to organs. *Science* **2012**, *335*, 1458–1462. [CrossRef] [PubMed]
3. Singh, M.S.; Jiang, H. Elastic property attributes to photoacoustic signals: An experimental phantom study. *Opt. Lett.* **2014**, *39*, 3970–3973. [CrossRef] [PubMed]
4. Singh, M.S.; Jiang, H. Ultrasound (US) transducer of higher operating frequency detects photoacoustic (PA) signals due to the contrast in elastic property. *AIP Adv.* **2016**, *6*, 1–9. [CrossRef]
5. Singh, M.S.; Jiang, H. Estimating both direction and magnitude of flow velocity using photoacoustic microscopy. *Appl. Phys. Lett.* **2014**, *104*, 1–5. [CrossRef]

6. Jiang, H.; Yuan, Z.; Gu, X. Spatially varying optical and acoustic property reconstruction using finite element-based photoacoustic tomography. *J. Opt. Soc.* **2006**, *23*, 878–888. [CrossRef]
7. Dean-Ben, X.L.; Fehm, T.F.; Ford, S.J.; Gottschalk, S.; Razansky, D. Spiral volumetric optoacoustic tomography visualizes multi-scale dynamics in mice. *Light Sci. Appl.* **2017**, *6*, 1–8. [CrossRef]
8. Singh, M.K.A.; Agano, T.; Sato, N.; Shigeta, Y.; Uemura, T. Real-time in vivo imaging of human lymphatic system using an LED-based photoacoustic/ultrasound imaging system. In Proceedings of the Photons Plus Ultrasound: Imaging and Sensing 2018, San Francisco, CA, USA, 19 February 2018.
9. Zhu, Y.; Feng, T.; Cheng, Q.; Wang, X.; Du, S.; Sato, N.; Yuan, J.; Singh, M.K.A. Towards clinical translation of LED-based Photoacoustic imaging: A review. *Sensors* **2020**, *20*, 2484. [CrossRef]
10. Singh, M.K.A. *LED-Based Photoacoustic Imaging from Bench to Bedside*; Springer nature Singapore Pte Ltd.: Singapore, 2020.
11. Zhong, H.; Duan, T.; Lan, H.; Zhou, M.; Gao, F. Review of low-cost photoacoustic sensing and imaging based on laser diode and light-emitting diode. *Sensors* **2018**, *18*, 2264. [CrossRef]
12. Larina, I.V.; Larin, K.V.; Esenaliev, R.O. Real-time optoacostic monitoring of temperature in tissues. *J. Phys. D* **2005**, *38*, 2633–2639. [CrossRef]
13. Petrova, E.; Anton, L.; Alexander, A.O.; Sergey, A.E. Temperature-dependent optoacoustic response and transient through zero Grüneisen parameter in optically contrasted media. *Photoacoustics* **2017**, *7*, 36–46. [CrossRef] [PubMed]
14. Nikitin, S.M.; Khokhlova, T.D.; Pelivanov, I.M. In-vitro study of the temperature dependence of the optoacoustic conversion efficiency in biological tissues. *IEEE J. Quantum Electron.* **2012**, *46*, 269–276. [CrossRef]
15. Nikitin, S.M.; Khokhlova, T.D.; Pelivanov, I.M. Temperature dependence of the optoacoustic transformation efficiency in ex-vivo tissues for application in monitoring thermal therapies. *J. Biomed. Opt.* **2012**, *46*, 061214. [CrossRef] [PubMed]
16. Larin, K.V.; Larina, I.V.; Esenaliev, R.O. Monitoring of tissue coagulation during thermotherapy using optoacoustic technique. *J. Phys. D Appl. Phys.* **2005**, *38*, 2645–2653. [CrossRef]
17. Petrova, E.; Ermilov, S.; Su, R.; Nadvoretskiy, V.; Conjusteau, A.; Oraevsky, A. Using optoacoustic imaging for measuring the temperature dependence of Grüeisenparameter in optically absorbing solutions. *Opt. Exp.* **2013**, *21*, 195733. [CrossRef]
18. Wang, L.; Zang, C.; Wang, L.V. Grueneisen relaxation photoacoustic microscopy. *Phys. Rev. Lett.* **2014**, *113*, 174301. [CrossRef]
19. Tian, C.; Xie, Z.; Fabiilli, M.L.; Liu, S.; Wang, C.; Cheng, Q.; Wang, X. Dual-pulse nonlinear photoacoustic technique: A practical investigation. *Biomed. Opt. Exp.* **2015**, *6*, 241695. [CrossRef]
20. Mahmoodkalayeh, S.; Jooya, H.Z.; Hariri, A.; Zhou, Y.; Xu, Q.; Ansari, M.A.; Avanaki, M.R.N. Low temperature-mediated enhancement of photoacoustic imaging depth. *Sci. Rep.* **2018**, *8*, 1–9. [CrossRef]
21. Olivier, S.; Amaury, P.; Jerome, G.; Bossy, E. Influence of nanoscale temperature rises on photoacoustic generation: Discrimination between optical absorbers based on thermal nonlinearity at high frequency. *Photoaocustics* **2015**, *3*, 20–25.
22. Zharov Vladimir, P. Ultrasharp nonlinear photothermal and photoacoustic resonances and holes beyond the spectral limit. *Nat. Photonics* **2011**, *5*, 110–116. [CrossRef]
23. Esenaliev, R.O. Optoacoustic monitoring of physiologic variables. *Front. Physiol.* **2017**, *8*, 1030. [CrossRef] [PubMed]
24. Wu, D.; Huang, L.; Jiang, M.S.; Jiang, H. Contrast agents for photoacoustic and thermoacoustic imaging: A review. *Int. J. Mol. Sci.* **2014**, *15*, 23616–23639. [CrossRef] [PubMed]
25. Dean-Ben, X.L.; Fehm, T.F.; Ford, S.J.; Gottschalk, S.; Razansky, D. Contrast agents for molecular photoacoustic imaging. *Nat. Methods* **2016**, *13*, 639–650.
26. Amaury, P.; Florian, P.; Emmanuel, B. Photoacoustic generation by a gold nanosphere: From the linear to the nonlinear thermoelastic regime. *Phys. Rev. B* **2015**, *92*, 115450.
27. Irio, G.C.; Craig, W.; Diebold, G.J. Photoacoustic point source. *Phys. Rev. Lett.* **2001**, *86*, 3550–3553.
28. Shah, J.; Park, S.; Aglyamov, S.; Larson, T.; Ma, L.; Sokolov, K.; Johnston, K.; Milner, T.; Emelianov, S.Y. Photoacoustic imaging and temperature measurement for photothermal cancer therapy. *J. Biomed. Opt.* **2008**, *13*, 034024. [CrossRef] [PubMed]
29. Larin, K.V.; Larina, I.V.; Motamedi, M.; Esenaliev, R.O. Optoacoustic laser monitoring of cooling and freezing of tissuess. *IEEE J. Quantum Electron.* **2002**, *32*, 953–958. [CrossRef]
30. Esenaliev, R.O.; Petrov, Y.Y.; Cicenaite, I.; Chumakova, O.V.; Petrova, I.Y.; Patrikeev, I.; Liopo, A. Real-time noninvasive optoacoustic monitoring of nanoparticle-mediated photothermal therapy of tumors. In *Photons Plus Ultrasound: Imaging and Sensing 2007: The Eighth Conference on Biomedical Thermoacoustics, Optoacoustics, and Acousto-Optics*; International Society for Optics and Photonics: San Jose, CA, USA, 19 March 2007; Volume 6437, p. 64370Q-1. [CrossRef]
31. Wang, Y.; Xie, X.; Wang, X.; Ku, G.; Gill, K.L.; O'Neal, D.P.; Stoica, G.; Wang, L.V. Photoacoustic tomography of a nanoshell contrast agent in the in vivo rat brain. *Nano Lett.* **2004**, *4*, 1689–1692. [CrossRef]
32. Copland, J.A.; Eghtedari, M.; Popov, V.L.; Kotov, N.; Mamedova, N.; Motamedi, M.; Oraevsky, A.A. Bioconjugated gold nanoparticles as a molecular based contrast agent: Implications for imaging of deep tumors using optoacoustic tomography. *Mol. Imaging Biol.* **2004**, *6*, 341–349. [CrossRef]
33. Mallidi, S.; Larson, T.; Aaron, J.; Sokolov, K.; Emelianov, S. Molecular specific optoacoustic imaging with plasmonic nanoparticles. *Opt. Exp.* **2007**, *15*, 6583–6588. [CrossRef]
34. Pramanik, M.; Wang, L.V. Thermoacoustic and photoacoustic sensing of temperature. *J. Biomed. Opt.* **2009**, *14*, 054024. [CrossRef] [PubMed]

35. Cox, B.T.; Laufer, J.; Kostli, K.; Beard, P. Experimental validation of photoacoustic k-space propagation models. In Proceedings of the Photons Plus Ultrasound: Imaging and Sensing, International Society for Optics and Photonics, San Jose, CA, USA, 25–26 January 2004; Volume 5320, pp. 238–248.
36. Treeby, B.E.; Cox, B.T. k-Wave: MATLAB toolbox for the simulation and reconstruction of photoacoustic wave fields. *J. Biomed. Opt.* **2010**, *15*, 021314. [CrossRef] [PubMed]
37. Beard, P. Biomedical photoacoustic imaging. *Interf. Focus* **2011**, *1*, 602–631. [CrossRef] [PubMed]
38. Singh, M.S.; Paul, S.; Thomas, A. Fundamentals of photoacoustic imaging: A theoretical tutorial. In *LED-Based Photoacoustic Imaging*; Progress in Optical Science and, Photonics; Kuniyil Ajith Singh, M., Ed.; Springer: Singapore, 2020; pp. 3–21.
39. Chen, Y.S.; Frey, W.; Aglyamov, S.; Emelianov, S. Environment-dependent generation of photoacoustic waves from plasmonic nanoparticles. *Small* **2012**, *8*, 47–52. [CrossRef]
40. Zemansky, M.W.; Dittman, R.H. *Heat and Thermodynamics*, 7th ed.; McGraw-Hill Companies Inc.: New York, NY, USA, 1997; pp. 154–196.
41. Hunter, S.D.; Jones, W.V.; Malbrough, D.J.; Van Buren, A.L.; Liboff, A.; Bowen, T.; Jones, J.J.; Learned, J.G.; Bradner, H.; Pfeffer, L.; et al. Acoustic-signals of nonthermal origin from high-energy protons in water. *J. Acoust. Soc. Am.* **1981**, *69*, 1557–1562. [CrossRef]
42. Gusev, V.E.; Karabutov, A.A. *Laser Optoacoustics*; AIP: New York, NY, USA, 1993.
43. Haemmerich, D.; Dos Santos, I.; Schutt, D.J.; Webster, J.G.; Mahvi, D.M. In vitro measurements of temperature-dependent specific heat of liver tissue. *Communication* **2006**, *28*, 194–197.
44. Burmistrova, L.V.; Karabutov, A.A.; Rudenko, O.V.; Cherepetskaya, E.B. Influence of thermal nonlinearity on the thermooptical generation of sound. *Sov. Phys. Acoust. USSR* **1979**, *25*, 348–350.
45. Singh, M.S.; Yalavarthy, P.K. Born-ratio type data normalization improves quantitation in photoacoustic tomography. In Proceedings of the Medical Imaging 2014: Ultrasonic Imaging and Tomography, International Society for Optics and Photonics, San Diego, CA, USA, 20 March 2014; Volume 9040. [CrossRef]
46. Ultrasonic Transducers Technical Notes. Available online: https://ctac.mbi.ufl.edu/files/2017/02/ultrasound-basics.pdf (accessed on 22 November 2020).
47. Wang, L.V. Tutorial on Photoacoustic Microscopy and Computed Tomography. *IEEE J. Sel. Top. Quant. Elect.* **2008**, *11*, 171–179. [CrossRef]
48. *American National Standard for Safe Use of Lasers*; ANSI Z136.1-2007; Laser Institute of America: Orlando, FL, USA, 2007; pp. 76–77. Available online: https://webstore.ansi.org/standards/lia/ansiz1362007 (accessed on 27 January 2021).
49. Ramachandran, T.; Sreenivasan, K.; Sivakumar, R. Water vaporization from heated tissue: An in vitro study by differential scanning calorime-try. *Lasers Surg. Med.* **1996**, *19*, 413–415. [CrossRef]
50. Durduran, T.; Choe, R.; Culver, J.P.; Zubkov, L.; Holboke, M.J.; Giammarco, J.; Chance, B.; Yodh, A.G. Bulk optical properties of healthy female breast tissue. *Phys. Med. Biol.* **2002**, *47*, 2847–2861. [CrossRef]
51. Rinaldo, C.; Antonio, P.; Paola, T.; Alessandro, T.; Gianluca, V. A solid tissue phantom for photon migration studies. *Phys. Med. Biol.* **1997**, *42*, 197–199.

sensors

Communication

Tomographic Ultrasound and LED-Based Photoacoustic System for Preclinical Imaging

Kalloor Joseph Francis [1,*], Richell Booijink [2], Ruchi Bansal [2] and Wiendelt Steenbergen [1,*]

[1] Biomedical Photonics Imaging (BMPI), Technical Medical Center, University of Twente, 7522 NH Enschede, The Netherlands

[2] Translational Liver Research, Medical Cell BioPhysics (MCBP), Technical Medical Center, University of Twente, 7522 NH Enschede, The Netherlands; r.bansal@utwente.nl (R.B.); r.s.booijink@utwente.nl (R.B.)

* Correspondence: f.kalloorjoseph@utwente.nl (K.J.F.); w.steenbergen@utwente.nl (W.S.)

Received: 17 April 2020; Accepted: 11 May 2020; Published: 14 May 2020

Abstract: Small animals are widely used as disease models in medical research. Noninvasive imaging modalities with functional capability play an important role in studying the disease state and treatment progress. Photoacoustics, being a noninvasive and functional modality, has the potential for small-animal imaging. However, the conventional photoacoustic tomographic systems use pulsed lasers, making it expensive, bulky, and require long acquisition time. In this work, we propose the use of photoacoustic and ultrasound tomographic imaging with LEDs as the light source and acoustic detection using a linear transducer array. We have demonstrated full-view tomographic imaging of a euthanized mouse and a potential application in liver fibrosis research.

Keywords: tomography; photoacoustic; ultrasound; small animal; liver; fibrosis

1. Introduction

Studying different diseases and developing new drugs in a controlled environment is vital in biomedical research. Small animals are widely used for this purpose, as they allow for controlled disease staging and evaluating the performance of drugs through histopathological validation [1]. Longitudinal monitoring of disease progression and treatment response to drugs can improve the outcome of preclinical studies and can reduce the number of laboratory animal deaths. Imaging modalities can be used for longitudinal monitoring of small animal models. However, there are limitations in using conventional imaging modalities such as MRI, CT, and ultrasound for small animal imaging [2]. Micro MRI is costly and has a slow data acquisition. Micro CT and PET, on the other hand, use ionizing radiation, which hinders longitudinal observations [2]. Ultrasound (US) is a noninvasive and real-time imaging modality but, being a structural imaging modality, in most cases it cannot quantify disease state. Photoacoustic (PA) imaging is a new modality that has functional and molecular capability while being noninvasive and real-time. Thus, PA is considered to be ideal for small animal imaging [3].

PA imaging utilizes pulsed light excitation to induce a temperature rise in optical absorbing structures inside the tissue resulting in thermoelastic expansion and acoustic wave generation. These acoustic waves are detected for imaging [4]. The advantage of PA imaging is that with optical excitation and acoustic detection it combines optical contrast at ultrasound resolution. Additionally, the use of ultrasound transducers enables us to combine PA imaging with conventional US imaging providing co-registered structural and functional imaging of the tissue. Several PA and US imaging systems were successfully demonstrated for small-animal whole-body imaging [3,5,6]. However, the use of a pulsed laser source in these systems not only makes them expensive but demands laser safe small animal labs, eye safety goggles, and additional manpower to operate the system. Therefore, for the wide use of PA imaging in small animal labs, there is a requirement for low cost, compact, safe to use tomographic systems which can be operated by a non-expert. Recent developments in LED-based PA imaging, being compact, low-cost, and safe to use, offer promising avenues to fill this gap [7]. LED-based handheld PA systems were used previously for small animal studies for imaging superficial structures such as tumors [8], wounds [9], and knee joints [7,10]. A limitation of the hand-held PA system using a linear transducer array is the limited view of the target tissue due to the directional sensitivity of the transducer. Additionally, with a small number of LED elements arranged on either side of the transducer, the imaging depth is shallow. We have recently developed a tomographic imaging configuration using a linear transducer array and four LED arrays, to overcome the limited view and to improve the imaging depth [11,12]. The system was originally developed for imaging finger joints for diagnosis and monitoring of rheumatoid arthritis [11].

In this study, we propose the application of our tomographic US and LED-based PA system for preclinical research. First, we demonstrate full-view tomographic imaging of the abdominal region of a mouse. We also compare the results with B-scan images obtained using a handheld probe. Further, we present a potential application of the system in liver fibrosis research. A large number of preclinical studies are currently being performed in small animals to develop antifibrotic therapies. However, the outcome of the preclinical study relies on endpoint histopathological analysis. A noninvasive imaging technique can provide longitudinal monitoring of animals and can improve the study outcome. We present the use of noninvasive and low-cost US and PA tomographic imaging system for liver imaging and compared the outcome with histology images.

2. Materials and Methods

An LED-based photoacoustic and ultrasound imaging system, AcousticX (Cyberdyne Inc., Japan), was used in this work. Four LED arrays having 576 elements (36×4 array) were used as the light source. We used LEDs having a wavelength of 850 nm, and each array has a pulse energy of 200 μJ with a pulse duration of 70 ns. We used a linear transducer array (128 elements) with a center frequency 7 MHz with 80% bandwidth for acoustic detection. A tomographic imaging configuration with four LED arrays and transducer scanning around the sample was used in this study.

PA imaging using a linear transducer array suffers from low image quality due to incomplete acoustic detection. To overcome the limited view problem in linear array-based imaging, we recently developed a tomographic imaging configuration [11]. The system consists of an imaging probe with a linear transducer array and four LED arrays as shown in Figure 1a. Two LED arrays were placed parallel to the long axis of the transducer, geometrically pointing towards the focus of the transducer (20 mm depth). The other two LED arrays were placed on either side of the transducer along its short axis at an angle of 105° with the transducer surface. These two LED arrays were also placed at an angle of 5° with the imaging plane to minimize the acoustic reflections being detected. In this way, the object to be imaged is illuminated from three sides. The probe is then scanned around the object for tomographic imaging. Based on a simulation study, we have estimated that 16 angular views with an angular step of 22.5° is sufficient to form a full-view tomographic image [11]. The illumination configuration was also developed to provide a significant overlapping region between two angular scans. Interested readers can refer to the work in [11,12] for more details. A 3D printed holder encompassing the transducer and LED arrays to form the imaging probe is shown in Figure 1b. For small animal imaging, the probe was mounted on a motorized linear and circular scanning system, installed on top of a water tank as shown in Figure 1c. A 3D printed small animal holder was used to keep the mouse in place for imaging (Figure 1c).

Male Balb/c mice (8 weeks old) were used in this study. All the animals used in the study received ad libitum normal chow diet and normal water, and were housed with 12h-light/12h-dark cycle. All the animal experiments were carried out strictly according to the ethical guidelines for the Care and Use of Laboratory Animals (Utrecht University, The Netherlands). A proof-of-concept study was performed whereby liver fibrosis was induced in one mouse by a single injection of 1ml/kg of carbon tetrachloride (CCl_4), and the control animal received olive oil. Mice were euthanized and imaged immediately using the PA and US tomographic system. The hair with the outer skin layer of the animals was removed around the abdominal region before imaging. After the imaging experiments, the livers of the animals were collected and fixed with cryomatrix in isopentane. Fixed livers were sectioned into 5 μm sections and immunohistological analysis using collagen I (fibrosis marker) antibody was performed.

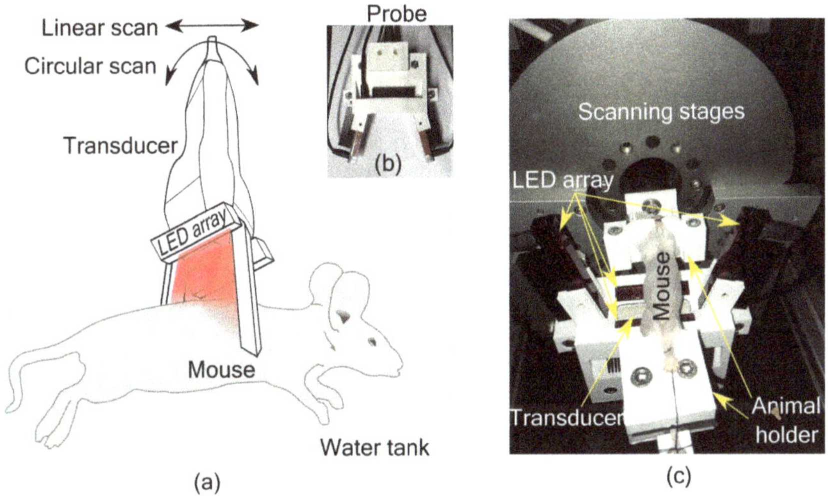

Figure 1. Small animal tomographic imaging system. (**a**) Schematic of the set-up with the imaging probe scanning around the mouse in a water tank. (**b**) Imaging probe with linear transducer array and four LED arrays in a 3D printed holder. (**c**) Photograph of the imaging set-up showing mouse holder, imaging probe and the scanning stages.

Two scanning techniques were used in this study, involving a combination of circular and linear scans. To obtain a complete tomographic image, a circular scanning of the entire 360° was performed. Combined PA and US imaging were performed at each angular step. Typically, a circular scan takes 42.5 secs. However, due to memory limitations of the system, the RF data was saved after each 180° scan and combined in the postprocessing stage. Multiple tomographic images of the abdominal area were obtained by translating the probe to a different location and repeating the circular scan. In the second experiment, the liver region was considered as the region of interest. Here, a limited view tomographic imaging was performed by spatial compounding of B-scan images from three angles (20° steps). In this case, linear scans were performed for the entire length (15 mm length) of the liver at each angular step. Each linear scan took 8.5 secs and the 20° circular scan took 2.5 secs. The US and PA B-scan images were formed by delay and sum algorithm. For speckle reduction and to improve the contrast of the US image, Bayesian mean filtering using an open source software was used [13]. To form tomographic images, the B-scan images from both US and PA imaging were rotated to the acquired angle and spatially compounded using a custom developed MATLAB program [14].

3. Results and Discussion

In Figure 2a–c, co-registered ultrasound and photoacoustic B-scan images of a mouse abdomen, obtained from three different angles, have been presented. The images demonstrate the limited view problem in small animal imaging using a linear transducer array. The limited view arises from the directional sensitivity of the transducer resulting in structures oriented away from the transducer being undetected. The tomographic images obtained from spatial compounding of the B-scan images are shown in Figure 2d–f. The ultrasound image shows the capability of the tomographic system to image major abdominal organs such as the liver, kidney, spleen, stomach, and intestines. Additionally, anatomical structures such as skin and spinal cord are also imaged in this mode. Kidney and lobes of the liver can be seen as hypoechogenic regions, while the stomach, spleen, and intestine are visible as hyperechogenic regions. In PA images major blood vessels are visible beneath the skin and around the spinal cord. In addition to that, high PA signals were also observed from the liver, kidney, and spleen. For the imaging configuration used in this work, it was estimated that PA signal strength drops to 30% at a depth of 9.7 mm in soft tissue [11]. This finding also holds in this study, as most of the structures visible in PA images are within the 10 mm depth from the skin surface. Illumination from three sides and spatial compounding from several angles enable PA tomographic imaging to visualize most of the absorbing structures in the small animal. The abdominal diameter of Balb/c mice is mostly less than 30 mm and most of the organs of interest are within the 10 mm depth from the skin surface, making the system useful for small animal studies. However, there is indeed a region at the center where the signal to noise ratio is low, which can only be improved with better illumination methods. Another challenge in PA imaging is reflection artifacts from multiple structures inside the small animal body. There are several methods proposed to solve reflection artifacts in PA images [15]. Methods for improving the imaging depth and removing artifacts will be considered in the future. In this study, we have imaged a euthanized animal; however, imaging a live animal in this set-up is also possible by positioning the scanning stages such that the snout of the animal is above the water while the rest of the body immersed and/or using a breathing mask. Imaging speed is another factor that needs to be considered while imaging a live animal. The current scanning time is 42.5 secs, which is longer when compared with the state-of-the-art system using a full ring transducer [3]. However, it is possible to synchronize the breathing cycle of the animal with the acquisition and by using postprocessing to reduce image distortion due to motion while acquisition [16].

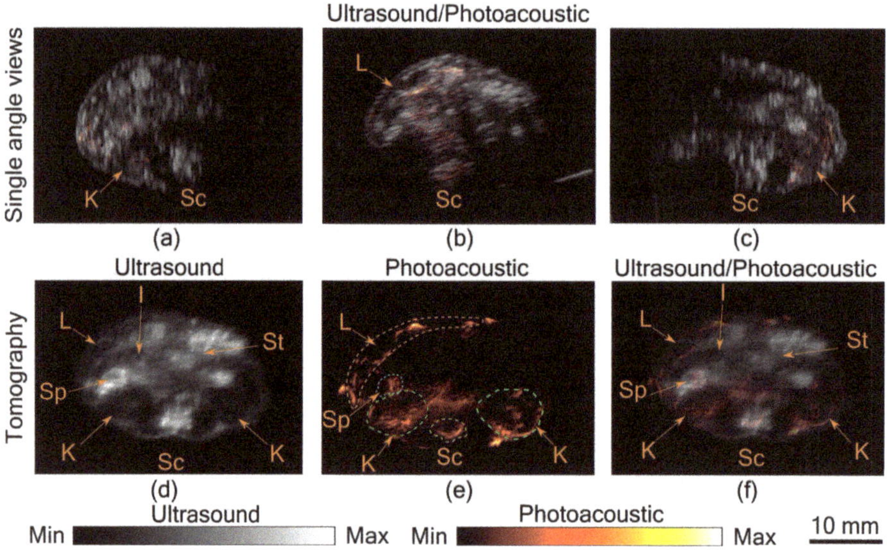

Figure 2. Tomographic photoacoustic and ultrasound imaging. (**a–c**) Co-registered ultrasound and photoacoustic B-Scan images of mouse abdomen, acquired from three different angles. Tomographic (**d**) ultrasound, (**e**) photoacoustic, and (**f**) co-registered ultrasound and photoacoustic image. Several organs in the abdominal region indicated by Sc-spinal cord, K-kidney, L-liver, Sp -spleen, St-stomach, and I-intestine. Dashed lines are used to mark the organs in the photoacoustic image.

In Figure 2, the liver is visible in both US and PA. Being a superficial organ, small animal liver imaging is one of the potential applications where the proposed system can be used. To demonstrate this aspect, tomographic imaging of a control and a fibrotic mouse was performed. Figure 3a,h shows five tomographic slices of control and fibrotic mouse, respectively, with the liver in the top region of the image. Remarkable differences can be observed between control and fibrotic liver in both US and PA images. To analyze this difference, let us consider a pair of US and PA images of the liver from both control and fibrotic mouse shown in Figure 3b,c,e,f. The liver is marked with a green boundary. Two lobes of the liver are visible in the US images. In the US image, the difference in control and fibrotic liver is evident with the difference in echogenicity. Echogenicity can be characterized by the brightness level or the mean value of the region of interest. The mean pixel value was calculated with the liver as the region of interest. For the control animal, the calculated mean value was 0.13 ± 0.02, while it was 0.25 ± 0.03 for the fibrotic case. Additionally, heterogeneity computed using the variance was found to be 0.52 ± 0.11 for the control liver and 0.85 ± 0.10 for the fibrotic liver. These metrics were computed over the liver region with multiple tomographic slices stacked and normalized with the maximum value. The difference in structure evidenced in the US image is due to the hepatic nodularity developed due to scar formation in fibrotic liver [17]. In addition to the differences in US images, the PA images in Figure 3c,f show a difference in contrast between control and fibrotic liver. The contrast value computed for the liver region for the control animal was 0.48 ± 0.08 and 0.69 ± 0.08 for the fibrotic case. The primary reason for the high PA contrast in the fibrotic liver is due to the neovasculature in hypoxic liver cells [18]. Angiogenesis or neovascularization plays a crucial role in the progression of liver fibrosis and is also considered as an early indicator of fibrosis thus PA imaging can be used for monitoring the disease progression, while heterogeneity observed with US imaging reflects scar formation at a later stage of the disease [18]. The liver fibrosis was further confirmed using histopathological analysis. Figure 3d,g show histology images of control and fibrotic liver respectively with Collagen I immunostainings. The fibrotic liver (Figure 3g) shows an increased level of collagen bridges resulting from an accumulation of the extracellular matrix protein. These results are also in line with our previous small animal liver fibrosis study [19]. The main improvement we achieved in this work is the superior image quality from the tomographic system and the use of LED-based PA and US imaging. Further works will focus on including more animals in the study and developing a facility to image live animals.

The advantage of using an LED-based PA system is primarily the cost-effectiveness of the system. Nanosecond pulsed lasers and laser diodes are commonly used for tomographic imaging. The cost for a pulsed laser is approximately 70–200k$, and for laser diodes, with drivers, the cost ranges from 10–25k$. For LEDs with driving electronics, the cost is as lower as 10–15k$ [20]. Another benefit is that LED arrays can be integrated within the imaging probe, hence the system can be compact and suitable for small animal labs. Additionally, there is no requirement for a laser-safe imaging facility. However, the limitations are the low pulse energy resulting in low imaging depth, to an extent, this can be addressed using LED arrays with a large number of elements and frame averaging. The longer pulse duration (30–100 ns) achieved with driving circuits for LEDs is sufficient for stress confinement in tomographic imaging applications where the expected resolution is 1mm. However, unlike lasers, wavelength tuning is not possible with LED-based illumination. LED arrays with elements having multiple wavelengths were tested for oxygen saturation imaging compromising pulse energy and imaging depth. Importantly, for different imaging applications, a wide choice of LEDs is available ranging from visible to near-infrared wavelengths (470–980nm). Custom-developed power LED arrays arranged around the small animal along with ring US transducer can be an ideal configuration for small animal imagers in the future.

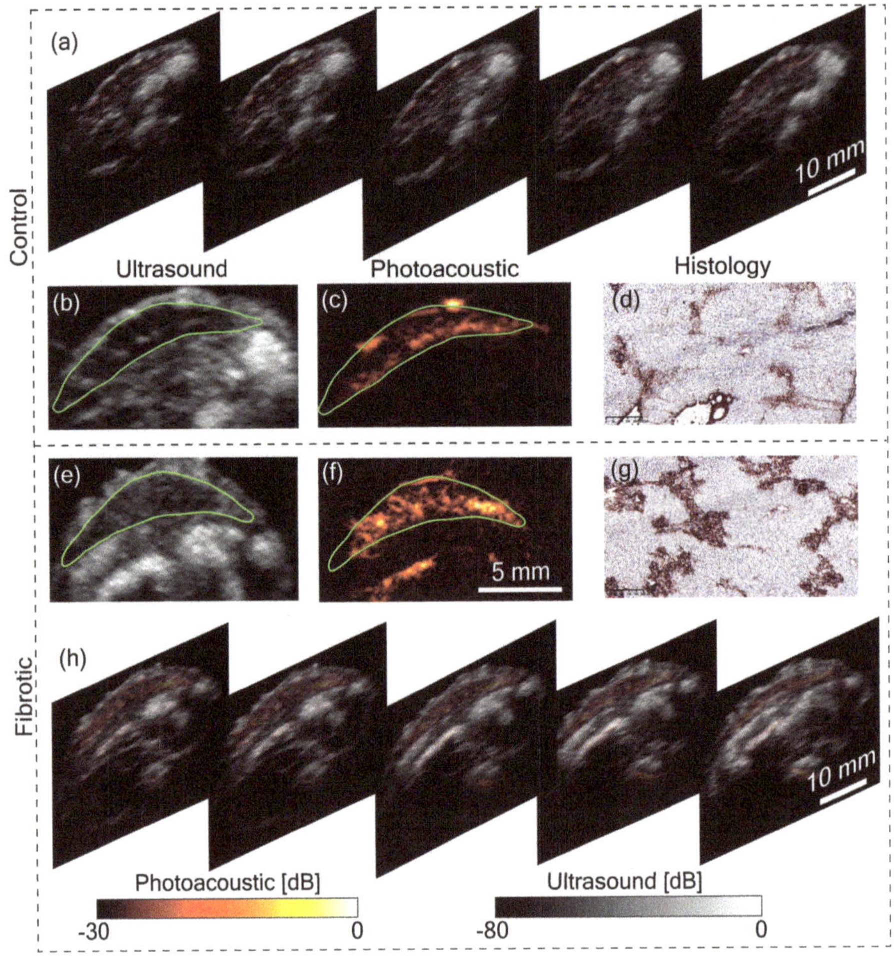

Figure 3. Tomographic imaging of the liver. (**a**) Tomographic photoacoustic and ultrasound images of a control mouse showing the liver at a scanning step of 2 mm. (**b**) Ultrasound image of the liver and (**c**) corresponding photoacoustic image. The green line markers the liver region. (**d**) Histology image of the control liver stained with Collagen I (fibrosis marker). (**e**) Ultrasound image of a mouse liver with fibrosis (CCl_4-treated) and (**f**) corresponding photoacoustic image. (**g**) Histology image of the fibrotic liver stained with collagen I (fibrosis marker). (**h**) Tomographic photoacoustic and ultrasound images of the fibrotic mouse showing liver at a scanning step of 2 mm.

4. Conclusions

In this work, a tomographic ultrasound and LED-based photoacoustic system was demonstrated for small animal imaging. Results show that the tomographic ultrasound imaging can be used to distinguish major organs in the abdominal region of a small animal. The tomographic photoacoustic imaging is capable of imaging an approximate depth of 10 mm from the surface of the skin, allowing imaging of blood vessels and major organs. Limited imaging depth and artifacts are some of the hurdles for photoacoustic imaging in this proposed configuration. The applicability of the proposed imaging system is demonstrated in small animal liver imaging in a fibrotic mouse model. An increase in photoacoustic contrast and ultrasound echogenicity was observed in the fibrotic liver compared to that of the control mouse. The proposed tomographic ultrasound and photoacoustic imaging system offer a cost-effective, compact, and eye-safe alternative for the expensive laser-based system for small animal research.

Author Contributions: K.J.F.- conceptualization, methodology, imaging experiments, data processing, analysis, draft preparation; R.B. (Richell Booijink) - histopathological analysis; R.B. (Ruchi Bansal) - conceptualization, small animal handling, review and editing, W.S. - supervision, project administration, funding acquisition, review and editing. All authors have read and agreed to the published version of the manuscript.

Funding: This work was funded by the National Centre for the Replacement Refinement and Reduction of Animals in Research (CRACKITRT-P1-3)

Conflicts of Interest: The authors declare no conflicts of interest.

References

1. Andersen, M.L.; Winter, L.M. Animal models in biological and biomedical research-experimental and ethical concerns. *An. Acad. Bras. CiÊNcias* **2019**, *91*, e20170238.
2. Xia, J.; Wang, L.V. Small-animal whole-body photoacoustic tomography: A review. *IEEE Trans. Biomed. Eng.* **2013**, *61*, 1380–1389.
3. Li, L.; Zhu, L.; Ma, C.; Lin, L.; Yao, J.; Wang, L.; Maslov, K.; Zhang, R.; Chen, W.; Shi, J.; others. Single-impulse panoramic photoacoustic computed tomography of small-animal whole-body dynamics at high spatiotemporal resolution. *Nat. Biomed. Eng.* **2017**, *1*, 1–11.
4. Ntziachristos, V.; Ripoll, J.; Wang, L.V.; Weissleder, R. Looking and listening to light: The evolution of whole-body photonic imaging. *Nat. Biotechnol.* **2005**, *23*, 313–320.
5. Jeon, M.; Kim, J.; Kim, C. Multiplane spectroscopic whole-body photoacoustic imaging of small animals in vivo. *Med Biol. Eng. Comput.* **2016**, *54*, 283–294.
6. Merčep, E.; Herraiz, J.L.; Deán-Ben, X.L.; Razansky, D. Transmission–reflection optoacoustic ultrasound (TROPUS) computed tomography of small animals. *Light. Sci. Appl.* **2019**, *8*, 1–12.
7. Zhu, Y.; Xu, G.; Yuan, J.; Jo, J.; Gandikota, G.; Demirci, H.; Agano, T.; Sato, N.; Shigeta, Y.; Wang, X. Light emitting diodes based photoacoustic imaging and potential clinical applications. *Sci. Rep.* **2018**, *8*, 1–12.
8. Xavierselvan, M.; Mallidi, S. LED-Based Functional Photoacoustics—Portable and Affordable Solution for Preclinical Cancer Imaging. In *LED-Based Photoacoustic Imaging*; Springer: Singapore, 2020; pp. 303–319.
9. Hariri, A.; Chen, F.; Moore, C.; Jokerst, J.V. Noninvasive staging of pressure ulcers using photoacoustic imaging. *Wound Repair Regen.* **2019**, *27*, 488–496.
10. Francis, K.J.; Xavierselvan, M.; Singh, M.K.A.; Mallidi, S.; Van Der Laken, C.; Van De Loo, F.; Steenbergen, W. LED-based photoacoustic imaging for early detection of joint inflammation in rodents: towards achieving 3Rs in rheumatoid arthritis research. In *Photons Plus Ultrasound: Imaging and Sensing 2020. International Society for Optics and Photonics*; SPIE: San Francisco, CA, USA, 2020; Vol. 11240, p. 112400M.
11. Francis, K.J.; Boink, Y.E.; Dantuma, M.; Singh, M.K.A.; Manohar, S.; Steenbergen, W. Tomographic imaging with an ultrasound and LED-based photoacoustic system. *Biomed. Opt. Express* **2020**, *11*, 2152–2165.
12. Francis, K.J.; Boink, Y.E.; Dantuma, M.; Singh, M.K.A.; Manohar, S.; Steenbergen, W. Light Emitting Diodes Based Photoacoustic and Ultrasound Tomography: Imaging Aspects and Applications. In *LED-Based Photoacoustic Imaging*; Springer: Berlin/Heidelberg, Germany, 2020; pp. 245–266.
13. Coupé, P.; Hellier, P.; Kervrann, C.; Barillot, C. Nonlocal means-based speckle filtering for ultrasound images. *IEEE Trans. Image Process.* **2009**, *18*, 2221–2229.
14. Francis, K.J.; Chinni, B.; Channappayya, S.S.; Pachamuthu, R.; Dogra, V.S.; Rao, N. Multiview spatial compounding using lens-based photoacoustic imaging system. *Photoacoustics* **2019**, *13*, 85–94.
15. Nguyen, H.N.Y.; Steenbergen, W. Reducing artifacts in photoacoustic imaging by using multi-wavelength excitation and transducer displacement. *Biomed. Opt. Express* **2019**, *10*, 3124–3138.
16. Ron, A.; Davoudi, N.; Deán-Ben, X.L.; Razansky, D. Self-gated respiratory motion rejection for optoacoustic tomography. *Appl. Sci.* **2019**, *9*, 2737.
17. Viganò, M.; Visentin, S.; Aghemo, A.; Rumi, M.G.; Ronchi, G.; Colli, A.; Colucci, A.; Paggi, S.; Fraquelli, M.; Conte, D. US features of liver surface nodularity as a predictor of severe fibrosis in chronic hepatitis C. *Radiology* **2005**, *234*, 641.

18. Bocca, C.; Novo, E.; Miglietta, A.; Parola, M. Angiogenesis and fibrogenesis in chronic liver diseases. *Cell. Mol. Gastroenterol. Hepatol.* **2015**, *1*, 477–488.
19. van den Berg, P.J.; Bansal, R.; Daoudi, K.; Steenbergen, W.; Prakash, J. Preclinical detection of liver fibrosis using dual-modality photoacoustic/ultrasound system. *Biomed. Opt. Express* **2016**, *7*, 5081–5091.
20. Zhu, Y.; Feng, T.; Cheng, Q.; Wang, X.; Du, S.; Sato, N.; Kuniyil Ajith Singh, M.; Yuan, J. Towards Clinical Translation of LED-Based Photoacoustic Imaging: A Review. *Sensors* **2020**, *20*, 2484.

sensors

MDPI

Article

In Vivo Tumor Vascular Imaging with Light Emitting Diode-Based Photoacoustic Imaging System

Marvin Xavierselvan [1], **Mithun Kuniyil Ajith Singh** [2] and **Srivalleesha Mallidi** [1,3,*]

[1] Department of Biomedical Engineering, Tufts University, Medford, MA 02155, USA; marvin.xavierselvan@tufts.edu

[2] Research & Business Development Division, Cyberdyne INC, Cambridge Innovation Center, 3013 Rotterdam, The Netherlands; mithun_ajith@cyberdyne.jp

[3] Wellman Center for Photomedicine, Massachusetts General Hospital, Harvard Medical School, Boston, MA 02114, USA

* Correspondence: Srivalleesha.mallidi@tufts.edu

Received: 23 July 2020; Accepted: 8 August 2020; Published: 12 August 2020

Abstract: Photoacoustic (PA) imaging has shown tremendous promise for imaging tumor vasculature and its function at deeper penetration depths without the use of exogenous contrast agents. Traditional PA imaging systems employ expensive and bulky class IV lasers with low pulse repetition rate, due to which its availability for preclinical cancer research is hampered. In this study, we evaluated the capability of a Light-Emitting Diode (LED)-based PA and ultrasound (US) imaging system for monitoring heterogeneous microvasculature in tumors (up to 10 mm in depth) and quantitatively compared the PA images with gold standard histology images. We used a combination of a 7 MHz linear array US transducer and 850 nm excitation wavelength LED arrays to image blood vessels in a subcutaneous tumor model. After imaging, the tumors were sectioned and stained for endothelial cells to correlate with PA images across similar cross-sections. Analysis of 30 regions of interest in tumors from different mice showed a statistically significant R-value of 0.84 where the areas with high blood vessel density had high PA response while low blood vessel density regions had low PA response. Our results confirm that LED-based PA and US imaging can provide 2D and 3D images of tumor vasculature and the potential it has as a valuable tool for preclinical cancer research.

Keywords: LED; photoacoustic imaging; ultrasound; tumor imaging

1. Introduction

The critical role of vasculature for tumor growth and metastasis is undisputed. Tumor aggressiveness and the extent of the aberrant tumor vascular structure and function are highly correlated. To meet the increasing needs for oxygen and nutrients, cancers induce neovascularization, one of the hallmarks of cancer [1]. The heterogeneity in tumor vasculature [2] affects drug distribution [3,4] and, as a result, impact treatment outcomes. Indeed, many anti-angiogenic agents or therapies that can prune these blood vessels and prevent nutrients from reaching the tumors have been developed [5]. Monitoring the dynamic changes in tumor vasculature post therapy is key for prognosis [6]. Several imaging modalities have been used to study changes in tumor vasculature both at structural and functional level. More recently, photoacoustic (PA) imaging has gained tremendous popularity for imaging tumor vasculature [7–10]. PA imaging involves detection of acoustic waves generated when the optical absorbers such as hemoglobin in the blood are irradiated with nanosecond pulsed light and undergo thermo-elastic expansion and contraction [11]. The acoustic waves are picked up by the ultrasound (US) transducer and processed to generate a PA image [11,12]. Notably, two features of PA imaging that make it highly suitable to image vasculature are: 1. ability to image blood vessels without exogenous contrast agents and 2. ability to complement US imaging that provides structural information of the tumor.

Traditionally PA imaging systems use lasers (typically Nd: YAG pumped optical parametric oscillator lasers) for exciting the optical absorbers and employ a US transducer (array or single element) to receive the generated acoustic wave. These lasers are expensive, bulky with large spatial occupancy, and often have low repetition rate which leads to low image frame rate and acquisition speed. These limitations hinder the wide adaption of PA imaging systems in both preclinical and clinical applications. Advances in semiconductor industry have led to the creation of powerful Light-Emitting Diodes (LED) which are compact, portable, affordable, and energy-efficient. AcousticX by Cyberdyne Inc. (Tsukuba, Japan) is a LED-based system that can perform PA and US imaging at high frame rates. These LEDs can be fired with a pulse repetition frequency (PRF) from 1 to 4 kHz. Despite generating low optical output (400 μJ/pulse when using two 850 nm arrays), the high PRF of LEDs offers the possibility of averaging several frames which results in generating real-time PA images with reasonable signal-to-noise ratio (SNR) [13–17]. Furthermore, the spatial resolution and SNR provided by LED-based PA imaging is very similar and comparable to laser-based PA systems [13]. Due to the inherent advantages of the LED-based PA system, several research groups have explored its ability for various potential applications [15,16,18–21]. None of these studies provide validation of the generated PA image with histology. In this study, for the first time to the best of our knowledge, we report the LED-based PA imaging system's ability to image microvasculature in subcutaneous murine tumor xenografts and validate the results using histology.

2. Materials and Methods

2.1. LED-Based PA Imaging System

An AcousticX system integrated with a 7 MHz US transducer (128 elements, −6 dB bandwidth is 80%) and two 850 nm LED arrays (for deeper light penetration) attached on either side of the transducer at a 30° angle was used for in vivo imaging (Figure 1A–C). The mean lateral and axial resolutions of AcousticX when using a 7 MHz transducer are 460 and 220 μm, respectively [14,19]. Each LED array had height, length and width dimensions of 12.4, 86.5, and 10.2 mm, respectively, and had 144 individual LEDs arranged in 4 rows (36 elements/row). The PRF of the LED array could be varied between 1, 2, 3, and 4 kHz. In this study, we operated the LED arrays at a PRF of 4 kHz to acquire and save PA images of the tumor vasculature at a frame rate of approximately 10 Hz i.e., 384 PA frames were continuously averaged to generate one display image. For offline MATLAB analysis, a higher number of frames (12,800) was averaged to enhance the PA image SNR. The SNR was defined as SNR = S/σ, where S is the mean of the image amplitude in a signal region, and σ is the standard deviation of the image amplitude in a noise region. The signal region was empirically chosen as a rectangular region (4 by 4 pixels) near a feeding blood vessel on the tumor surface where the PA signal was apparent; the noise region was manually chosen to correspond to a rectangular region outside the tumor region.

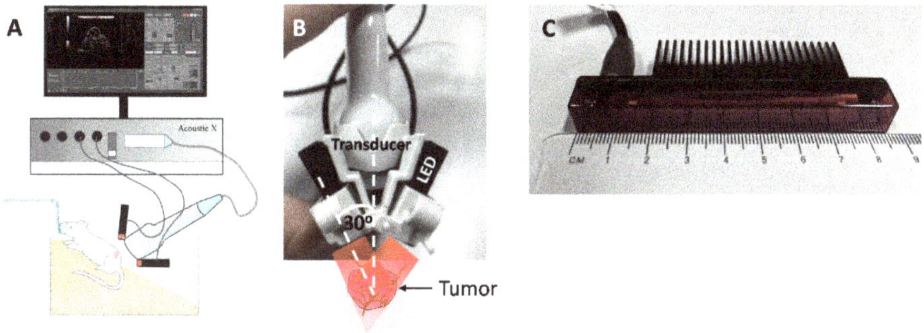

Figure 1. (**A**) Schematic of our experimental setup in which the Light-Emitting Diode (LED)-based photoacoustic (PA) system was utilized for imaging subcutaneous tumors in a mouse. (**B**,**C**) LED array used for illumination and the angle at which the LEDs are connected to the ultrasound transducer. (Scale bar = 2 mm).

2.2. *Animal Model*

All animal experiments were performed in compliance with the Institutional Animal Care and Use Committee (IACUC) of Massachusetts General Hospital (MGH). Swiss nu/nu female mice (aged 6–8 weeks, body weight 20–25 g) were raised in aseptic conditions in the institution's animal facility with a 12 h light and dark cycles. We used human hypopharyngeal squamous cell carcinoma (FaDu) xenograft model because these are highly vascularized tumors [22,23]. FaDu cells were cultured in Eagle's Minimum Essential Medium supplemented with 10% fetal bovine serum and 1% antibiotics (Penicillin and Streptomycin 1:1 *v*/*v*) in a T75 flask and maintained at 37 °C and 5% CO_2. At 80–90% confluency, cells were trypsinized and centrifuged to remove the supernatant. The cells were then suspended in a mixture of matrigel (BD Bioscience, CA) and phosphate buffered saline in 1:1 *v*/*v* at a density of 1 million cells per 100 µL and injected subcutaneously into mice.

2.3. *In-Vivo Photoacoustic Imaging*

A total of 3 mice were imaged using the AcousticX system when the tumors reached 6–8 mm in diameter as measured with a digital caliper (VWR 62379-531, Radnor, PA, USA). The mice were anesthetized using 2% isoflurane and placed in a warm water bath to achieve optical coupling with the US transducer as shown in Figure 1A. The LED arrays were positioned at a 30° angle (Figure 1B) to the US transducer to colocalize the light with the transducer focus. The US transducer was then vertically positioned such that the tumor was approximately at the focal region of the transducer (~20 mm depth). Radio frequency (RF) data and US/PA images for several cross-sections of the tumors were acquired and saved for further analysis. Motion artifacts in the tumor region due to mouse breathing were minimal as shown in Video S1 (captured at 10 Hz frame rate, see Supplementary materials). In the post-acquisition image analysis, frames with breathing motion were not included. The RF data was processed with previously reported algorithms [24–27] using custom written MATLAB code to obtain US/PA images of tumor vasculature. Three-dimensional imaging was performed by moving the US/PA transducer linearly across the XY plane with the aid of an in-built translation stage and the sequence was rendered in ImageJ.

2.4. *Histology*

Post imaging mice were sacrificed, and tumors were surgically removed without skin. The tumors were embedded and frozen in optical cutting temperature (OCT) compound in the same orientation as the US/PA image with the aid of fiducial markers placed prior to euthanasia. The frozen tumors were cryosectioned (5 μm) in the same plane as the US/PA images based on previously published methodology [28] and stained for platelet endothelial cell adhesion molecule-1 (or CD31). Tissue sections were incubated with anti-CD31 primary antibody (1:50 dilution; Rabbit polyclonal anti-CD31, Abcam) overnight at 4 °C, washed, and incubated with fluorescently tagged secondary antibody (1:100 dilution, Donkey Anti-Rabbit IgG NL637 Affinity Purified Ab, R&D systems) for one hour. The sections were mounted using slowfade gold antifade mountant with 4′,6-diamidino-2-phenylindole (DAPI, Invitrogen, Carlsbad, CA, USA) and sealed with coverslips. A whole-slide scanning fluorescence imaging system (Hamamatsu NanoZoomer 2.0-HT) was used to image the slides at 40 x magnification and images were saved in Hamamatsu Nanozoomer Digital Pathology Image (NDPI) format.

2.5. Image Processing

A 5x magnification of the fluorescence image was extracted from the NDPI file using ImageJ, rotated to match the PA image orientation and saved for further analysis in MATLAB (Figure 2). Several 1.25 × 1.85 mm^2 regions of interest (ROI) were chosen from the immunofluorescence (IF) and PA images for correlation analysis. This corresponds to 679 × 1005 and 5 × 50 pixels in IF and PA images, respectively. The background from a similar ROI area was calculated from a region outside the tumor area for both IF and PA images and subtracted from the tumor ROI values. The CD31 intensity from the ROI in IF images were summed and compared to the total PA signal intensity from the corresponding ROI in the PA image. Linear regression analysis was performed using GraphPad Prism software.

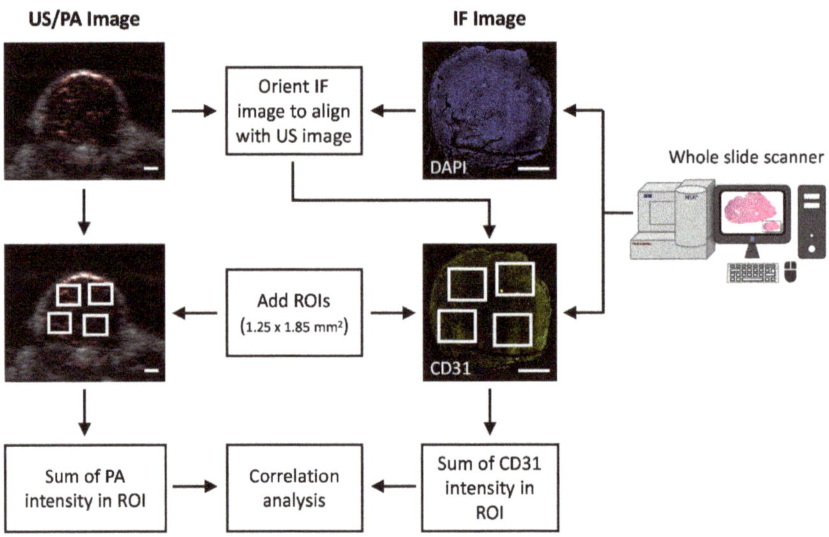

Figure 2. Flowchart describing the image processing methodology to correlate CD31 signal intensity with PA signals. (Scale bar = 2 mm).

3. Results

3.1. SNR Enhances with Frame Averaging

The PA image of a representative tumor acquired and processed at different frame rates (0.31–60 Hz) is shown in Figure 3A. To enhance the SNR and improve image quality, subsequent PA frames were averaged resulting in lower frame rate. This relationship between the frame rate and SNR of the PA image is displayed in Figure 3B, where low frame rate offers better SNR as expected. The low SNR of the image at higher frame rates may be attributed to the low output power of the LED, however, due to the high pulse repetition rate of LED excitation for PA data acquisition, more PA frames can be acquired and averaged to increase the SNR (Figure 3C). A frame rate of 10 Hz was chosen for data acquisition (2D and 3D acquisition) during the experiment as it provided the best possible SNR for near real-time combined US/PA imaging. To quantitatively compare the 2D images with IF images, the acquired PA images were further averaged to a final frame rate of 0.31 Hz to enhance the SNR by 20 dB, i.e., overall 12,800 frames were averaged to obtain an SNR of ~40 dB (Figure 3C). The enhanced SNR facilitated accurate comparison with histology images without the involvement of background noise.

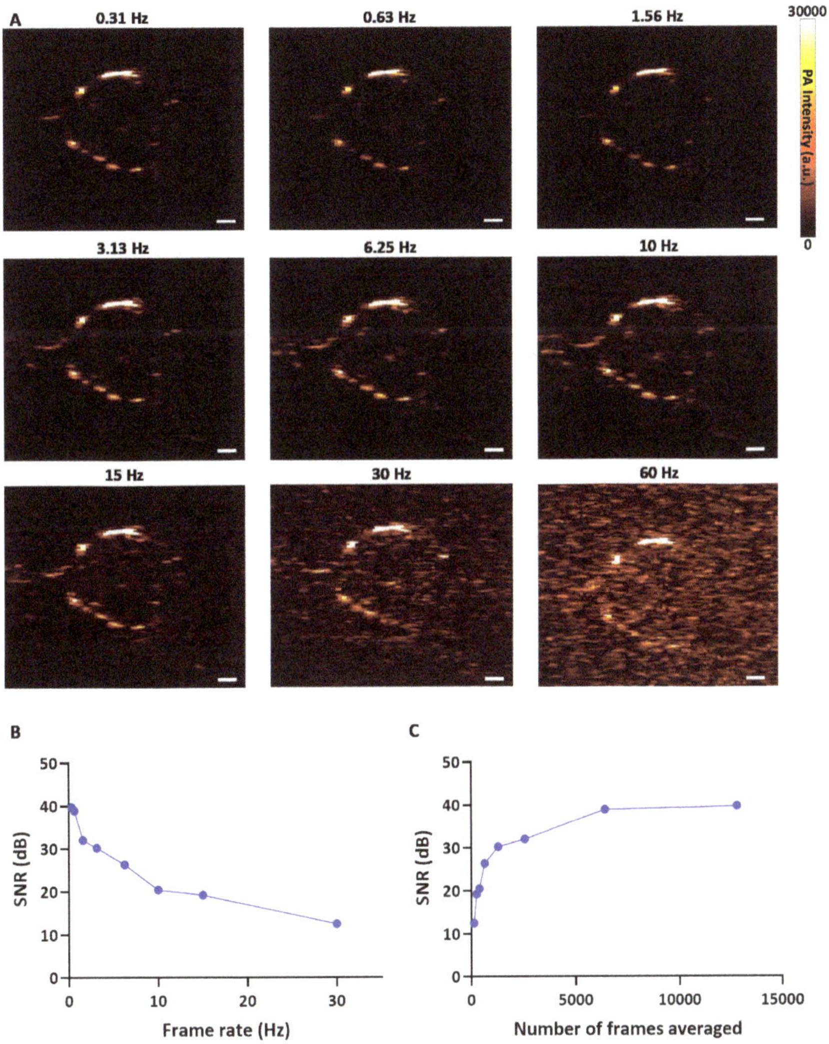

Figure 3. The signal-to-noise ratio (SNR) is enhanced by averaging several frames in the LED-based PA imaging system. (**A**) PA images of a tumor acquired and processed at different frame rates. (**B**) Relationship between the frame rate and SNR. (**C**) Plot detailing the SNR change with respect to the averaging of frames. (Scale bar = 2 mm).

3.2. Tumor Vascular Imaging with AcousticX

Figure 4A shows the overlay of the US image (in grayscale) with the PA image of two representative tumors (offline reconstructed). The B-mode US image depicts the shape and structure of the tumor (~6 to 10 mm in diameter and depth), and the PA image provides the blood vessels location inside and around the tumor. The PA image is displayed on a pseudo-colored hot map where low and high signals are represented by black and white, respectively. Subcutaneous tumors are fed by large blood vessels usually originating from the skin tissue above or the muscle tissue lying beneath the tumor, as visualized in Figure 4A. AcousticX can detect the PA signal not only from the surface of the tumor but also from the deep edge (~10 mm) of the tumor. The Hematoxylin and Eosin (H&E) stain is a standard histology stain to showcase tissue microanatomy. Figure 4B showcases the H&E image of the tumor at the same cross-section as the US/PA image where the tumor structure is analogous to the shape and size observed in the US image. Figure 4C shows the IF image of the tumor cross-section with CD31 stain for blood vessels (yellow) and DAPI for cell nuclei (blue). DAPI staining was performed to visualize the tumor structure and aided in selecting corresponding ROIs as the US/PA image. Qualitatively we noticed regions of the high PA signal have high vessel density on the IF images as displayed by the yellow arrows in Figure 4. The blood vessels on the top of the tumor (particularly Figure 4A top panel) are not strongly visualized in the IF image as these vessels were part of the skin above the tumors which was removed during the histological processing of the tissue.

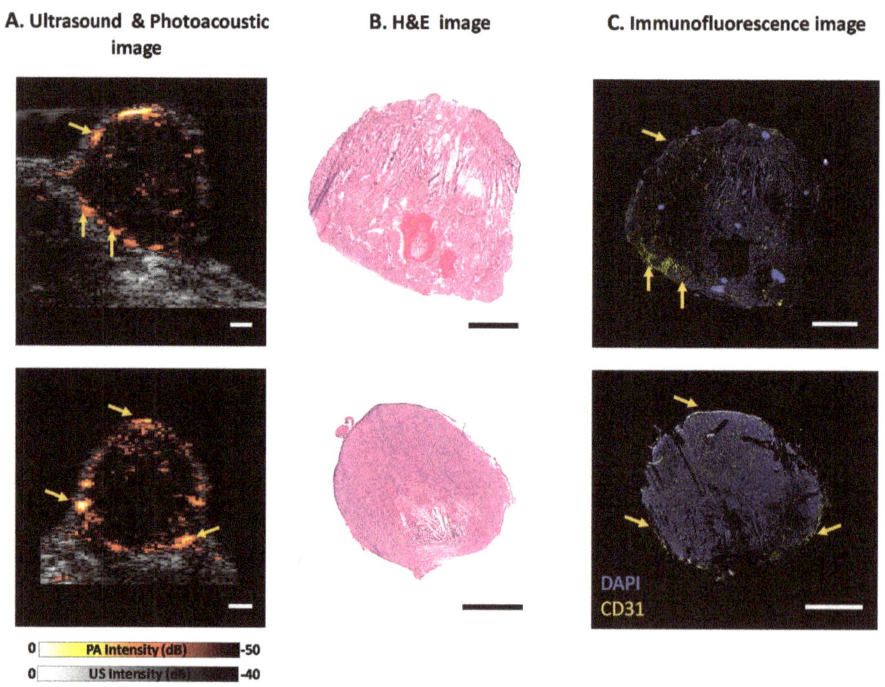

Figure 4. Evaluation of AcousticX on subcutaneous tumor xenografts in mice. (**A**) Ultrasound and photoacoustic image overlay of the tumor vasculature of two mice. Images of the Mouse 1 tumor acquired at a frame rate of 10 Hz are shown in Video 1. Motion due to breathing was minimal and only frames that did not have motion artifacts were used for image analysis. (**B**) H&E stain of the tumor cross-section. (**C**) Immunofluorescence stain for blood vessels in yellow and cell nuclei in blue. (Scale bar on all images = 2 mm, Video S1, MP4, 1.7 MB).

3.3. Correlation of PA Signal with Histology

The correlation plot between the PA signal from blood vessels and CD31 intensity from the IF image is displayed in Figure 5. As expected, we observed that larger blood vessels or areas with high blood vessel density had a high PA signal while the micro vessels or low blood vessel density regions had a low signal. The overall R-value for correlation was 0.84 and individual R-values for the three tumors used in the study were 0.65, 0.91 and 0.91. The differences in the R-values among the tumors were due to the inter-tumor heterogeneity in the vasculature and biological variations across different animals. The results obtained here with the LED-based PA imaging system are on par with reports using traditional laser-based PA imaging systems. For example, Mallidi et al. [28] and Bar-Zion et al. [29] qualitatively demonstrated good spatial co-registration between tumor vascular oxygenation maps obtained from PA imaging and hypoxia markers such as pimonidazole or carbonic anhydrase IX. Gerling et al. [30] observed an excellent R-value of 0.95 when quantitatively correlating oxygenation saturation measurements from PA imaging to the intensity of pimonidazole in the tumor, however, there were only six points in this correlation. Furthermore, the PA signals were averaged over the entire tumor underplaying the intra-tumor heterogeneity. In our study we divided the tumor region into several ROIs to consider both high and low vascular regions. In addition, these studies [28–30] utilized higher frequency transducers with limitations in imaging depth. It is obvious that some microvascular structures inside the tumors may be not within the detection limits of the 7 MHz transducer used in our study, which would have resulted in slightly lower correlation between PA signals and CD31 intensity in certain regions of the tumor. Nevertheless, we clearly demonstrate a good correlation between histology and PA images generated using an LED source with comparatively less optical output energy.

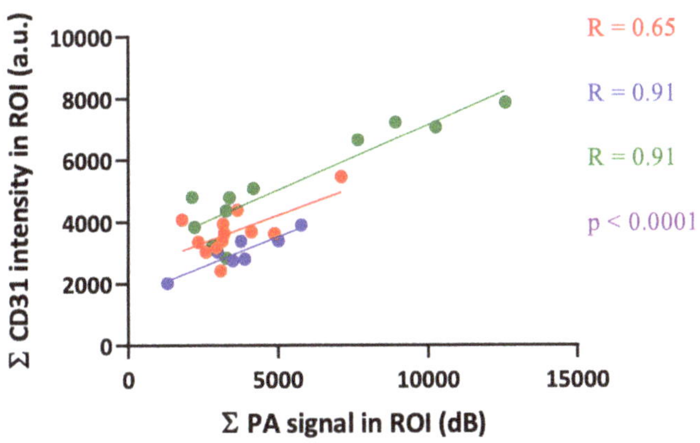

Figure 5. Correlation between background subtracted signal intensity from PA images and CD31 immunofluorescence (IF) image. Each color represents data from different mice. A total of 30 ROIs from three mice were used in our analysis. The R-value for correlation based on Pearson's coefficient was calculated between the CD31 intensity from the IF image and PA signals from blood vessels.

3.4. 3D Tumor Imaging with AcousticX

An LED-based photoacoustic imaging system can also be used to visualize vasculature in 3D. Figure 6 shows the overlay of a US/PA image of tumor acquired by AcousticX at a frame rate of 10 Hz and reconstructed in 3D using ImageJ. A corresponding video showcasing the 3D rendering of US/PA images of the tumor is shown in Video S2. Figure 6C shows the structure of the tumor in different cross-sectional planes. The measurements obtained from AcousticX images have good agreement with the caliper measurements for the tumor size (length, width, and depth). Two LED arrays on either side of the US transducer delivered a total optical energy output of 400 μJ/pulse on the skin surface. In the absence of optical scattering, the irradiation area at the US focus will be approximately 50 × 7 mm in the YX-plane, with a maximum radiant exposure of 0.11 mJ/cm^2, which is orders of magnitude lower than the ANSI safety limit and the radiant exposure in laser-based PA imaging systems. It is encouraging that the system used in this study could visualize tumor vasculature in vivo at an imaging depth of 10 mm, considering the low pulse energies provided by the LED arrays.

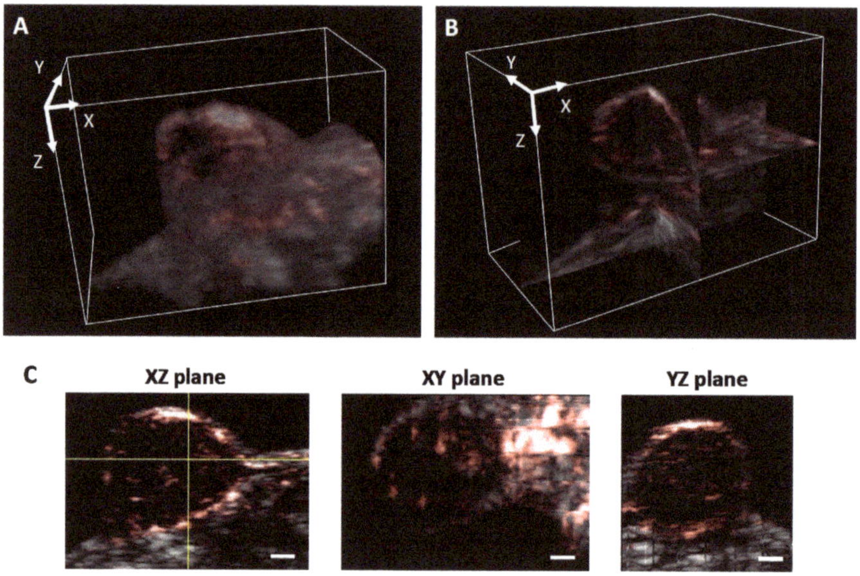

Figure 6. (**A**) Overlay of ultrasound and photoacoustic images of tumors reconstructed in 3D (Video 2) and (**B**) in orthoslice view. (**C**) Tumor images displayed in different 2D planes. (Scale bar = 2 mm, Video S2, MOV, 779 kB).

4. Discussion

In preclinical cancer research, rodent models are widely used to study the complex physiological process involved in tumor formation and therapy response. The tumor is grown either subcutaneously or orthotopically at the site of the origin of tumor cells. The maximum size for subcutaneous tumors is less than 1–2 cm in diameter, and orthotopic tumors depending on the organ can vary between 3 and 6 mm in diameter. The length of the LED array in AcousticX is 8.65 cm [14] and is significantly larger than that of the tumors generally seen in preclinical research. Hence most of the light from the LED arrays is delivered outside of the tumor. Considering this, the imaging depth of ~10 mm we achieved demonstrated the high sensitivity of the system, which is commendable. Better light delivery strategies to illuminate tumors efficiently may help in improving the imaging depth and SNR further.

Currently the AcousticX system is integrated with a 7 MHz ultrasound transducer that has limitations in spatial resolution. Higher frequency transducers offer better resolution and our future work involves integrating these transducers with LED arrays to enhance image resolution. In addition to improvements in spatial resolution, image SNR can be enhanced through novel beamforming algorithms. Traditional beamforming algorithms are based on "delay and sum" and "delay-multiply and sum" methodology. Although they are simple to implement in imaging systems, they generate low quality images with low contrast, especially when the data is noisy. Recent studies have shown that the "double stage delay-multiply and sum" algorithm improves lateral resolution and offers high contrast for an LED-based PA imaging system [31,32]. Moreover, in this study we used a single wavelength

(850 nm) LED array that provides deeper light penetration to image tumor vasculature. The availability of high-power multi-wavelength LED arrays will enable imaging blood oxygen saturation levels in the tumor vasculature [33,34]. This information is key for monitoring treatment response and predicting recurrence. These multi-wavelength LED arrays can also enable the imaging of tumor via receptor targeted exogenous contrast agents [15,18,25,27], as previously demonstrated by us and others with traditional pulsed laser sources [9,35–41]. With the availability of low-cost multi-wavelength PA imaging systems, our future studies will aim at developing and monitoring various targeted therapies guided by multi-wavelength PA imaging that can provide vascular information with information on the localization of targeted contrast agents.

5. Conclusions

In summary, we demonstrate the capability of an LED-based PA imaging system for monitoring tumor vasculature in vivo. The PA images obtained are in good correlation with the gold standard IF images. With the widespread rise of PA imaging technology, new improved reconstruction algorithms, high frequency transducers, better illumination strategies, and multi-wavelength LED arrays, we can expect LED-based PA imaging to become a promising tool and play a highly significant role in both preclinical research and clinical applications.

Supplementary Materials: The following are available online at http://www.mdpi.com/1424-8220/20/16/4503/s1, Video S1: Visualization demonstrates that the motion artifacts in the tumor region caused due to mouse breathing, while acquiring the RF data for Figure 4 in the manuscript, is minimal. Video S2: Visualization demonstrating the 3D rendering of US/PA images of the tumor acquired by AcousticX.

Author Contributions: Conceptualization, M.X., M.K.A.S., and S.M.; methodology, M.X. and S.M.; software, M.X. and M.K.A.S.; validation, M.X. and S.M.; formal analysis, M.X.; investigation, M.X. and S.M.; resources, S.M.; data curation, M.X. and M.K.A.S.; writing—original draft preparation, M.X. and S.M.; writing—review and editing, M.X., M.K.A.S., and S.M.; visualization, M.X. and S.M.; supervision, S.M.; project administration, S.M.; funding acquisition, S.M. All authors have read and agreed to the published version of the manuscript.

Funding: The work was supported by funds from the Tufts School of Engineering and NIH R41CA221420.

Acknowledgments: The authors would like to thank the Pathology core at the Wellman Center for Photomedicine, MGH for sectioning the tissues and access to the NanoZoomer imaging system. We also would like to thank Annika Schaad at Tufts University for help with the Figure 1 schematic drawing.

Conflicts of Interest: M.K.A.S. is employed by Cyberdyne INC. The authors have no relevant financial interests or potential conflicts of interest to disclose.

References

1. Hanahan, D.; Weinberg, R.A. Hallmarks of cancer: The next generation. *Cell* **2011**, *144*, 646–674. [CrossRef] [PubMed]
2. Eberhard, A.; Kahlert, S.; Goede, V.; Hemmerlein, B.; Plate, K.H.; Augustin, H.G. Heterogeneity of angiogenesis and blood vessel maturation in human tumors: Implications for antiangiogenic tumor therapies. *Cancer Res.* **2000**, *60*, 1388–1393. [PubMed]
3. Garattini, S.; Fuso Nerini, I.; D'Incalci, M. Not only tumor but also therapy heterogeneity. *Ann. Oncol.* **2018**, *29*, 13–19. [CrossRef] [PubMed]
4. Stylianopoulos, T.; Munn, L.L.; Jain, R.K. Reengineering the Tumor Vasculature: Improving Drug Delivery and Efficacy. *Trends Cancer* **2018**, *4*, 258–259. [CrossRef] [PubMed]
5. Vasudev, N.S.; Reynolds, A.R. Anti-angiogenic therapy for cancer: Current progress, unresolved questions and future directions. *Angiogenesis* **2014**, *17*, 471–494. [CrossRef] [PubMed]
6. Cai, W.; Gambhir, S.S.; Chen, X. Methods in Enzymology. In *Angiogenesis: In Vivo Systems, Part B*; Academic Press: Cambridge, MA, USA, 2008; Volume 445, pp. 141–176.
7. Laufer, J.; Johnson, P.; Zhang, E.; Treeby, B.; Cox, B.; Pedley, B.; Beard, P. In vivo preclinical photoacoustic imaging of tumor vasculature development and therapy. *J. Biomed. Opt.* **2012**, *17*, 56016. [CrossRef]
8. Bohndiek, S.E.; Sasportas, L.S.; Machtaler, S.; Jokerst, J.V.; Hori, S.; Gambhir, S.S. Photoacoustic Tomography Detects Early Vessel Regression and Normalization During Ovarian Tumor Response to the Antiangiogenic Therapy Trebananib. *J. Nucl. Med.* **2015**, *56*, 1942–1947. [CrossRef]

9. Mallidi, S.; Luke, G.P.; Emelianov, S. Photoacoustic imaging in cancer detection, diagnosis, and treatment guidance. *Trends Biotechnol.* **2011**, *29*, 213–221. [CrossRef]
10. Brown, E.; Brunker, J.; Bohndiek, S.E. Photoacoustic imaging as a tool to probe the tumour microenvironment. *Dis. Model. Mech.* **2019**, *12*, 039636. [CrossRef]
11. Xu, M.; Wang, L.V. Photoacoustic imaging in biomedicine. *Rev. Sci. Instrum.* **2006**, *77*, 41101. [CrossRef]
12. Gigan, S. Optical microscopy aims deep. *Nat. Photonics* **2017**, *11*, 14–16. [CrossRef]
13. Agrawal, S.; Kuniyil Ajith Singh, M.; Yang, X.; Albahrani, H.; Dangi, A.; Kothapalli, S.-R. *Photoacoustic Imaging Capabilities of Light Emitting Diodes (LED) and Laser Sources: A Comparison Study*; Oraevsky, A.A., Wang, L.V., Eds.; Photons Plus Ultrasound: Imaging and Sensing: San Francisco, CA, USA, 2020.
14. Hariri, A.; Lemaster, J.; Wang, J.; Jeevarathinam, A.S.; Chao, D.L.; Jokerst, J.V. The characterization of an economic and portable LED-based photoacoustic imaging system to facilitate molecular imaging. *Photoacoustics* **2018**, *9*, 10–20. [CrossRef] [PubMed]
15. Zhu, Y.; Xu, G.; Yuan, J.; Jo, J.; Gandikota, G.; Demirci, H.; Agano, T.; Sato, N.; Shigeta, Y.; Wang, X. Light Emitting Diodes based Photoacoustic Imaging and Potential Clinical Applications. *Sci. Rep.* **2018**, *8*, 9885. [CrossRef] [PubMed]
16. Joseph Francis, K.; Boink, Y.E.; Dantuma, M.; Ajith Singh, M.K.; Manohar, S.; Steenbergen, W. Tomographic imaging with an ultrasound and LED-based photoacoustic system. *Biomed. Opt. Express* **2020**, *11*, 2152–2165. [CrossRef] [PubMed]
17. Agrawal, S.; Fadden, C.; Dangi, A.; Yang, X.; Albahrani, H.; Frings, N.; Heidari Zadi, S.; Kothapalli, S.R. Light-Emitting-Diode-Based Multispectral Photoacoustic Computed Tomography System. *Sensors* **2019**, *19*, 4861. [CrossRef] [PubMed]
18. Jo, J.; Xu, G.; Zhu, Y.; Burton, M.; Sarazin, J.; Schiopu, E.; Gandikota, G.; Wang, X. Detecting joint inflammation by an LED-based photoacoustic imaging system: A feasibility study. *J. Biomed. Opt.* **2018**, *23*, 1–4. [CrossRef]
19. Xia, W.; Kuniyil Ajith Singh, M.; Maneas, E.; Sato, N.; Shigeta, Y.; Agano, T.; Ourselin, S.; West, S.J. Handheld Real-Time LED-Based Photoacoustic and Ultrasound Imaging System for Accurate Visualization of Clinical Metal Needles and Superficial Vasculature to Guide Minimally Invasive Procedures. *Sensors* **2018**, *18*, 1394. [CrossRef]
20. Zhu, Y.; Lu, X.; Dong, X.; Yuan, J.; Fabiilli, M.L.; Wang, X. LED-Based Photoacoustic Imaging for Monitoring Angiogenesis in Fibrin Scaffolds. *Tissue Eng. Part C Methods* **2019**, *25*, 523–531. [CrossRef]
21. Hariri, A.; Zhao, E.; Jeevarathinam, A.S.; Lemaster, J.; Zhang, J.; Jokerst, J.V. Molecular imaging of oxidative stress using an LED-based photoacoustic imaging system. *Sci. Rep.* **2019**, *9*, 1–10.
22. Lauk, S.; Zietman, A.; Skates, S.; Fabian, R.; Suit, H.D. Comparative Morphometric Study of Tumor Vasculature in Human Squamous Cell Carcinomas and Their Xenotransplants in Athymic Nude Mice. *Cancer Res.* **1989**, *49*, 4557–4561.
23. Seshadri, M.; Mazurchuk, R.; Spernyak, J.A.; Bhattacharya, A.; Rustum, Y.M.; Bellnier, D.A. Activity of the vascular-disrupting agent 5,6-dimethylxanthenone-4-acetic acid against human head and neck carcinoma xenografts. *Neoplasia* **2006**, *8*, 534–542. [CrossRef] [PubMed]
24. Jaeger, M.; Schüpbach, S.; Gertsch, A.; Kitz, M.; Frenz, M. Fourier reconstruction in optoacoustic imaging using truncated regularized inversek-space interpolation. *Inverse Probl.* **2007**, *23*, S51–S63. [CrossRef]
25. Xia, W.; Maneas, E.; Trung Huynh, N.; Kuniyil Ajith Singh, M.; Montaña Brown, N.; Ourselin, S.; Gilbert-Kawai, E.; West, S.J.; Desjardins, A.E. *Imaging of Human Peripheral Blood Vessels during Cuff Occlusion with a Compact LED-Based Photoacoustic and Ultrasound System*; Oraevsky, A.A., Wang, L.V., Eds.; Photons Plus Ultrasound: Imaging and Sensing: San Francisco, CA, USA, 2019.
26. Hussain, A.; Petersen, W.; Staley, J.; Hondebrink, E.; Steenbergen, W. Quantitative blood oxygen saturation imaging using combined photoacoustics and acousto-optics. *Opt. Lett.* **2016**, *41*, 1720–1723. [CrossRef] [PubMed]
27. Kuniyil Ajith Singh, M.; Sato, N.; Ichihashi, F.; Sankai, Y. In Vivo Demonstration of Real-Time Oxygen Saturation Imaging Using a Portable and Affordable LED-Based Multispectral Photoacoustic and Ultrasound Imaging System. In Proceedings of the Photons Plus Ultrasound: Imaging and Sensing 2019, San Francisco, CA, USA, 27 February 2019.
28. Mallidi, S.; Watanabe, K.; Timerman, D.; Schoenfeld, D.; Hasan, T. Prediction of tumor recurrence and therapy monitoring using ultrasound-guided photoacoustic imaging. *Theranostics* **2015**, *5*, 289–301. [CrossRef]

29. Bar-Zion, A.; Yin, M.; Adam, D.; Foster, F.S. Functional Flow Patterns and Static Blood Pooling in Tumors Revealed by Combined Contrast-Enhanced Ultrasound and Photoacoustic Imaging. *Cancer Res.* **2016**, *76*, 4320–4331. [CrossRef]
30. Gerling, M.; Zhao, Y.; Nania, S.; Norberg, K.J.; Verbeke, C.S.; Englert, B.; Kuiper, R.V.; Bergstrom, A.; Hassan, M.; Neesse, A.; et al. Real-time assessment of tissue hypoxia in vivo with combined photoacoustics and high-frequency ultrasound. *Theranostics* **2014**, *4*, 604–613. [CrossRef]
31. Miri Rostami, S.R.; Mozaffarzadeh, M.; Ghaffari-Miab, M.; Hariri, A.; Jokerst, J. GPU-accelerated Double-stage Delay-multiply-and-sum Algorithm for Fast Photoacoustic Tomography Using LED Excitation and Linear Arrays. *Ultrason. Imaging* **2019**, *41*, 301–316. [CrossRef]
32. Mozaffarzadeh, M.; Hariri, A.; Moore, C.; Jokerst, J.V. The double-stage delay-multiply-and-sum image reconstruction method improves imaging quality in a LED-based photoacoustic array scanner. *Photoacoustics* **2018**, *12*, 22–29. [CrossRef]
33. Xavierselvan, M.; Mallidi, S. LED-Based Functional Photoacoustics—Portable and Affordable Solution for Preclinical Cancer Imaging. In *LED-Based Photoacoustic Imaging: From Bench to Bedside*; Kuniyil Ajith Singh, M., Ed.; Springer: Singapore, 2020; pp. 303–319.
34. Zhu, Y.; Feng, T.; Cheng, Q.; Wang, X.; Du, S.; Sato, N.; Kuniyil Ajith Singh, M.; Yuan, J. Towards Clinical Translation of LED-Based Photoacoustic Imaging: A Review. *Sensors* **2020**, *20*, 2484. [CrossRef]
35. Mallidi, S.; Larson, T.; Tam, J.; Joshi, P.P.; Karpiouk, A.; Sokolov, K.; Emelianov, S. Multiwavelength Photoacoustic Imaging and Plasmon Resonance Coupling of Gold Nanoparticles for Selective Detection of Cancer. *Nano Lett.* **2009**, *9*, 2825–2831. [CrossRef]
36. Obaid, G.; Bano, S.; Mallidi, S.; Broekgaarden, M.; Kuriakose, J.; Silber, Z.; Bulin, A.L.; Wang, Y.; Mai, Z.; Jin, W.; et al. Impacting Pancreatic Cancer Therapy in Heterotypic in Vitro Organoids and in Vivo Tumors with Specificity-Tuned, NIR-Activable Photoimmunonanoconjugates: Towards Conquering Desmoplasia? *Nano Lett.* **2019**, *19*, 7573–7587. [CrossRef] [PubMed]
37. Li, P.-C.; Wang, C.-R.C.; Shieh, D.-B.; Wei, C.-W.; Liao, C.-K.; Poe, C.; Jhan, S.; Ding, A.-A.; Wu, Y.-N. In vivo Photoacoustic Molecular Imaging with Simultaneous Multiple Selective Targeting Using Antibody-Conjugated Gold Nanorods. *Opt. Express* **2008**, *16*, 18605–18615. [CrossRef] [PubMed]
38. Yang, H.W.; Liu, H.L.; Li, M.L.; Hsi, I.W.; Fan, C.T.; Huang, C.Y.; Lu, Y.J.; Hua, M.Y.; Chou, H.Y.; Liaw, J.W.; et al. Magnetic gold-nanorod/ PNIPAAmMA nanoparticles for dual magnetic resonance and photoacoustic imaging and targeted photothermal therapy. *Biomaterials* **2013**, *34*, 5651–5660. [CrossRef] [PubMed]
39. Pan, D.; Pramanik, M.; Senpan, A.; Allen, J.S.; Zhang, H.; Wickline, S.A.; Wang, L.V.; Lanza, G.M. Molecular photoacoustic imaging of angiogenesis with integrin-targeted gold nanobeacons. *FASEB J.* **2011**, *25*, 875–882. [CrossRef] [PubMed]
40. Wu, D.; Huang, L.; Jiang, M.S.; Jiang, H. Contrast agents for photoacoustic and thermoacoustic imaging: A review. *Int. J. Mol. Sci.* **2014**, *15*, 23616–23639. [CrossRef] [PubMed]
41. Pan, D.; Kim, B.; Wang, L.V.; Lanza, G.M. A brief account of nanoparticle contrast agents for photoacoustic imaging. *WIREs Nanomed. Nanobiotechnol.* **2013**, *5*, 517–543. [CrossRef]

sensors

Article

Oxygen Saturation Imaging Using LED-Based Photoacoustic System

Rianne Bulsink [1,†], Mithun Kuniyil Ajith Singh [2,†], Marvin Xavierselvan [3], Srivalleesha Mallidi [3,4], Wiendelt Steenbergen [1] and Kalloor Joseph Francis [1,*]

1 Biomedical Photonic Imaging (BMPI), Technical Medical Center, University of Twente, 7500 AE Enschede, The Netherlands; r.e.bulsink@student.utwente.nl (R.B.); w.steenbergen@utwente.nl (W.S.)
2 Research & Business Development Division, CYBERDYNE Inc., Cambridge Innovation Center, 3013 AK Rotterdam, The Netherlands; mithun_ajith@cyberdyne.jp
3 Department of Biomedical Engineering, Science and Technology Center, Tufts University, Medford, MA 02155, USA; marvin.xavierselvan@tufts.edu (M.X.); srivalleesha.mallidi@tufts.edu (S.M.)
4 Wellman Center for Photomedicine, Massachusetts General Hospital, Harvard Medical School, Boston, MA 02114, USA
* Correspondence: f.kalloorjoseph@utwente.nl
† These authors contributed equally.

Abstract: Oxygen saturation imaging has potential in several preclinical and clinical applications. Dual-wavelength LED array-based photoacoustic oxygen saturation imaging can be an affordable solution in this case. For the translation of this technology, there is a need to improve its accuracy and validate it against ground truth methods. We propose a fluence compensated oxygen saturation imaging method, utilizing structural information from the ultrasound image, and prior knowledge of the optical properties of the tissue with a Monte-Carlo based light propagation model for the dual-wavelength LED array configuration. We then validate the proposed method with oximeter measurements in tissue-mimicking phantoms. Further, we demonstrate in vivo imaging on small animal and a human subject. We conclude that the proposed oxygen saturation imaging can be used to image tissue at a depth of 6–8 mm in both preclinical and clinical applications.

Keywords: oxygen saturation imaging; LED; photoacoustics; ultrasound; fluence compensation; in vivo; hypoxia

Citation: Bulsink, R.; Kuniyil Ajith Singh, M.; Xavierselvan, M.; Mallidi, S.; Steenbergen, W.; Francis, K.J. Oxygen Saturation Imaging Using LED-Based Photoacoustic System. *Sensors* **2021**, *21*, 283. https://doi.org/10.3390/s21010283

Received: 15 December 2020
Accepted: 1 January 2021
Published: 4 January 2021

1. Introduction

Blood oxygen saturation (sO_2) is the ratio of oxygen saturated hemoglobin to the total hemoglobin concentration [1]. sO_2 is a key physiological marker for detection, diagnosis, and treatment of several devastating diseases like cancer [1,2]. Hypoxia, reduction in sO_2 is one of the early biomarkers for all tumors and inflammatory diseases [3,4]. A fast-growing tumor is expected to have abnormal vasculature growth (angiogenesis) in and around it, resulting in hypoxia [5]. Detecting hypoxia at an early stage has a profound impact in early diagnosis and better treatment planning. In addition, sO_2 deviations inside the brain and its detection is important in the diagnosis of cerebrovascular diseases [6,7]. Hence, sensitive detection of changes in sO_2 is useful in early diagnosis and can help plan better treatment strategies.

In both preclinical and clinical settings, it is required to measure the changes in sO_2 with high spatial and temporal resolution without compromising on the imaging depth. Currently, point sO_2 measurement using commercial pulse oximeter is widely used to obtain an average sO_2 value. However, the spatial distribution of sO_2 is important in several applications, including tumor analysis. Non-invasive functional Magnetic Resonance Imaging (fMRI) is capable of imaging sO_2 changes with spatial resolutions ranging from 1–6 mm [8,9]. However, fMRI systems are too expensive and are not suitable in a point-of-care resource-limited setting. Another imaging modality that is suitable for sO_2 imaging is positron emission tomography (PET). The PET involves ionizing radiations and can offer far lower spatial resolution when compared to other techniques [10]. Purely optical imaging techniques can detect changes in sO_2 at reasonable imaging depth. However, the spatial resolution offered by optical-techniques is not sufficient for most clinical applications [11]. Photoacoustic imaging (PAI) has shown potential in multiple preclinical and clinical applications [12,13]. PAI or optoacoustic imaging is a modality that combines the advantages of ultrasound (US) and optical imaging techniques [12]. In PAI, nanosecond pulsed light excitation of optical absorbers in the tissue results in US signal generation [12]. These US signals can be detected using US transducers. Further, the signal can be used to reconstruct an image proportional to the optical absorption of the tissue. PAI is one of the fastest-growing imaging modalities of recent times, offering high-resolution images and optical contrast in deep tissue. Since acoustic detection hardware can be shared between US and PAI, it is straightforward to implement a dual-mode US and PAI system capable of offering structural, functional, and molecular contrast from a single measurement [14]. Hemoglobin is an excellent optical absorber with well-defined absorption spectra in the near-infrared wavelengths. Hence, measurement of sO_2 is undoubtedly the most interesting application of PAI [15,16]. By carefully selecting the light wavelengths (NIR range is commonly used because of its tissue penetration and high absorption in hemoglobin), one can measure deep-tissue sO_2 with unprecedented spatio-temporal resolution using PAI [17]. Conventionally, PAI utilizes bulky, slow, and expensive class IV lasers for tissue illumination [18], a key factors hindering the clinical translation of this imaging modality [19]. In recent years, there have been advancements in solid-state device technology leading to the development of laser diodes (LD) and light-emitting diodes (LED), which are suitable for PAI [20–22]. High power LEDs have shown potential in PAI-based superficial and sub-surface (skin and <10 mm imaging depth) imaging in both preclinical and clinical applications [23–28]. LED-PAI is portable, affordable, and energy-efficient, for that reason it may have an easy clinical acceptance. These advantages of LED-PAI will be also useful in a preclinical setting, where it is of great importance to improve small animal study outcome by monitoring disease and treatment progress [29,30]. Further, we have reported an affordable LED-based tomographic system by imaging an object from multiple angles and demonstrated it in finger joint and small animal imaging [30,31]. In this work, we propose the use of US image information for fluence compensation to improve the accuracy of sO_2 in dual-wavelength LED-based handheld PAI.

Fluence variations in the tissue hinder the possibility of quantitative sO_2 imaging [16]. It is important to develop fluence-compensation methods that are accurate, fast, and easy-to-implement in real-time PAI systems. Guo et al. utilized the acoustic spectra of PA signals to compute sO_2 in the tissue [32]. Deep learning-based methods were also used for sO_2 [33,34]. Xia et al. utilized the dynamics in sO_2 to correct for the fluence. Tzoumas et al. modeled the fundamental optical absorption spectra (eigenspectra) in the tissue, with the consideration of unknown optical fluence and demonstrated improvement in sO_2 imaging [35]. Hussain et al. used acoustically tagged photons (acousto-optics) for fluence compensation in sO_2 imaging [36]. All the above-mentioned works were focused on improving the quantitative nature of sO_2 imaging in laser-based PAI, with either algorithmic or optical methods. As most PAI systems can also perform US imaging, using US image information and the prior knowledge of tissue for fluence compensation can be a potential direction for quantitative sO_2 imaging in practical applications. US information was previously used for fluence estimation using Beer's law in tissue [37]. Even though sO_2 imaging using LED-based PAI system was previously reported by the authors [29,38] and by Zhu et al. [25], depth-dependent fluence variation was not considered in these works. Assumption of homogeneous light distribution or simple fitting using Beer's law is not suitable in handheld LED-PAI probes, with a gap between tissue and US probe surface caused by LED arrays placed at an angle to maximize light in the imaging plane.

In this work, we utilized the information offered by conventional US imaging to segment the tissue and used Monte-Carlo simulations of the LED probe to estimate the fluence map in the imaging plane. For the first time, we characterized the sO_2 imaging capability of commercially available LED-based PAI system using human blood in vitro and compared the results with oximeter readings. Further, we thoroughly tested and validated our proposed US-assisted fluence compensation method and its efficacy in improving LED-PAI based sO_2 imaging using tissue-mimicking phantoms with human blood at different depths and mice in vivo. Finally, we applied our proposed method in a real-time experiment showing clear differentiation between a vein and pulsating artery in a human wrist.

2. Materials and Methods

In this work, we used a US and LED-based PA imaging system, AcousticX (Cyberdyne Inc., Tsukuba, Japan). For oxygen saturation imaging, we used a dual-wavelength approach with an LED array having 750 nm and 850 nm elements. Figure 1e shows the probe with LED array with elements having two wavelengths. Two LED units were used in the probe on either side of the transducer (Figure 1e). The 850 nm array has pulse energy of 200 μJ and 750 nm array has 100 μJ, with a pulse duration of 70 ns. The LED array unit consists of two rows of 36 element 850 nm arrays and two rows of 24 element 750 nm arrays arranged alternately. A 128 element, 7 MHz linear US transducer with a bandwidth of 80% was used in the probe (Figure 1e). In this section, we present the proposed fluence compensation method and the details of our experimental studies.

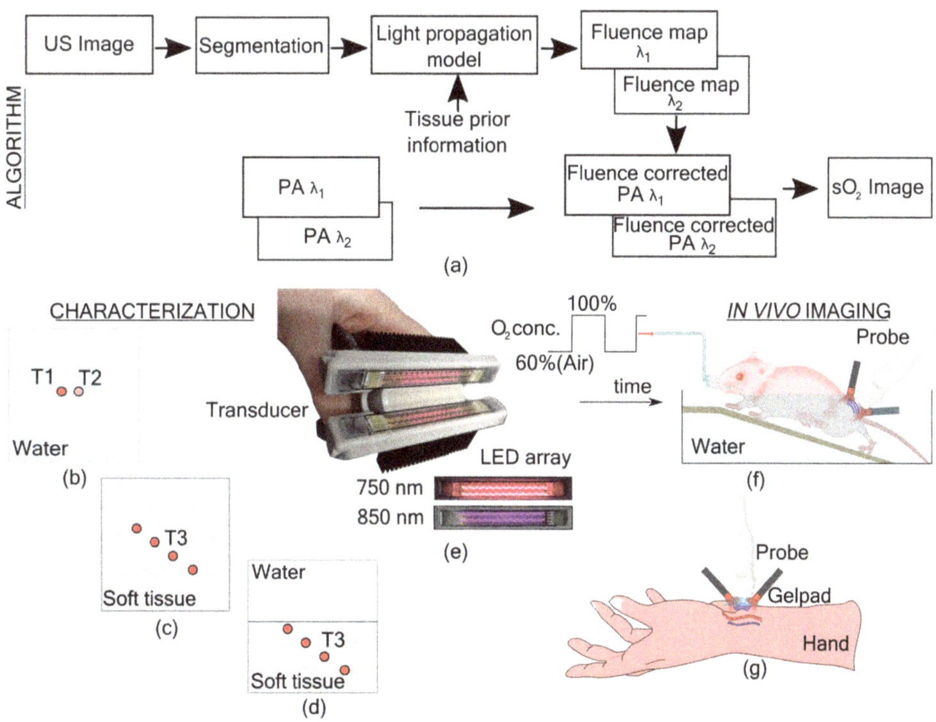

Figure 1. Algorithm and experimental setup. (**a**) Fluence compensation algorithm using ultrasound information. Schematic of the imaging plane in the photoacoustic oxygen saturation (PA sO_2) characterization experiments (**b–d**). (**b**) PA sO_2 imaging of tube 1 (T1) having constant sO_2 against T2 with varying sO_2. (**c**) PA sO_2 imaging in soft tissue-mimicking medium with tube T3 imaged at four different depths. (**d**) PA sO_2 imaging in medium with two optical properties, water on the top and soft tissue-mimicking medium at the bottom. (**e**) LED-based photoacoustic probe with an ultrasound transducer and LED arrays of two wavelengths (750 and 850 nm). (**f**) In vivo imaging of a mouse with cyclically changing oxygen concentration in the breathing air. (**g**) In vivo imaging of an artery and a vein in the wrist of a human volunteer.

2.1. Oxygen Saturation Imaging Using Linear Unmixing

An overview of PA oxygen saturation using linear unmixing with fluence compensation is presented in this section for completeness [17,39]. PA initial pressure resulting from pulsed light excitation of optical absorber with the assumption of stress confinement can be expressed as,

$$p_0(\boldsymbol{r}, \lambda_i) = \Gamma \mu_a(\boldsymbol{r}, \lambda_i) \Phi(\boldsymbol{r}, \lambda_i). \tag{1}$$

Grüneisen parameter Γ is the thermal to pressure conversion efficiency and μ_a is optical absorption coefficient and Φ is light fluence (integrated radiance). The optical absorption coefficient can be written as the product of the molar extinction coefficient (ϵ) and concentration (c). Thus the reconstructed PA image can be written as,

$$p(\boldsymbol{r}, \lambda_i) = \Phi(\boldsymbol{r}, \lambda_i)[\epsilon_{HbR}(\lambda_i) c_{HbR}(\boldsymbol{r}) + \epsilon_{HbO_2}(\lambda_i) c_{HbO_2}(\boldsymbol{r})] \tag{2}$$

Let the two wavelengths used in this study be λ_1 = 750 nm and λ_2 = 850 nm. Further, let $b(\boldsymbol{r}, \lambda_i = p(\boldsymbol{r}, \lambda_i)/\Phi(\boldsymbol{r}, \lambda_i)$ be the fluence compensated PA images given by

$$b = \begin{bmatrix} b(\boldsymbol{r}, \lambda_1) \\ b(\boldsymbol{r}, \lambda_2) \end{bmatrix}. \tag{3}$$

The molar extinction coefficient of HbR and HbO_2 for both wavelengths can be combined into a matrix,

$$A = \begin{bmatrix} \epsilon_{HbR}(\lambda_1) & \epsilon_{HbO_2}(\lambda_1) \\ \epsilon_{HbR}(\lambda_2) & \epsilon_{HbO_2}(\lambda_2) \end{bmatrix}. \tag{4}$$

The molar concentration of HbR and HbO_2,

$$x = \begin{bmatrix} c_{HbR}(\boldsymbol{r}) \\ c_{HbO_2}(\boldsymbol{r}) \end{bmatrix}, \tag{5}$$

can be retrieved by solving the linear equation $Ax = b$. The most common method to solve this linear equation is using the least square solution given by,

$$x = (A^T A)^{-1} A^T b. \tag{6}$$

Oxygen saturation image can then be obtained from the molar concentrations,

$$sO_2(\boldsymbol{r}) = \frac{c_{HbO_2}(\boldsymbol{r})}{c_{HbO_2}(\boldsymbol{r}) + c_{HbR}(\boldsymbol{r})} \times 100\% \tag{7}$$

2.2. Fluence Compensation

Using LED arrays in the PA probe introduces some constraints. First, it is near impossible to have the transducer touching the tissue surface due to the inclined position of the LED array. Thus, a medium such as water or a gel pad is required to fill this gap for acoustic coupling. Second, with multiple LED element arrays in the unit, the light source has several mm width and cannot be assumed as a line source with both wavelengths of light originating from the same location. Further, with the large opening angle and low power of LEDs, it is not possible to assume a homogeneous illumination in the imaging plane. In this section, we present a method to estimate the fluence map inside the tissue utilizing the information from the US image. The following steps were performed to obtain a fluence compensated sO_2 image (Figure 1a).

1. US and PA images were reconstructed offline using a Fourier based algorithm [40]
2. The US image was segmented to obtain a binary mask of the tissue boundary.
3. The binary mask and the optical properties of the tissue were used in the light propagation model to obtain fluence maps at the imaging plane for two wavelengths.
4. PA images at two wavelengths were normalized using the fluence maps.

5. Linear unmixing was used to obtain oxygen saturation images from the fluence normalized PA images.

Characterization of the proposed approach was performed in three sets of experiments. In the first experiment, a validation of oxygen saturation imaging using the linear unmixing algorithm using the two-wavelength LED array was performed. Next, the effect of fluence compensation on oxygen saturation imaging was studied by imaging tubes carrying blood at four different depths in a soft tissue-mimicking medium. Finally, the US-assisted fluence compensation was tested on a two slab phantom and in vivo imaging.

2.2.1. Ultrasound Segmentation

The tissue boundary was obtained by segmenting the US image. First, a median filter was applied to smoothen the speckles in the US image. Next, a binary image was obtained by thresholding the grayscale US image. The threshold value was manually selected for different tissue types. The binary image was then processed to obtain the tissue boundary using the Sobel edge detection operator. Further, the speckle created holes were coved using morphological operator, filling. The largest connected component was identified and used as the mask for the target tissue area. All image processing operations were performed using MATLAB (MathWorks, MA, USA) imaging processing toolbox. For the liquid phantom used in the experiment, the US information about the separating film was obtained and the water region was set to zeros.

2.2.2. Light Propagation Model

A binary mask from the US segmentation and prior optical properties of the tissue was used as input to the model. We used a Monte Carlo based light propagation model (MCXLAB) for the simulations [41]. The LED-based PA probe consists of a US transducer and LED units with two-wavelength arrays. The positions of LED elements in the array, its opening angle, and pulse energy were used to define the light source. Measured optical properties of the phantom and values from the literature for the in vivo experiments at the two wavelengths were used as medium properties. The imaging plane was defined as the center slice in a three-dimensional grid and the fluence map was retrieved after the simulation from the same location.

A three dimensional geometry 55 mm (x-axis) $\times$ 55 mm (y-axis) $\times$ 37.9 mm (z-axis) (744 $\times$ 744 $\times$ 512 pixels) with uniform grid size of 74 μm was used for the simulation. The acquired images (PA and US) had a dimension of 40.32 mm (x) $\times$37.9 mm (z) resulting from a transducer with 128 elements with a pitch of 0.315 mm and 1024 time samples at 40 MHz sampling rate. The PA and US images were then interpolated to 545 $\times$ 512 pixel images with a uniform spacing of 74 μm in both dimensions. The input binary mask to the model was also the same size (545 $\times$ 512). The LED units have a length of 50 mm and a width of 10 mm. The LED units were placed at an angle of 41.4° with respect to the transducer surface and 0.15 mm in front of it. The LED units were placed on either side of the transducer with a distance of 9.56 mm between them. Four LED element arrays, two arrays from two wavelengths (750 nm and 850 nm) are present in each LED unit. The 750 nm arrays have 24 elements and the 850 nm arrays have 36 elements. The elements were placed with a spacing of 1.4 mm within the array and the spacing between the arrays was 1.72 mm. An opening angle of 120 degree for each element was used in the model. The two wavelengths LED elements have a power ratio of 2:1 for 850 nm and 750 nm, respectively. This power ratio was incorporated in defining the light source.

The length of the input US mask (545 pixels) was smaller than the length of the LED array (744 pixels) along the x-axis. Hence, the first and the last pixels were replicated to fill the entire length of the grid, with the US mask at the center. Further, the mask is two dimensional, and to define the medium properties in the third dimension, 744 copies of the same mask were stacked along the y-axis. Two phantom experiments and two in vivo experiments were performed in this study (Figure 1). More details are provided in Sections 2.3. The optical properties used in the simulation are provided in Table 1 below.

The tubes used in the experiments were assumed to be transparent in the wavelengths used and hence not considered in the light propagation model.

Table 1. Optical properties used in the simulations, optical absorption coefficient (μ_a) and reduced scattering coefficient (μ_s') (anisotropy, g = 0.9).

Medium	Wavelength [nm]	μ_a [cm^{-1}]	μ_s' [cm^{-1}]
Soft tissue phantom [42]	750	0.101	10.5
	850	0.089	9
Mice thigh muscle [43]	850	0.76	5.3
	850	0.64	4.8
Human forearm [44]	750	0.41	7.2
	850	0.3	6.5

2.3. *In Vitro Characterization in Phantoms*

2.3.1. In Vitro Validation of Oxygen Saturation Imaging

The goal of the experiment is to check the accuracy of two-wavelength PA sO_2 imaging against the ground truth method, oximeter sO_2. Fluence compensation was not considered in this case. However, the pulse energy of both wavelengths were used to normalize the PA images. Figure 1b shows imaging plane, where two tubes were placed in a water tank at 15 mm from the transducer surface. Both the tubes were placed at the same depth with a small distance between them so that the fluence variations are minimal. Polythene tubes (Smiths Medical, USA) with an inner diameter of 0.5 mm and an outer diameter of 0.84 mm filled with human blood were used for imaging. In the first tube (T1), blood with a constant oxygen saturation of around 65% was maintained. In the second tube (T2), different levels of blood oxygen were introduced. In this way, we also aim to see the relative difference in PA estimated sO_2. The blood (9 mL with heparin as an anticoagulant) for the experiment was obtained by TechMed Centre donor services, University of Twente, from a human volunteer after completing an ethical approval, adhering to the Dutch Medical Research involving human subjects Act (WMO). One part (4 mL) of the blood was used for tube T1. To obtain deoxygenated blood, 17 mg of sodium hydrosulfite (Sigma Aldrich, Darmstadt, Germany) was added to 5 mL of blood. The deoxygenated blood was then exposed to atmospheric oxygen to obtain different levels of oxygen concentration. Blood oxygen saturation was measured in both tubes before and after the imaging using an oximeter (AVOXimeter 4000, ITC, Munich, Germany). PA images at both 750 nm and 850 nm were performed by toggling between two-wavelength arrays. PA sO_2 images were obtained using the linear unmixing method mentioned above. The oxygen saturation in the tube was estimated by averaging 20 PA frames, with each frame generated by averaging 64 frames on-board in the system. A region of interest of 0.6 × 0.6 mm was selected for each tube and the mean PA sO_2 value was used for the analysis. The PA estimated sO_2 values were then compared against the oximeter measured sO_2 values.

2.3.2. Imaging a Homogeneous Phantom

In the second experiment, the effect of fluence compensation on sO_2 imaging at different depths was studied. Figure 1c shows the imaging plane where the same tube (T3) carrying blood was arranged such that it crosses the imaging plane at four different depths. An absorbing and scattering medium was used with soft tissue optical properties [42]. The optical properties of the medium are given in Table 1. Intralipid and Indian ink were used to prepare the phantom. Five liters solution was prepared in which 275 mL of 20% intralipid (Fresenius Kabi, Bad Homburg Germany) was used and 71.8 µL Indian ink (Talens, The Netherlands) was added from a stock solution (dilution factor 69,642). Using the light propagation model, fluence maps for both the wavelengths were computed. The fluence maps were used to compensate for the non-uniform illumination in the PA images. The fluence compensated PA images were used to compute the PA sO_2. PA sO_2 images with and without fluence compensation were compared with the measured blood oxygen saturation from the oximeter.

2.3.3. Imaging a Two Slab Phantom

In a realistic scenario, the transducer cannot touch the tissue surface because of the LED arrays placed on both sides of it in an angle. In most cases, water or a gel pad is used as a coupling medium. Oxygen saturation imaging in such a scenario is studied in this experiment with a two slab phantom. Water was used for the top slab and a soft tissue-mimicking medium for the bottom slab Figure 1d. The two mediums were separated by polyurethane US film (Protection Cover Ultrasound B. V., The Netherlands) with a thickness of 140 μm. The optical properties of the soft tissue phantom are given in Table 1. The tube arrangement used in the previous experiment at four different depths was used in this experiment as well. With the same blood in the tube, constant oxygen saturation was expected at all four depths. B-mode US imaging was performed and used to identify the two mediums. Light propagation was modeled in the two slab medium and the fluence maps were used to correct the PA images before estimating the oxygen saturation.

2.4. In Vivo Imaging

Two in vivo imaging experiments were performed. In the first experiment, fluence compensated oxygen saturation imaging was performed on a mouse. In the second experiment, in vivo oxygen saturation imaging was performed in the wrist of a healthy volunteer.

2.4.1. Small Animal Imaging

For tracking the changes in the oxygen saturation of the blood in vivo, we conducted a small animal study. The animal (BALB/c nude mice) was anesthetized with Isoflurane vapor added to the breathing air through a nose cone (Figure 1f). The transducer as well as the lower part of the mice body were immersed in a warm water bath for acoustic coupling (Figure 1f). The flank region closer to the thigh was aligned to the imaging plane using US imaging. A combination LED array (750 nm and 850 nm) was used at a repetition rate of 4 kHz, and the resulting PA signal was averaged 640 times to display PA image with a frame rate of 6 Hz. At first, the animal was allowed to breathe in normal air (21% oxygen) and the oxygen saturation images were acquired. Then the breathing gas was changed to 100% oxygen and the animal was allowed to stabilize to the oxygen level for two minutes. US and PA raw RF files were saved for the entire duration. The US image was used to identify the tissue boundary. Optical properties of mouse tissue used in the light propagation model are given in Table 1. Fluence compensated oxygen saturation images were computed. The oxygen levels were compared for both levels by selecting three blood vessels as regions of interest.

2.4.2. Human Imaging

In the final experiment, the wrist of a human volunteer with brown skin was imaged. An artery and vein were identified and imaged in real-time (30 Hz) using both the B-mode US and two-wavelength PA imaging. US-assisted fluence compensation was performed, and oxygen saturation images were computed. From the PA sO_2 image, the artery and vein regions were analyzed to estimate the mean sO_2 values by computing the mean over a region of interest.

3. Results

3.1. In Vitro Validation of PA sO_2

Validation of PA sO_2 imaging is presented in this section to analyze the accuracy of the two-wavelength PA sO_2 against a ground truth method. Figure 2a shows two tubes having blood with three different levels of oxygen. Oximeter readings are marked next to the tubes, and the difference in oxygen saturation is visible in the PA sO_2 images. To quantify the changes in oxygen saturation and validate it with oximeter readings, PA sO_2 in a region of interest of the tubes are compared in Figure 2b. The graph shows the PA estimated sO_2 against the measured oxygen saturation. sO_2 readings before and after imaging are indicated by the error bar in the measured sO_2. A linear fit to the data is shown with the dashed line. The correlation coefficient (r) between measured sO_2 and PA estimated sO_2 is 0.893 ($p < 0.001$), which shows that the two measurements are linearly correlated. The error in the PA estimated sO_2, given by the difference between the two methods is plotted in Figure 2c. A mean error of 0.18% and a standard deviation of 10.33% was observed. Most of the error values are around the mean, and 94% of the values are within the Mean $\pm$ 2SD region, indicating that the variations are from the noise in the PA measurements.

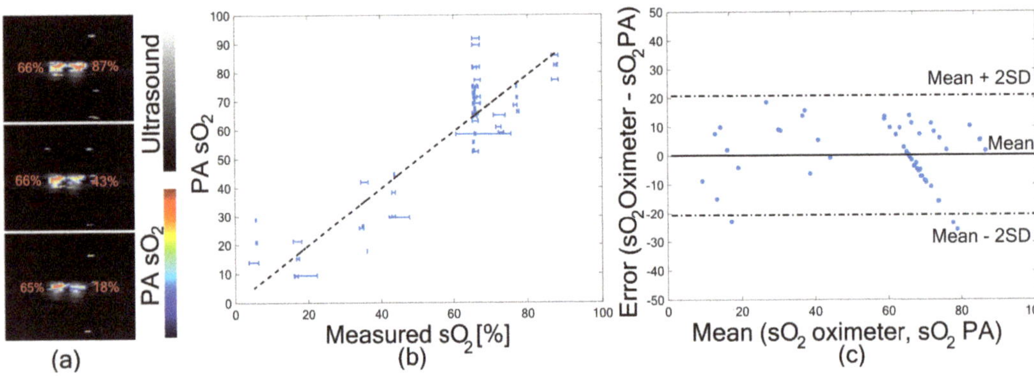

Figure 2. Photoacoustic Oxygen saturation (PA sO_2) validation. (**a**) PA sO_2 images from three blood oxygen levels with oximeter reading marked in the image. (**b**) PA sO_2 plotted against measured sO_2 from oximeter readings. (**c**) Error plot showing average sO_2 from oximeter values and PA sO_2 plotted against difference in sO_2 between the two methods.

3.2. Fluence Compensated PA sO_2

In the second experiment, a homogeneous phantom was considered with soft tissue optical properties (Table 1). Fluence for both 750 and 850 nm is shown in Figure 3. The peak fluence can be observed at the LED locations in Figure 3a,b. Due to the advanced position (9.2 mm) of LED arrays on the sides of the transducer, this region in the imaging plane cannot be used for imaging tissue. We only considered the imaging plane below the LED position as the region of interest (ROI). Figure 3c shows a normalized line profile through the fluence map for the entire imaging plane and in the ROI. Figure 3d,h shows fluence map at 750 and 850 nm, respectively. Figure 3e,i shows the PA images of the same tube with blood at different depths at 750 and 850 nm, respectively. Figure 3f,j shows fluence compensated PA images. PA images show the signal from the tube located at three depths. With fluence compensation an enhancement in the PA signal from deeper locations is visible. Figure 3g shows the PA sO_2 image before fluence compensation and Figure 3k shows PA sO_2 image after fluence compensation.

Fluence drop of $1/e$ was observed at a depth of 3.6 mm for 750 nm and 4.2 mm for 850 nm. The maximum depth obtained in the system with the denoising filters was 10.5 mm and using the offline program it was 6 mm. The measured oximeter readings of the blood in the tube before imaging was 95.9% and after imaging it was 96.6%. The PA estimated sO_2 at three tube locations (mean value) before fluence compensation were 85.8%, 83.9% and 81.7% and after fluence compensation were 99.1%, 99.3% and 99.6%. In the offline reconstruction, the fourth location of the tube was not visible. However, all four tubes were visible in the system.

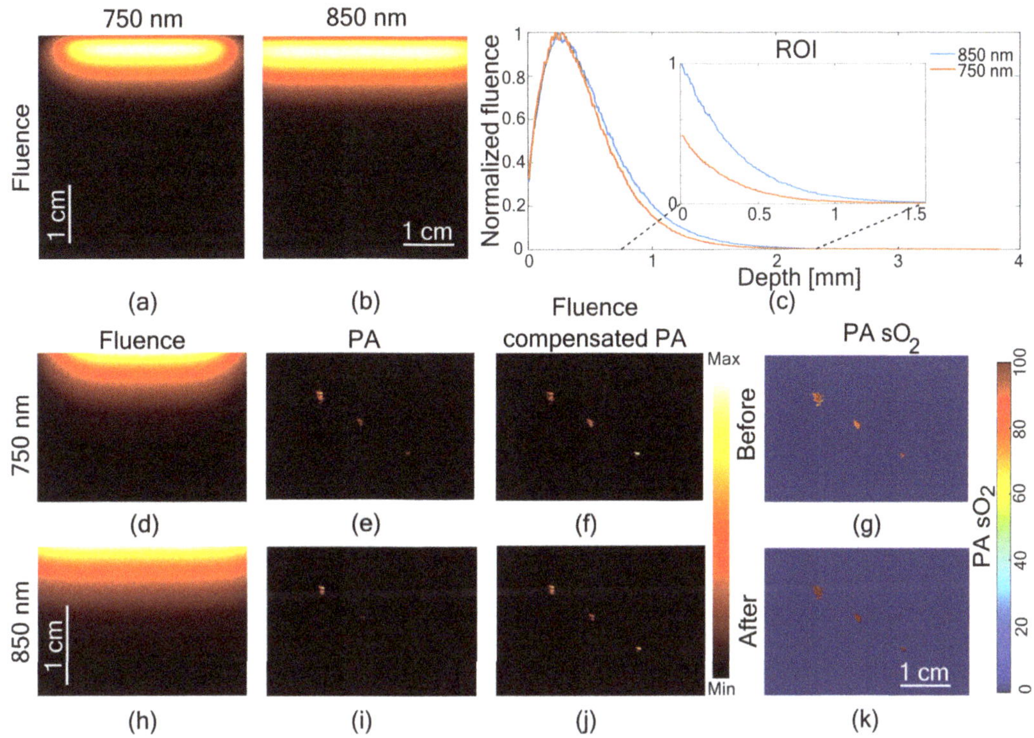

Figure 3. Fluence compensation in a homogeneous medium. Fluence map at the imaging plane from (**a**) 750 nm and (**b**) 850 nm. (**c**) Line profile showing normalized fluence in the depth direction and the zoomed-in region showing fluence decay in the Region Of Interest (ROI), marked with dashed lines. Fluence map in the ROI with (**d**) 750 nm and (**h**) 850 nm. Uncompensated photoacoustic (PA) images at (**e**) 750 nm and (**i**) 850 nm and corresponding fluence compensated PA images (**f**,**j**). Photoacoustic oxygen saturation images (PA sO_2) (**g**) before fluence compensation and (**k**) after fluence compensation.

In the third experiment, a two-slab phantom was considered. The upper region was water and the lower region a soft tissue-mimicking phantom. Figure 4a is the US image showing the film separating the two mediums. Figure 4b is the segmented image with ones for soft tissue region and zeros for water. Figure 4c shows the fluence in the medium at 850 nm. The PA image of the tube at three different depths is shown in Figure 4d. The fluence compensated PA image in Figure 4e shows the enhancement at deeper locations. Figure 4f shows the sO_2 image before fluence compensation overlaid on the US image and Figure 4g is sO_2 image after fluence compensation. The enhancement in the PA sO_2 is visible in Figure 4g. The PA sO_2 estimated by taking the mean value for the tube locations before fluence compensation were 82.4%, 76.9% and 77.7%, and fluence compensation it was 91.8%, 89.5%, and 92.1%. The oximeter readings before and after imaging were 96.8% and 97.4%, respectively.

3.3. *In Vivo PA sO_2 Imaging*

Results from in vivo PA sO_2 imaging on mice thigh muscle is shown in Figure 5. The animal was given two levels of oxygen, switching between normal air and 100% oxygen. Figure 5a shows the US image of the thigh muscle of the mice. The fluence map at 850 nm utilizing the US segmented image is shown in Figure 5d. Figure 5b,e show fluence compensated PA sO_2 images corresponding to the normal air (low cycle) and 100% oxygen (high cycle), respectively. The zoomed-in region from both low and high cycle is shown in Figure 5c,f, respectively. The difference in sO_2 images is clear between normal air and

100% oxygen. To quantify the change, three regions marked by 1, 2, and 3 in Figure 5c,f were analyzed for the mean PA sO_2 values. During the low cycle PA sO_2 values were 72.9%, 67.8% and 67.3%, respectively, for regions 1, 2, and 3 and it changed to 93.4%, 89.2% and 85.4% during high cycle.

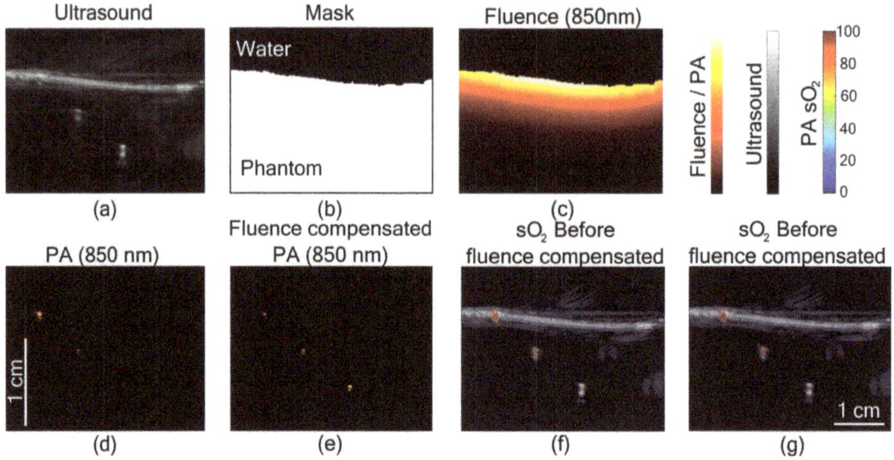

Figure 4. Ultrasound-assisted fluence compensation. (**a**) Ultrasound image of the two-layer phantom. (**b**) Binary mask obtained by segmenting ultrasound image. (**c**) Fluence map at 850 nm. (**d**) Photoacoustic (PA) images before fluence compensation at 850 nm and corresponding (**e**) fluence compensated image. Photoacoustic oxygen saturation (PA sO_2) images (**f**) before fluence compensation and (**g**) after fluence compensation overlaid on the ultrasound image.

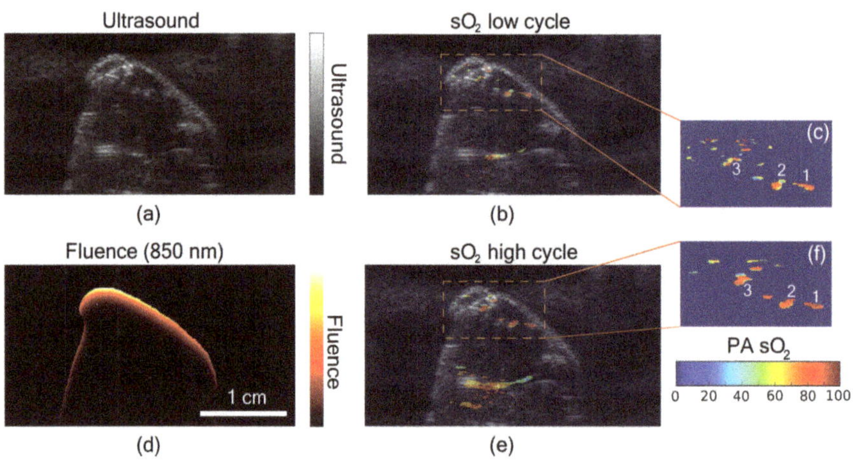

Figure 5. In vivo small animal imaging. (**a**) Ultrasound image of the thigh muscle of a mouse. (**b**) Fluence compensated photoacoustic oxygen saturation (PA sO_2) during the low sO_2 cycle overlaid on the ultrasound image. (**c**) Zoomed-in region of the PA sO_2 image from the low sO_2 cycle. (**d**) Fluence map at 850 nm obtained using segmented ultrasound image. (**e**) Fluence compensated PA sO_2 during the high sO_2 cycle, overlaid on the ultrasound image. (**f**) Zoomed-in region of the PA sO_2 image from the high sO_2 cycle.

In the final experiment, an in vivo imaging on the wrist of a healthy volunteer was performed. Figure 6a shows fluence compensated PA sO_2 image overlaid on US image. An artery on the right side and a vein on the left side of the PA sO_2 image is visible, along with the skin signal. An artery is characterized by higher sO_2 while a low sO_2 for the vein. The estimated PA sO_2 by selecting a region of interest in the artery was 92.6% and that of the vein was 78.7%. The difference in the sO_2 between artery and vein is evident in Figure 6a. Mean PA signal acquired real-time (30 Hz) from the artery at 850 nm is presented in Figure 6b, which shows the pulsation due to the heart rate. The estimated heart rate of the subject from the PA signal was 84 beats per minute. Figure 6c shows the real-time (without fluence compensation) PA sO_2 images from the system. Figure 6c shows the pulsating artery, demonstrating the real-time PA sO_2 imaging capability of LED-based PA sO_2 imaging.

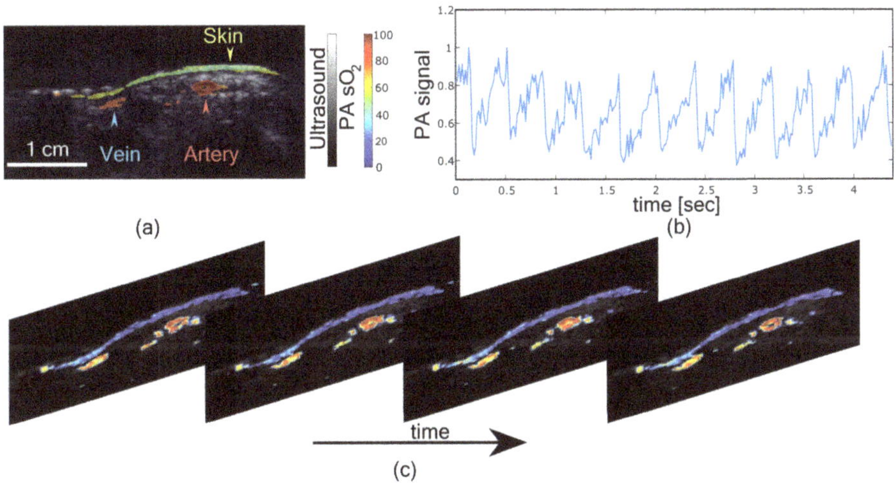

Figure 6. In vivo imaging on a human wrist. (**a**) Fluence compensated photoacoustic oxygen saturation (PA sO_2) image of the human wrist overlaid on an ultrasound image. (**b**) Normalized photoacoustic (PA) signal from artery showing pulsation. (**c**) PA sO_2 frames showing pulsating artery. (Video S1).

4. Discussion

In recent years, LED-PAI has shown potential in various clinical and preclinical applications. However, depth-dependent fluence variation was not considered for PA sO_2 calculation. An accurate fluence compensated PA sO_2 for LED-PAI is expected to accelerate the clinical translation of LED-PAI. In this work, we characterized the accuracy of PA sO_2 imaging by using in vitro human blood and compared the PA sO_2 with conventional oximeter readings. A good correlation of 0.893 ($p < 0.001$) was achieved between measured sO_2 and PA sO_2. We utilized the structural information offered by US images for tissue segmentation. Further, the segmented tissue information was used in the light propagation model using Monte-Carlo simulations to compensate for fluence variations to improve the sO_2 imaging accuracy. The fluence compensation was validated using two different tissue-mimicking phantoms with human blood-filled tubes at different depths. The sO_2 values were compared before and after the fluence compensation step. In both phantoms, blood with the same oxygenation level was present at different depths. It is evident from our results (decreasing sO_2 values with depth) that it is important to account for light fluence variations to estimate sO_2 accurately. After applying the fluence compensation algorithm, the PA sO_2 imaging accuracy improved by 12% in the two slab phantom and 15% in the homogeneous phantom studies. The best enhancement was achieved for deeper locations in comparison to the uncompensated PA sO_2 imaging. An error of 15% (when no fluence compensation was performed) is not acceptable in a clinical scenario, especially when one has to differentiate an artery and a vein (difference in sO_2 for arterial and venous blood is less than 15–20%). To demonstrate the potential of fluence compensated PA sO_2 imaging in vivo, we performed a small animal study in which the thigh muscle of a live mouse was imaged. First, the mouse was set to breathe normal air (21%) and then switched to (100%) oxygen and PA sO_2 was monitored to see the dynamic imaging capability. Our approach yielded expected results showing differences in sO_2 values of different blood vessels (67–73% sO_2 during normal air-breathing and around 85–94% during 100% oxygen breathing) with an imaging depth of approximately 6 mm. Finally, we also performed a fluence-compensated PA sO_2 measurement on the wrist of a human volunteer and imaged a vein and a pulsating artery for which a difference of 14% sO_2 was observed.

We foresee that the implementation of this idea in the imaging system will open up the possibility of improving the sO_2 imaging capability. Since US imaging is already available in the system, this data can be used for tissue segmentation and consequently compute a fluence map specific for the tissue and compensate the PA data for sO_2 estimation. We believe that our approach is suitable for any PAI system with US imaging capability. In this work, we were able to achieve a penetration depth of around 10 mm in phantom experiments when sO_2 imaging was performed along with conventional US imaging in the AcousticX system. For the in vivo experiments, we achieved an imaging depth of 6 mm, which is good enough for small animal imaging and human wrist imaging that we have demonstrated. However, it is critical to improve the imaging depth to explore a wide range of clinical applications. Improving the LED array packaging, pulse repetition rate, and coded-excitation schemes may be some aspects to be considered in this direction [45]. Further, novel deep learning and image processing methods are also expected to have an impact in improving spatial resolution and imaging depth of LED-PAI [46–48]. In this work, we used a combination LED array of 750 nm and 850 nm with a pulse energy of around 200 μJ and 100 μJ per pulse, respectively. As one can see, pulse energy of 750 nm is just 100μJ, which is half of that of 850 nm, and this is the main factor hindering the imaging depth in this work. The higher amount of background noise present in the 750 nm images (when compared to 850 nm) may very well be the reason for the slight inaccuracies (especially in deeper features) in our sO_2 estimates [49]. Our future work will include the design and development of new LED arrays with optimal wavelengths and improved optical energy.

Even though we achieved promising results in this proof-of-concept study, there is a lot of scope in improving the quantitative nature of PA sO_2 imaging. In our proposed fluence compensation model, we segmented the phantom and tissue as two layers (water and soft tissue layer in phantom, water and muscle in mouse imaging, and water and soft tissue in human wrist imaging), which is not an optimal approach. In our future work, we consider using the US images to segment different tissue types [50] and then perform a more accurate fluence compensation for quantitative sO_2 imaging. In the Monte-Carlo light propagation model, the reflection of light at different tissue layers was not considered, which will be considered in our future work. Another factor to consider in future studies is the band-limited frequency response of the detectors and its impact on the accuracy of sO_2 imaging. PA sO_2 is limited in predicting the absolute sO_2 values, as imaging depends on multiple factors, and it is difficult to calibrate the method in live tissue. However, relative sO_2 imaging is still useful in clinical applications. We foresee several applications including monitoring hypoxia in the tumor microenvironment in the future.

In summary, for the first time, we characterized and validated fluence-compensated LED-PAI based sO_2 imaging using in vitro, phantom and in vivo small animal and human volunteer studies. Our results show the importance of accounting for fluence variations in tissue when performing sO_2 using an LED-PAI system. Further, our results demonstrate the potential of LED-PAI in obtaining sO_2 information dynamically along with conventional pulse-echo imaging. We believe that combined US and sO_2 information in such high temporal and spatial resolution may be useful for diagnosis and treatment monitoring of several inflammatory and cancer-related diseases.

5. Conclusions

Through our results, we have demonstrated that oxygen saturation imaging using a dual-wavelength LED-based photoacoustic system has potential in pre-clinical and clinical applications. The proposed fluence compensation method utilizing structural information of the tissue from the US images was found to improve the oxygen saturation imaging accuracy by 12–15%. Within the limited power of the LEDs, an imaging depth of 6–8 mm was achieved in soft tissue. We have demonstrated imaging different levels of oxygenated blood in vivo in both small animal and a healthy human subject and found it to be within the expected range, showing the accuracy of the proposed approach. Further research on improving the pulse energy of LED sources to enhance imaging depth and accurate segmentation of different tissue types for fluence compensation can potentially translate this technology for pre-clinical and clinical applications.

Supplementary Materials: The following are available online at https://www.mdpi.com/1424-8220/21/1/283/s1, Video S1: Real-time ultrasound and photoacoustic oxygen saturation imaging of a human wrist, showing a pulsating artery, vein, and skin layer.

Author Contributions: Conceptualization: M.K.A.S. and K.J.F.; Formal analysis: K.J.F.; Funding acquisition: W.S. and M.K.A.S., Investigation: K.J.F.; Methodology: R.B., M.K.A.S., M.X. and K.J.F.; Project administration: W.S.; Supervision: W.S., S.M. and K.J.F., Writing—original draft: R.B., M.K.A.S. and K.J.F.; Review and editing: M.K.A.S., W.S., R.B., M.X., S.M. and K.J.F. All authors have read and agreed to the published version of the manuscript.

Funding: We would like to acknowledge funding from National Centre for the Replacement Refinement and Reduction of Animals in Research (CRACKITRT-P1-3) and 4TU Precision Medicine program.

Institutional Review Board Statement: All animal experiments were performed in compliance with the Institutional Animal Care and Use Committee (IACUC) of Tufts University.

Informed Consent Statement: Informed consent was obtained from all subjects involved in the study.

Data Availability Statement: Data available on request from the authors.

Acknowledgments: We would like to thank Maura Dantuma for inputs on the light propagation model.

Conflicts of Interest: M.K.A.S. is employed by Cyberdyne Inc. The authors have no financial interests or conflict of interest to disclose.

References

1. Vaupel, P.; Kallinowski, F.; Okunieff, P. Blood flow, oxygen and nutrient supply, and metabolic microenvironment of human tumors: A review. *Cancer Res.* **1989**, *49*, 6449–6465. [PubMed]
2. Cairns, R.A.; Harris, I.S.; Mak, T.W. Regulation of cancer cell metabolism. *Nat. Rev. Cancer* **2011**, *11*, 85–95. [CrossRef] [PubMed]
3. Mallidi, S.; Luke, G.P.; Emelianov, S. Photoacoustic imaging in cancer detection, diagnosis, and treatment guidance. *Trends Biotechnol.* **2011**, *29*, 213–221. [CrossRef] [PubMed]
4. Szekanecz, Z.; Koch, A.E. Mechanisms of disease: Angiogenesis in inflammatory diseases. *Nat. Clin. Pract. Rheumatol.* **2007**, *3*, 635–643. [CrossRef]
5. Diot, G.; Metz, S.; Noske, A.; Liapis, E.; Schroeder, B.; Ovsepian, S.V.; Meier, R.; Rummeny, E.; Ntziachristos, V. Multispectral optoacoustic tomography (MSOT) of human breast cancer. *Clin. Cancer Res.* **2017**, *23*, 6912–6922. [CrossRef]
6. Petrova, I.; Petrov, Y.; Esenaliev, R.; Deyo, D.; Cicenaite, I.; Prough, D. Noninvasive monitoring of cerebral blood oxygenation in ovine superior sagittal sinus with novel multi-wavelength optoacoustic system. *Opt. Express* **2009**, *17*, 7285–7294. [CrossRef]
7. Zauner, A.; Daugherty, W.P.; Bullock, M.R.; Warner, D.S. Brain oxygenation and energy metabolism: Part I—biological function and pathophysiology. *Neurosurgery* **2002**, *51*, 289–302.
8. Tang, J.; Coleman, J.E.; Dai, X.; Jiang, H. Wearable 3-D photoacoustic tomography for functional brain imaging in behaving rats. *Sci. Rep.* **2016**, *6*, 25470. [CrossRef]
9. Dale, A.M.; Halgren, E. Spatiotemporal mapping of brain activity by integration of multiple imaging modalities. *Curr. Opin. Neurobiol.* **2001**, *11*, 202–208. [CrossRef]
10. Halmos, G.B.; de Bruin, L.B.; Langendijk, J.A.; van der Laan, B.F.; Pruim, J.; Steenbakkers, R.J. Head and neck tumor hypoxia imaging by 18F-fluoroazomycin-arabinoside (18F-FAZA)-PET: A review. *Clin. Nucl. Med.* **2014**, *39*, 44–48. [CrossRef]
11. Wang, L.V. Prospects of photoacoustic tomography. *Med. Phys.* **2008**, *35*, 5758–5767. [CrossRef] [PubMed]

12. Xu, M.; Wang, L.V. Photoacoustic imaging in biomedicine. *Rev. Sci. Instrum.* **2006**, *77*, 041101. [CrossRef]
13. Attia, A.B.E.; Balasundaram, G.; Moothanchery, M.; Dinish, U.; Bi, R.; Ntziachristos, V.; Olivo, M. A review of clinical photoacoustic imaging: Current and future trends. *Photoacoustics* **2019**, *16*, 100144. [CrossRef] [PubMed]
14. Singh, M.K.A.; Steenbergen, W.; Manohar, S. Handheld probe-based dual mode ultrasound/photoacoustics for biomedical imaging. In *Frontiers in Biophotonics for Translational Medicine*; Springer: Berlin, Germany, 2016; pp. 209–247.
15. Vogt, W.C.; Zhou, X.; Andriani, R.; Wear, K.A.; Pfefer, T.J.; Garra, B.S. Photoacoustic oximetry imaging performance evaluation using dynamic blood flow phantoms with tunable oxygen saturation. *Biomed. Opt. Express* **2019**, *10*, 449–464. [CrossRef] [PubMed]
16. Li, M.; Tang, Y.; Yao, J. Photoacoustic tomography of blood oxygenation: A mini review. *Photoacoustics* **2018**, *10*, 65–73. [CrossRef]
17. Cox, B.T.; Laufer, J.G.; Beard, P.C.; Arridge, S.R. Quantitative spectroscopic photoacoustic imaging: A review. *J. Biomed. Opt.* **2012**, *17*, 061202. [CrossRef]
18. Manohar, S.; Razansky, D. Photoacoustics: A historical review. *Adv. Opt. Photonics* **2016**, *8*, 586–617. [CrossRef]
19. Kuniyil Ajith Singh, M.; Xia, W. Portable and Affordable Light Source-Based Photoacoustic Tomography. *Sensors* **2020**, *20*, 6173. [CrossRef]
20. Upputuri, P.K.; Pramanik, M. Fast photoacoustic imaging systems using pulsed laser diodes: A review. *Biomed. Eng. Lett.* **2018**, *8*, 167–181. [CrossRef]
21. Allen, T.J. High-Power Light Emitting Diodes; An Alternative Excitation Source for Photoacoustic Tomography. In *LED-Based Photoacoustic Imaging*; Springer Nature: Gateway East, Singapore, 2020; pp. 23–43.
22. Sato, N.; Singh, M.K.A.; Shigeta, Y.; Hanaoka, T.; Agano, T. High-speed photoacoustic imaging using an LED-based photoacoustic imaging system. In *Photons Plus Ultrasound: Imaging and Sensing 2018*; International Society for Optics and Photonics: Bellingham, WA, USA, 2018; Volume 10494, p. 104943N.
23. Zhu, Y.; Feng, T.; Cheng, Q.; Wang, X.; Du, S.; Sato, N.; Yuan, J.; Kuniyil Ajith Singh, M. Towards Clinical Translation of LED-Based Photoacoustic Imaging: A Review. *Sensors* **2020**, *20*, 2484. [CrossRef]
24. Mackle, E.; Maneas, E.; Xia, W.; West, S.; Desjardins, A. LED-Based Photoacoustic Imaging for Guiding Peripheral Minimally Invasive Procedures. In *LED-Based Photoacoustic Imaging*; Springer Nature: Gateway East, Singapore 2020; pp. 321–334.
25. Zhu, Y.; Xu, G.; Yuan, J.; Jo, J.; Gandikota, G.; Demirci, H.; Agano, T.; Sato, N.; Shigeta, Y.; Wang, X. Light emitting diodes based photoacoustic imaging and potential clinical applications. *Sci. Rep.* **2018**, *8*, 1–12. [CrossRef]
26. Xavierselvan, M.; Singh, M.K.A.; Mallidi, S. In Vivo Tumor Vascular Imaging with Light Emitting Diode-Based Photoacoustic Imaging System. *Sensors* **2020**, *20*, 4503. [CrossRef]
27. Singh, M.K.A.; Agano, T.; Sato, N.; Shigeta, Y.; Uemura, T. Real-time in vivo imaging of human lymphatic system using an LED-based photoacoustic/ultrasound imaging system. In *Photons Plus Ultrasound: Imaging and Sensing 2018*; International Society for Optics and Photonics: Bellingham, WA, USA, 2018; p. 1049404.
28. Hariri, A.; Chen, F.; Moore, C.; Jokerst, J.V. Noninvasive staging of pressure ulcers using photoacoustic imaging. *Wound Repair Regen.* **2019**, *27*, 488–496. [CrossRef]
29. Joseph, F.K.; Xavierselvan, M.; Singh, M.K.A.; Mallidi, S.; Van Der Laken, C.; Van De Loo, F.; Steenbergen, W. LED-based photoacoustic imaging for early detection of joint inflammation in rodents: Towards achieving 3Rs in rheumatoid arthritis research. In *Photons Plus Ultrasound: Imaging and Sensing 2020*; International Society for Optics and Photonics: Bellingham, WA, USA, 2020; Volume 11240, p. 112400M.
30. Francis, K.J.; Booijink, R.; Bansal, R.; Steenbergen, W. Tomographic Ultrasound and LED-Based Photoacoustic System for Preclinical Imaging. *Sensors* **2020**, *20*, 2793. [CrossRef]
31. Francis, K.J.; Boink, Y.E.; Dantuma, M.; Singh, M.K.A.; Manohar, S.; Steenbergen, W. Tomographic imaging with an ultrasound and LED-based photoacoustic system. *Biomed. Opt. Express* **2020**, *11*, 2152–2165. [CrossRef]
32. Guo, Z.; Hu, S.; Wang, L.V. Calibration-free absolute quantification of optical absorption coefficients using acoustic spectra in 3D photoacoustic microscopy of biological tissue. *Opt. Lett.* **2010**, *35*, 2067–2069. [CrossRef]
33. Kirchner, T.; Gröhl, J.; Maier-Hein, L. Context encoding enables machine learning-based quantitative photoacoustics. *J. Biomed. Opt.* **2018**, *23*, 056008. [CrossRef]
34. Bench, C.; Hauptmann, A.; Cox, B.T. Toward accurate quantitative photoacoustic imaging: Learning vascular blood oxygen saturation in three dimensions. *J. Biomed. Opt.* **2020**, *25*, 085003. [CrossRef]
35. Tzoumas, S.; Nunes, A.; Olefir, I.; Stangl, S.; Symvoulidis, P.; Glasl, S.; Bayer, C.; Multhoff, G.; Ntziachristos, V. Eigenspectra optoacoustic tomography achieves quantitative blood oxygenation imaging deep in tissues. *Nat. Commun.* **2016**, *7*, 12121. [CrossRef]
36. Hussain, A.; Petersen, W.; Staley, J.; Hondebrink, E.; Steenbergen, W. Quantitative blood oxygen saturation imaging using combined photoacoustics and acousto-optics. *Opt. Lett.* **2016**, *41*, 1720–1723. [CrossRef]
37. Kim, S.; Chen, Y.S.; Luke, G.P.; Emelianov, S.Y. In vivo three-dimensional spectroscopic photoacoustic imaging for monitoring nanoparticle delivery. *Biomed. Opt. Express* **2011**, *2*, 2540–2550. [CrossRef] [PubMed]
38. Singh, M.K.A.; Sato, N.; Ichihashi, F.; Sankai, Y. In vivo demonstration of real-time oxygen saturation imaging using a portable and affordable LED-based multispectral photoacoustic and ultrasound imaging system. In *Photons Plus Ultrasound: Imaging and Sensing 2019*; International Society for Optics and Photonics: Bellingham, WA, USA, 2019; Volume 10878, p. 108785N.

39. Li, M.L.; Oh, J.T.; Xie, X.; Ku, G.; Wang, W.; Li, C.; Lungu, G.; Stoica, G.; Wang, L.V. Simultaneous molecular and hypoxia imaging of brain tumors in vivo using spectroscopic photoacoustic tomography. *Proc. IEEE* **2008**, *96*, 481–489.
40. Jaeger, M.; Schüpbach, S.; Gertsch, A.; Kitz, M.; Frenz, M. Fourier reconstruction in optoacoustic imaging using truncated regularized inverse k-space interpolation. *Inverse Probl.* **2007**, *23*, S51. [CrossRef]
41. Fang, Q.; Boas, D.A. Monte Carlo simulation of photon migration in 3D turbid media accelerated by graphics processing units. *Opt. Express* **2009**, *17*, 20178–20190. [CrossRef]
42. Jacques, S.L. Optical properties of biological tissues: A review. *Phys. Med. Biol.* **2013**, *58*, R37. [CrossRef]
43. Krainov, A.; Mokeeva, A.; Sergeeva, E.; Agrba, P.; Kirillin, M.Y. Optical properties of mouse biotissues and their optical phantoms. *Opt. Spectrosc.* **2013**, *115*, 193–200. [CrossRef]
44. Bashkatov, A.N.; Genina, E.A.; Tuchin, V.V. Optical properties of skin, subcutaneous, and muscle tissues: A review. *J. Innov. Opt. Health Sci.* **2011**, *4*, 9–38. [CrossRef]
45. Allen, T.J.; Beard, P.C. High power visible light emitting diodes as pulsed excitation sources for biomedical photoacoustics. *Biomed. Opt. Express* **2016**, *7*, 1260–1270. [CrossRef]
46. Hauptmann, A.; Cox, B. Deep Learning in Photoacoustic Tomography: Current approaches and future directions. *J. Biomed. Opt.* **2020**, *25*, 112903. [CrossRef]
47. Anas, E.M.A.; Zhang, H.K.; Kang, J.; Boctor, E. Enabling fast and high quality LED photoacoustic imaging: A recurrent neural networks based approach. *Biomed. Opt. Express* **2018**, *9*, 3852–3866. [CrossRef]
48. Farnia, P.; Najafzadeh, E.; Hariri, A.; Lavasani, S.N.; Makkiabadi, B.; Ahmadian, A.; Jokerst, J.V. Dictionary learning technique enhances signal in LED-based photoacoustic imaging. *Biomed. Opt. Express* **2020**, *11*, 2533–2547. [CrossRef] [PubMed]
49. Hochuli, R.; An, L.; Beard, P.C.; Cox, B.T. Estimating blood oxygenation from photoacoustic images: Can a simple linear spectroscopic inversion ever work? *J. Biomed. Opt.* **2019**, *24*, 121914. [CrossRef] [PubMed]
50. Han, T.; Yang, M.; Yang, F.; Zhao, L.; Jiang, Y.; Li, C. A three-dimensional modeling method for quantitative photoacoustic breast imaging with handheld probe. *Photoacoustics* **2020**, *2020*, 100222.

sensors

MDPI

Letter

Photoacoustic Imaging as a Tool for Assessing Hair Follicular Organization

Ali Hariri [1], Colman Moore [1], Yash Mantri [2] and Jesse V. Jokerst [1,3,4,*]

1 Nanoengineering Department, University of California-San Diego, La Jolla, CA 92093, USA; a1hariri@ucsd.edu (A.H.); cam081@eng.ucsd.edu (C.M.)
2 Bioengineering Department, University of California-San Diego, La Jolla, CA 92093, USA; ymantri@eng.ucsd.edu
3 Material Science and Engineering Program, University of California-San Diego, La Jolla, CA 92093, USA
4 Radiology Department, University of California-San Diego, La Jolla, CA 92093, USA
* Correspondence: jjokerst@ucsd.edu

Received: 21 September 2020; Accepted: 11 October 2020; Published: 16 October 2020

Abstract: Follicular unit extraction (FUE) and follicular unit transplantation (FUT) account for 99% of hair transplant procedures. In both cases, it is important for clinicians to characterize follicle density for treatment planning and evaluation. The existing gold-standard is photographic examination. However, this approach is insensitive to subdermal hair and cannot identify follicle orientation. Here, we introduce a fast and non-invasive imaging technique to measure follicle density and angles across regions of varying density. We first showed that hair is a significant source of photoacoustic signal. We then selected regions of low, medium, and high follicle density and showed that photoacoustic imaging can measure the density of follicles even when they are not visible by eye. We performed handheld imaging by sweeping the transducer across the imaging area to generate 3D images via maximum intensity projection. Background signal from the dermis was removed using a skin tracing method. Measurement of follicle density using photoacoustic imaging was highly correlated with photographic determination (R^2 = 0.96). Finally, we measured subdermal follicular angles—a key parameter influencing transection rates in FUE.

Keywords: LED-based photoacoustic imaging; hair follicles; FUE; FUT

1. Introduction

Follicular unit extraction (FUE) and follicular unit transplantation (FUT) are the gold standard surgical interventions for androgenic alopecia and account for >99% of transplant procedures [1,2]. Both of these techniques involve transplantation of healthy hair follicles from a safe donor area to a low-density region of thinning/balding [3,4]. The procedures differ in how the donor follicles are collected. In FUE, follicles are individually extracted using a handheld punching device. In FUT, a strip of scalp is resected and then dissected to obtain individual follicles [5,6]. Consequently, FUE results in less severe scarring than FUT, and is more commonly requested by patients for this reason [7]. However, FUE is a more technically demanding procedure and has a higher risk of follicle transection; in some cases, it can also lead to cyst formation or a "moth-eaten" appearance of the donor region [8].

One important element for both of these procedures is characterization of the follicular units both at the donor and implant site (pre- and post-operatively). Factors such as graft size, angle, and grafting density have a significant effect on follicular unit survival rates [9,10]. Beyond implant survival, these metrics also have implications for potential complications such as the development of subdermal cysts or telogen effluvium ("shock loss") [11,12]. Finally, variations in graft density and angle can have major impacts on cosmetic satisfaction.

Photography and/or visual inspection is the current gold-standard means of characterization during treatment planning and evaluation of results [1]. However, this practice neglects the subdermal orientation/depth of follicles, which is the primary factor governing undesirable transection rates in FUE [13]. Furthermore, these optics-based inspections cannot evaluate the follicle below the skin surface or during the "shock loss" period after transplant [14]. Indeed, subdermal characterization of the follicles at the donor and implant sites could inform the clinician on the best donor sites as well as the risk and status of downstream complications such as telogen effluvium and cysts. Therefore, we propose the use of a non-invasive imaging modality for gleaning both supra- and subdermal information about the density and orientation of follicles.

Ultrasound is a widely deployed imaging modality across medical specialties with applications in dermatology for melanoma, inflammatory diseases, and lipoablation [15]. These applications are currently being expanded as photoacoustic imaging (PAI)—an augmented form of ultrasound—continues to gain clinical traction [16,17]. PAI uses pulsed light to generate ultrasound waves from the imaging target. This shifts the mechanism of imaging contrast from differences in acoustic refraction to optical absorption. When performed simultaneously with ultrasound (routine practice for commercially available systems), the modality is capable of real-time imaging with anatomical and molecular contrast through many centimeters of tissue. PAI has experienced tremendous research growth in the past decade as a diagnostic platform and is currently being investigated for a number of dermatological conditions [18–24]. Recently, a laser-based PAI system was proposed for volumetric imaging of a single follicular unit [25]. In this work, we introduce an LED-based photoacoustic imaging method for fast and non-invasive characterization of follicle density and subdermal angle with an emphasis on its diagnostic value in FUE and FUT.

2. Materials and Methods

2.1. Photoacoustic Imaging System

In this work, we used a AcousticX CYBEDYNE LED-based photoacoustic imaging system from CYBERDYNE Inc. (formerly Prexion) (Tokyo, Japan) [26]. The system is equipped with a 128-element linear array ultrasound transducer with a central frequency of 10 MHz and a bandwidth of 80.9% fitted with two LED arrays. The imaging equipment could be used with a variety of wavelengths. We used both 690 and 850 nm LED arrays in this study. We found that 690 nm provides more information about the skin layer and 850 nm shows more penetration depth.

The repetition rate of these LEDs is tunable between 1, 2, 3, and 4 KHz. The pulse width can be changed from 50 to 150 ns with a 5-ns step size. The transducer can be scanned to generate three-dimensional (3D) data using a maximum intensity projection (MIP) algorithm. The lateral and axial resolution of this imaging system is ~550 and 260 μm, respectively. This system can detect blood vessels to a depth of ~1.5 cm [26].

2.2. Imaging and Data Collection

The study enrolled a single adult healthy Caucasian male; the subject provided written informed consent. All work was conducted with approval from the UCSD Institutional Review Board and was in accordance with the ethical guidelines for human subject research set forth by the Helsinki Declaration of 1975. In order to maintain a sterile imaging environment, we used sterile CIV-Flex probe cover (CIVCO Medical Solution, Coraville, IA, USA) on the photoacoustic transducer.

The images were acquired using acoustically transparent coupling gel (clear image singles ultrasound scanning gel, Next medical product company, Somerville, NJ, USA), and we evaluated several regions of the body with varying follicle density. First, we imaged the nape of the subject's neck to include both skin and hair in the same field of view. We then imaged the parietal region of the scalp after trimming the hair with clippers (Wahl Clipper Corporation, Sterling, IL, USA). We used both ultrasound and photoacoustic mode to image the follicle density on the scalp.

In order to show that this technique can image hairs that are not visible (subdermal hairs), we imaged the abdomen at baseline, trimmed to leave a residual ~1 mm of hair (Wahl trimmer), and shaved with a razor (Gillette Mach 3, Boston, MA, USA). To further evaluate the capacity of the imaging technique to evaluate follicle density, we also imaged the subject's arm (little hair; no trimming or shaving) and face (freshly shaved). These various regions of the body were also photographed while protecting the subject's anonymity.

The imaging experiments used both 690 and 850 nm excitation wavelengths. The scanner was swept by hand at ~1 cm/s across the skin collecting both ultrasound and PA data for subsequent 3D visualization via maximum intensity projection (MIP)—a volume rendering that projects the voxels with maximum intensity in the correspondence plane [27,28]. Photographs were also taken at each imaging site for comparison of hair density.

2.3. Data Analysis

All imaging data were recorded as rf data and reconstructed using Fourier transform analysis (FTA) [29]. Images were exported and analyzed as bmp files types. The ImageJ 1.48 v toolbox was used to analyze all the data [30]. We used both B-mode cross-sections [15,31] and MIP volumes [23] to evaluate the capabilities of our LED-based photoacoustic imaging system for this application. In all images, ultrasound pixel intensities are shown in 8-bit grayscale, and photoacoustic pixel intensities are shown as hot color maps. In order to measure follicle densities, we compared the region of interest (ROI) for a photoacoustic MIP image to a photograph (2×2 cm^2 for both). To identify the follicles, we combined two metrics: photoacoustic intensity and morphology of the image. We evaluated the background photoacoustic intensity histogram, and we found that the background had a mean 8-bit pixel intensity lower than 35. Therefore, if intensity values are higher than 35, we set a threshold to define those data to be a feature—i.e., vein or follicle. Next, we could visually distinguish between vein and follicle structures. Each photoacoustic image and photograph were divided into four quadrants, and the number of follicles was counted manually. Subdermal follicular angles were measured with respect to the dermis using the angle tool function in ImageJ.

2.4. Statistical Analysis

Follicle densities for each region were quantified by averaging the counts for each quadrant of the ROI, and error bars represent the standard deviation across these four areas. We used a Student's t-test to assess statistical differences between the measurement techniques.

3. Results

The main goal of this study was to evaluate photoacoustic ultrasound as a diagnostic and staging tool for hair transplant procedures including FUE. Figure 1 shows the current and potential clinical workflow for the implementation of photoacoustic imaging to evaluate follicular organization. Figure 1A illustrates the value of photoacoustic imaging in pre-surgical staging in FUE: the images could be used to measure subdermal follicular angles (alpha and beta). Since these angles affect transection rate and can lead to more hair loss [32], a threshold angle could be established such that follicles below this value are not chosen for extraction. Follicle characterization steps are detailed in Figure 1B–E. These steps demonstrate that hair trimming and shaving followed by photoacoustic imaging can detect individual hair follicles, subdermal roots, and follicular angles. Future applications in follicular unit extraction and transplantation (FUE and FUT) are detailed in Figure 1F–J highlighting the value of imaging at the short and long-term post-implantation stages. Specifically, the implanted follicles can be visualized prior to eruption from the dermis immediately post-implantation (Figure 1G), during telogen effluvium to verify the presence of the residual papilla (Figure 1E), or periodically long-term (over 12–24 months) to monitor the formation of cysts from improper implantation (Figure 1J).

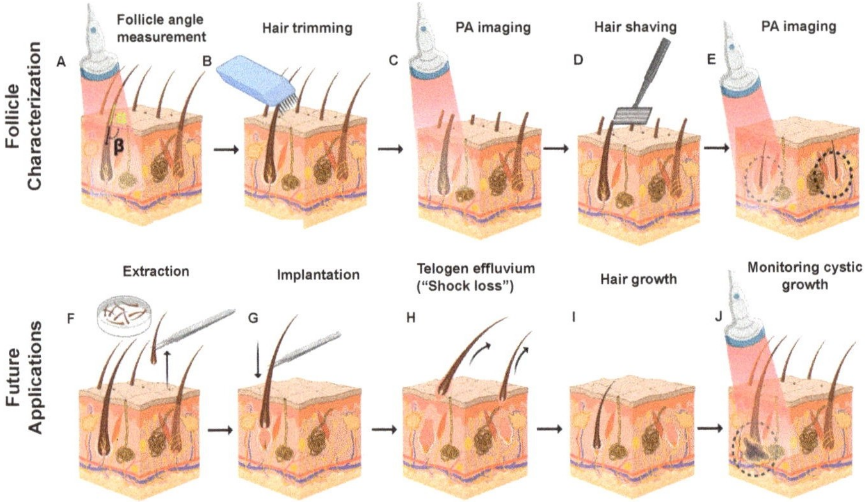

Figure 1. Photoacoustic imaging of follicular organization. The upper panel shows current work. (**A**) The follicular angle under the skin can be determined using photoacoustic imaging (PAI). (**B**) Hair trimming, (**D**) shaving, and subsequent (**C**,**E**) PA imaging to detect individual hair follicles, subdermal roots, and follicular angles. Future applications in follicular unit extraction and transplantation (FUE and FUT). (**F**) Hair follicles are excised from a safe donor area. (**G**) Extracted follicles are implanted in the donor spot. (**H**) Original hair shaft sheds while the dermal papilla is retained in a process called telogen effluvium ("shock loss"). PAI can quickly confirm successful implantation by imaging hair growth within the papilla before the follicle is visible on the skin surface. (**I**) Hair growth from successfully transplanted follicles. (**J**) Follicles implanted too deep can result in cyst formations 1–2 years after the procedure. PA imaging can monitor this cystic growth.

3.1. Hair Follicle Imaging Using PAI

We first used 690 nm LED light source and did some positive (hair area) and negative (skin area) control experiments to confirm that hair produced PA signal (Figure 2). We then used PAI to image below the skin, measure follicle angle, and determine follicle density (Figures 3–6). Figure 2A shows a photograph of the nape of the neck with unshaved and shaved hair. The corresponding PA image is presented in Figure 2B taken along the dashed line in Figure 2A. The PA intensity of unshaved hair in this area was 4.23-fold higher than shaved scalp. These data show that the PA signal corresponds to hair volume. Our subject had gray-brown hair color. Ford et al. demonstrated that gray and blond color generate significantly lower photoacoustic signal than black hair due to the higher melanin content in black hair [25].

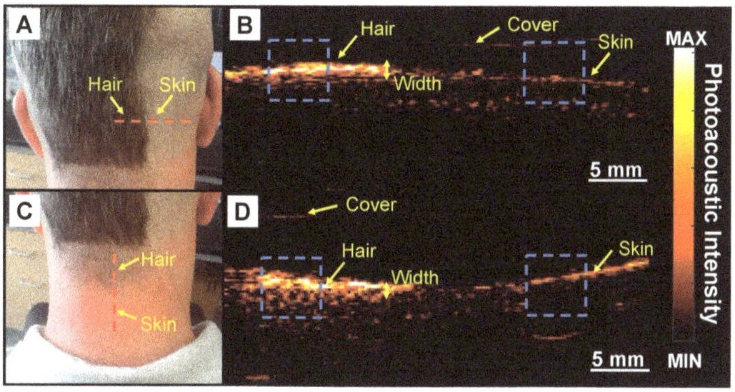

Figure 2. Hair imaging using PAI. This experiment contains positive and negative controls to validate the imaging technique using 690 nm light source. (**A**) Photograph of unshaved and shaved portions of the scalp. Red dashed line represents the imaging plane. (**B**) Photoacoustic image of dashed line in A which includes the skin and full hair. Photoacoustic intensity of hair in full hair area is 4.23-fold higher than shaved skin. (**C**) Photograph of the nape with lower hair density than the scalp. Red dashed line represents the imaging plane in D. (**D**) Photoacoustic image along red dashed line in C. Photoacoustic intensity in this area was 2.8-fold higher than the surrounding skin. Blue dashed squares are the region of interest (ROI) used to measure the PA intensity on the skin and hair.

We further confirmed this approach by imaging an area with varying hair density. The nape of the neck has decreasing hair density along a plane moving down the spine (Figure 2C). This subject also had recent sun exposure and with red skin in areas not covered by hair. Figure 2D shows a photoacoustic image collected in the red dashed line in Figure 2C. We note the low signal for the bare skin relative to the area with hair. The freshly shaved skin has a lower signal than the nape of the neck because the nape had been exposed to sun and had more pigmentation (Figure 2D). The intensity of hair on the nape was 2.8-fold higher than the surrounding skin. We used the photoacoustic intensity line profile of the hair and measured the average full width at half maximum (FWHM) for the width of the PA profile in Figure 2B,D to be 1.02 ± 0.19 and 0.37 ± 0.09 mm, respectively.

This shows that the width of PA signal can also be an acceptable metric to quantify the hair density. Scanning a 2 × 2 cm^2 area took 10 s to collect the raw data, 15 s to reconstruct the data, and up to 5 min to process. Future work can integrate automated image-processing tools to streamline the workflow [33]. It is also important to mention that, since the width of transducer is small (4 cm), we are easily able to perform multiple scans from multiple directions and surfaces such as flat and curved.

Figure 3A shows the absorption spectra of melanin, oxyhemoglobin, deoxyhemoglobin, and water. All data were downloaded from http://omlc.ogi.edu/spectra/. For hemoglobin, we used molar extinction coefficient (e) from [34], and the absorption coefficient was measured using the following: $\mu_a = \frac{(2.303)e\left(150\frac{gHb}{liter}\right)}{66{,}500\ gHb/mole}$. For melanin's μ_a correspondence to skin, we used $\mu_a = 1.70 \times 10^{12}\ \lambda^{-3.48}$, where, λ is the wavelength [35]. Using the same imaging setup, we performed PAI and ultrasound on the trimmed (with no guard) region of the scalp (see Figure 5G for exact region imaged). Figure 3B–D show the photoacoustic, ultrasound, and overlay of photoacoustic and ultrasound, respectively. With PAI, individual hair follicles are seen as discrete spots on the scalp Figure 3B. Ultrasound imaging was used to measure skin thickness of 3.78 ± 0.28 mm which is within reported values of 2–6 mm for normal adult humans [36]. The skull thickness was measured to be 6.05 ± 0.64 mm which is consistent with literature values ~6.5 mm for adult men [37,38].

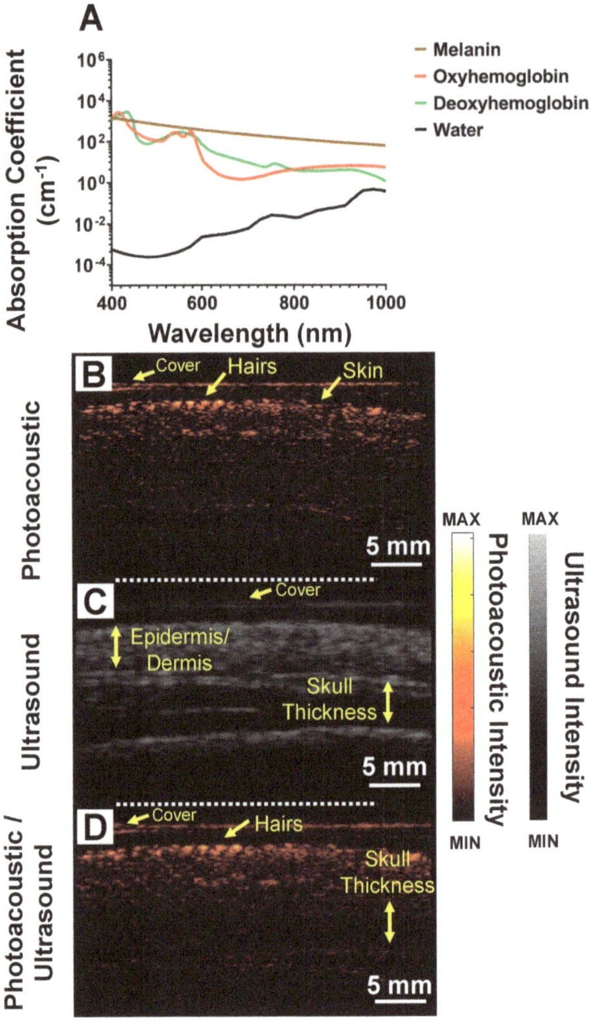

Figure 3. Trimmed scalp imaging. (**A**) Absorption spectra of melanin, oxyhemoglobin, deoxyhemoglobin, and water. Data from http://omlc.ogi.edu/spectra/. (**B**) Photoacoustic image of trimmed scalp. Individual hair follicles are represented by discrete spots on the scalp. (**C**) Ultrasound image from same plane of A including measurements of skull thickness from the ultrasound data. The skull thickness for this subject was 6.05 ± 0.64 mm. (**D**) Photoacoustic and ultrasound overlay allows us to locate hair follicles with respect to the anatomical features of the skull.

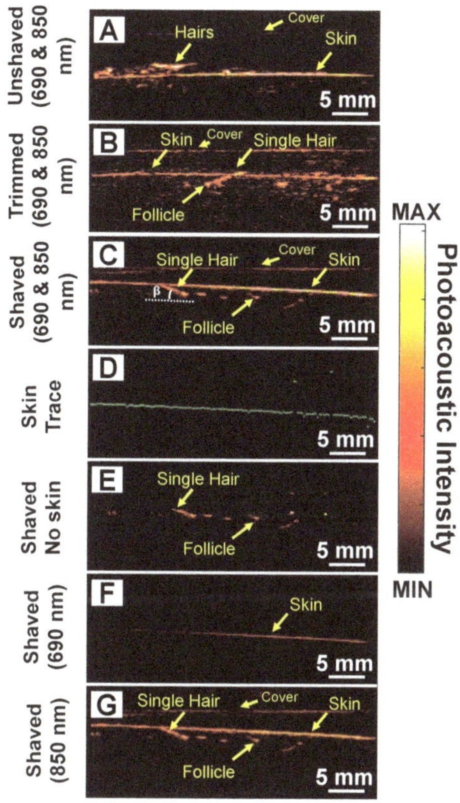

Figure 4. Subdermal imaging. Photoacoustic image of (**A**) unshaved and untrimmed, (**B**) trimmed, and (**C**) shaved abdominal skin using both wavelengths (690 and 850 nm). We could detect individual roots and subdermal hair strands using PAI after trimming and shaving. (**D**) We used a skin tracer algorithm to remove the photoacoustic contribution from the skin. (**E**) Photoacoustic image of shaved area after removing the photoacoustic signal of the skin via digital image processing. Photoacoustic image of the shaved region using (**F**) 690 and (**G**) 850 nm. Melanin in the skin absorbs strongly at 690 nm limiting depth penetration. Imaging at 850 nm allows deeper penetration allowing visualization of both the roots and subdermal strands.

counting from photographs. These results show that PAI can quantify hair follicle density across different regions—including in regions where the hair is not visible above the skin surface.

3.4. Follicle Angle

The transection rate in FUE can impact outcomes, and one key variable here is the follicle angle: The rate of transection increases when the angles point toward the nuchal lines [32]. We utilized PAI to measure the angle of multiple follicles (n = 25) relative to the skin surface on the shaved stomach. The follicular angle (β) was measured to be 18.62 ± 5.281° in this area (Figure 4C). To the best of our knowledge, no other study has reported the follicle angle on the stomach; values reported on the scalp vary from 10° to 40° depending on the location on the scalp [50–52].

4. Discussion

There are several error sources unique to photoacoustic imaging in this application. First, hair and skin color are key parameters here. Lighter hair and lighter skin color can generate less photoacoustic signal relative to darker skin. Darker skin has higher melanin content, absorbing more light and reducing depth penetration. Second, scanning is performed manually using a hand-held transducer, and the risk of shaking and instability can vary between operators but could be minimized with training. Image processing algorithms can also resolve and remove these artifacts [53,54]. Third, the resolution of our photoacoustic system could limit scanning of very fine hair. The resolution of our imaging system depends on the center frequency of the ultrasound transducer. As we mentioned before, our LED-based imaging system has an axial resolution of 260 µm. If the hair width is too small (lower than resolution range), then our imaging system may not able to resolve it. Similarly, if two follicles are separated by less than ~550 µm (lateral resolution) we cannot resolve both follicles. We could overcome these limitations by increasing the center frequency of the ultrasound transducer.

Previously, Ford et al. demonstrated the structural and functional analysis of hair follicles using volumetric multispectral optoacoustic tomography [25]. That work offered higher resolution images than this work, but we believe that our LED-based photoacoustic imaging system has important advantages in terms of a ~7-fold larger field of view and the use of LED light sources that are more compact, safe, cheap, and rugged.

In future work, we will image follicular features such as density and subdermal angle with the LED-based photoacoustic imaging technique and validate their prognostic value for in vivo models of FUE and FUT. We will also evaluate the technique in patients with a range of skin tones to better understand and study the limitations of the technique. We will also develop an image-processing algorithm to segment the photoacoustic images and measure the follicle density in a more automatic fashion.

5. Conclusions

In this study, we evaluated the application of photoacoustic imaging for the fast and non-invasive characterization of follicular density and subdermal angles with an emphasis on its diagnostic value for FUE and FUT. We showed that a portable, inexpensive, and low fluence LED-based imaging system has potential value for these procedures.

Author Contributions: A.H. conducted the experiments, data analysis, and prepared the manuscript. C.M. and Y.M. both helped to perform the literature research. J.V.J. had the idea of this article, helped in data collection, and critically revised the manuscript. All authors have read and agreed to the published version of the manuscript.

Funding: This study was funded by National Institutes of Health under grants R21 AG065776, R21 DE029025, and DP2 HL137187. We acknowledge NSF funding under grants 1842387 and 1937674.

Conflicts of Interest: The authors declare no conflict of interest.

24. Hindelang, B.; Aguirre, J.; Schwarz, M.; Berezhnoi, A.; Eyerich, K.; Ntziachristos, V.; Biedermann, T.; Darsow, U. Non-invasive imaging in dermatology and the unique potential of raster-scan optoacoustic mesoscopy. *J. Eur. Acad. Dermatol. Venereol.* **2019**, *33*, 1051–1061. [CrossRef]
25. Ford, S.J.; Bigliardi, P.L.; Sardella, T.C.P.; Urich, A.; Burton, N.C.; Kacprowicz, M.; Bigliardi, M.; Olivo, M.; Razansky, D. Structural and Functional Analysis of Intact Hair Follicles and Pilosebaceous Units by Volumetric Multispectral Optoacoustic Tomography. *J. Investig. Dermatol.* **2016**, *136*, 753–761. [CrossRef] [PubMed]
26. Hariri, A.; Lemaster, J.; Wang, J.; Jeevarathinam, A.S.; Chao, D.L.; Jokerst, J.V. The characterization of an economic and portable LED-based photoacoustic imaging system to facilitate molecular imaging. *Photoacoustics* **2018**, *9*, 10–20. [CrossRef] [PubMed]
27. Raman, R.; Napel, S.; Rubin, G.D. Curved-slab maximum intensity projection: Method and evaluation. *Radiology* **2003**, *229*, 255–260. [CrossRef] [PubMed]
28. Sakas, G.; Grimm, M.; Savopoulos, A. Optimized maximum intensity projection (MIP). In *Rendering Techniques' 95*; Springer: Berlin/Heidelberg, Germany, 1995; pp. 51–63.
29. Köstli, K.P.; Beard, P.C. Two-dimensional photoacoustic imaging by use of Fourier-transform image reconstruction and a detector with an anisotropic response. *Appl. Opt.* **2003**, *42*, 1899–1908. [CrossRef] [PubMed]
30. Rasband, W.S. *ImageJ 1.48v*; Bethesda: Rockville, MD, USA, 1997.
31. Schmid-Wendtner, M.-H.; Dill-Müller, D. Ultrasound technology in dermatology. *Semin. Cutan. Med. Surg.* **2008**, *27*, 44–51. [CrossRef]
32. Mohmand, M.H.; Ahmad, M. Effect of follicular unit extraction on the donor area. *World J. Plast. Surg.* **2018**, *7*, 193.
33. Miri Rostami, S.R.; Mozaffarzadeh, M.; Ghaffari-Miab, M.; Hariri, A.; Jokerst, J. GPU-accelerated double-stage delay-multiply-and-sum algorithm for fast photoacoustic tomography using LED excitation and linear arrays. *Ultrason. Imaging* **2019**, *41*, 301–316. [CrossRef] [PubMed]
34. Moaveni, M.K. A Multiple Scattering Field Theory Applied to Whole Blood. Ph.D. Thesis, University of Washington, Seattle, WA, USA, 1970.
35. Jacques, S.L.; McAuliffe, D.J. The melanosome: Threshold temperature for explosive vaporization and internal absorption coefficient during pulsed laser irradiation. *Photochem. Photobiol.* **1991**, *53*, 769–775. [CrossRef]
36. Oltulu, P.; Ince, B.; Kökbudak, N.; Kılıç, F. Measurement of epidermis, dermis, and total skin thicknesses from six different body regions with a new ethical histometric technique. *Türk Plastik Rekonstrüktif Estet. Cerrahi Derg. (Turk. J. Plast. Surg.)* **2018**, *26*, 56–61. [CrossRef]
37. Law, S.K. Thickness and resistivity variations over the upper surface of the human skull. *Brain Topogr.* **1993**, *6*, 99–109. [CrossRef]
38. Lillie, E.M.; Urban, J.E.; Weaver, A.A.; Powers, A.K.; Stitzel, J.D. Estimation of skull table thickness with clinical CT and validation with microCT. *J. Anat.* **2015**, *226*, 73–80. [CrossRef] [PubMed]
39. Kim, J.; Kim, J.Y.; Jeon, S.; Baik, J.W.; Cho, S.H.; Kim, C. Super-resolution localization photoacoustic microscopy using intrinsic red blood cells as contrast absorbers. *Light Sci. Appl.* **2019**, *8*, 1–11. [CrossRef] [PubMed]
40. Weber, J.; Beard, P.C.; Bohndiek, S.E. Contrast agents for molecular photoacoustic imaging. *Nat. Methods* **2016**, *13*, 639–650. [CrossRef]
41. Zhang, H.F.; Maslov, K.; Stoica, G.; Wang, L.V. Functional photoacoustic microscopy for high-resolution and noninvasive in vivo imaging. *Nat. Biotechnol.* **2006**, *24*, 848–851. [CrossRef] [PubMed]
42. Singh, M.K.A.; Sato, N.; Ichihashi, F.; Sankai, Y. Real-time improvement of LED-based photoacoustic image quality using intermittent pulse echo acquisitions. In Proceedings of the Photons Plus Ultrasound: Imaging and Sensing 2020, San Francisco, CA, USA, 17 February 2020; p. 1124051.
43. Hariri, A.; Alipour, K.; Mantri, Y.; Schulze, J.P.; Jokerst, J.V. Deep learning improves contrast in low-fluence photoacoustic imaging. *Biomed. Opt. Express* **2020**, *11*, 3360–3373. [CrossRef] [PubMed]
44. Crippa, P.; Cristofoletti, V.t.; Romeo, N. A band model for melanin deduced from optical absorption and photoconductivity experiments. *Biochim. Biophys. Acta (BBA)-Gen. Subj.* **1978**, *538*, 164–170. [CrossRef]
45. Hasegawa, H.; de Korte, C.L. Impact of element pitch on synthetic aperture ultrasound imaging. *J. Med Ultrason.* **2016**, *43*, 317–325. [CrossRef]

References

1. Vogel, J.E.; Jimenez, F.; Cole, J.; Keene, S.A.; Harris, J.A.; Barrera, A.; Rose, P.T. Hair Restoration Surgery: The State of the Art. *Aesthetic Surg. J.* **2013**, *33*, 128–151. [CrossRef] [PubMed]
2. International Society of Hair Restoration Surgery. *2017 Practice Census Results*; International Society of Hair Restoration Surgery: Geneva, IL, USA, 2017.
3. Unger, W.P.; Shapiro, R.; Unger, R.; Unger, M. *Hair Transplantation*; CRC Press: Boca Raton, FL, USA, 2010.
4. Paus, R.; Cotsarelis, G. The biology of hair follicles. *N. Engl. J. Med.* **1999**, *341*, 491–497. [CrossRef] [PubMed]
5. Rassman, W.R.; Bernstein, R.M.; McClellan, R.; Jones, R.; Worton, E.; Uyttendaele, H. Follicular Unit Extraction: Minimally Invasive Surgery for Hair Transplantation. *Dermatol. Surg.* **2002**, *28*, 720–728. [CrossRef] [PubMed]
6. Bernstein, R.M.; Rassman, W.R. Follicular unit transplantation: 2005. *Dermatol. Clin.* **2005**, *23*, 393–414. [CrossRef]
7. Nirmal, B.; Somiah, S.; Sacchidanand, S.A. A study of donor area in follicular unit hair transplantation. *J. Cutan. Aesthet. Surg.* **2013**, *6*, 210.
8. Avram, M.R.; Rogers, N.; Watkins, S. Side-effects from follicular unit extraction in hair transplantation. *J. Cutan. Aesthet. Surg.* **2014**, *7*, 177–179. [CrossRef]
9. Alhaddab, M.; Kohn, T.; Sidloi, M. Effect of Graft Size, Angle, and Intergraft Distance on Dense Packing in Hair Transplant. *Dermatol. Surg.* **2005**, *31*, 650–654. [CrossRef]
10. Jimenez, F.; Ruifernández, J.M. Distribution of human hair in follicular units: A mathematical model for estimating the donor size in follicular unit transplantation. *Dermatol. Surg.* **1999**, *25*, 294–298. [CrossRef]
11. Loh, S.-H.; Lew, B.-L.; Sim, W.-Y. Localized Telogen Effluvium Following Hair Transplantation. *Ann. Dermatol.* **2018**, *30*, 214–217. [CrossRef]
12. Poswal, A.; Bhutia, S.; Mehta, R. When fue goes wrong! *Indian J. Dermatol.* **2011**, *56*, 517–519. [CrossRef]
13. Harris, J.A. New Methodology and Instrumentation for Follicular Unit Extraction: Lower Follicle Transection Rates and Expanded Patient Candidacy. *Dermatol. Surg.* **2006**, *32*, 56–62. [CrossRef]
14. Zito, P.M.; Raggio, B.S. Hair Transplantation. In *StatPearls [Internet]*; StatPearls Publishing: Treasure Island, FL, USA, 2019.
15. Kleinerman, R.; Whang, T.B.; Bard, R.L.; Marmur, E.S. Ultrasound in dermatology: Principles and applications. *J. Am. Acad. Dermatol.* **2012**, *67*, 478–487. [CrossRef] [PubMed]
16. Attia, A.B.E.; Balasundaram, G.; Moothanchery, M.; Dinish, U.S.; Bi, R.; Ntziachristos, V.; Olivo, M. A review of clinical photoacoustic imaging: Current and future trends. *Photoacoustics* **2019**, *16*, 100144. [CrossRef] [PubMed]
17. Mantri, Y.; Davidi, B.; Lemaster, J.E.; Hariri, A.; Jokerst, J.V. Iodide-doped precious metal nanoparticles: Measuring oxidative stress in vivo via photoacoustic imaging. *Nanoscale* **2020**, *12*, 10511–10520. [CrossRef] [PubMed]
18. Choi, W.; Park, E.-Y.; Jeon, S.; Kim, C. Clinical photoacoustic imaging platforms. *Biomed. Eng. Lett.* **2018**, *8*, 139–155. [CrossRef] [PubMed]
19. Moore, C.; Jokerst, J.V. Strategies for Image-Guided therapy, surgery, and drug delivery using photoacoustic imaging. *Theranostics* **2019**, *9*, 1550. [CrossRef]
20. Kim, J.; Kim, Y.; Park, B.; Seo, H.M.; Bang, C.; Park, G.; Park, Y.; Rhie, J.; Lee, J.; Kim, C. Multispectral ex vivo photoacoustic imaging of cutaneous melanoma for better selection of the excision margin. *Br. J. Dermatol.* **2018**, *179*, 780–782. [CrossRef]
21. Petri, M.; Stoffels, I.; Jose, J.; Leyh, J.; Schulz, A.; Dissemond, J.; Schadendorf, D.; Klode, J. Photoacoustic imaging of real-time oxygen changes in chronic leg ulcers after topical application of a haemoglobin spray: A pilot study. *J. Wound Care* **2016**, *25*, 87–91. [CrossRef] [PubMed]
22. Schwarz, M.; Aguirre, J.; Omar, M.; Ntziachristos, V. Chapter 26—Optoacoustic Imaging of Skin. In *Imaging in Dermatology*; Hamblin, M.R., Avci, P., Gupta, G.K., Eds.; Academic Press: Boston, MA, USA, 2016; pp. 375–385. [CrossRef]
23. Zabihian, B.; Weingast, J.; Liu, M.; Zhang, E.; Beard, P.; Pehamberger, H.; Drexler, W.; Hermann, B. In vivo dual-modality photoacoustic and optical coherence tomography imaging of human dermatological pathologies. *Biomed. Opt. Express* **2015**, *6*, 3163–3178. [CrossRef]

46. Cheong, W.-F.; Prahl, S.A.; Welch, A.J. A review of the optical properties of biological tissues. *IEEE J. Quantum Electron.* **1990**, *26*, 2166–2185. [CrossRef]
47. Maneas, E.; Aughwane, R.; Huynh, N.; Xia, W.; Ansari, R.; Kuniyil Ajith Singh, M.; Hutchinson, J.C.; Sebire, N.J.; Arthurs, O.J.; Deprest, J. Photoacoustic imaging of the human placental vasculature. *J. Biophotonics* **2020**, *13*, e201900167. [CrossRef]
48. Favazza, C.P.; Wang, L.V.; Cornelius, L.A. In vivo functional photoacoustic microscopy of cutaneous microvasculature in human skin. *J. Biomed. Opt.* **2011**, *16*, 026004. [CrossRef]
49. Taruttis, A.; Timmermans, A.C.; Wouters, P.C.; Kacprowicz, M.; van Dam, G.M.; Ntziachristos, V. Optoacoustic imaging of human vasculature: Feasibility by using a handheld probe. *Radiology* **2016**, *281*, 256–263. [CrossRef] [PubMed]
50. Ahmad, M.; Mohmand, M.H. Analysis of the changes in scalp hair angles: In vivo and in vitro comparison before and after tumescence. *J. Cosmet. Dermatol.* **2019**, *18*, 390–394. [CrossRef] [PubMed]
51. Lam, S.; Williams, K. *Hair Transplant 360 Follicular Unit Extraction (FUE) Jaypee Medical Inc*; The Health Sciences Publisher: New Delhi, India, 2016.
52. Lindelöf, B.; Forslind, B.; Hedblad, M.-A. Human hair form: Morphology revealed by light and scanning electron microscopy and computer aided three-dimensional reconstruction. *Arch. Dermatol.* **1988**, *124*, 1359–1363. [CrossRef] [PubMed]
53. Fergus, R.; Singh, B.; Hertzmann, A.; Roweis, S.T.; Freeman, W.T. *Removing Camera Shake from a Single Photograph*; ACM Siggraph 2006 Papers: Boston, MA, USA, 2006; pp. 787–794.
54. Yan, W.-Q.; Kankanhalli, M.S. Detection and removal of lighting & shaking artifacts in home videos. In Proceedings of the Tenth ACM International Conference on Multimedia, Juan les Pins, France, 1–6 December 2002; pp. 107–116.

MDPI
St. Alban-Anlage 66
4052 Basel
Switzerland
Tel. +41 61 683 77 34
Fax +41 61 302 89 18
www.mdpi.com

Sensors Editorial Office
E-mail: sensors@mdpi.com
www.mdpi.com/journal/sensors

www.ingramcontent.com/pod-product-compliance
Ingram Content Group UK Ltd.
Pitfield, Milton Keynes, MK11 3LW, UK
UKHW060124010826
14089UKWH00006B/139

9 783036 511993